35th Oil Shale Symposium 2015

Salt Lake City, Utah, USA
5-8 October 2015

ISBN: 978-1-5108-3065-3

Printed from e-media with permission by:

Curran Associates, Inc.
57 Morehouse Lane
Red Hook, NY 12571

Some format issues inherent in the e-media version may also appear in this print version.

Printed by Curran Associates, Inc. (2016)

For permission requests, please contact Colorado School of Mines
at the address below.

Colorado School of Mines
1500 Illinois St.
Golden, CO 80401
USA

Phone: 303-273-3000 or 800-446-9488

www.mines.edu

Additional copies of this publication are available from:

Curran Associates, Inc.
57 Morehouse Lane
Red Hook, NY 12571 USA
Phone: 845-758-0400
Fax: 845-758-2633
Email: curran@proceedings.com
Web: www.proceedings.com

TABLE OF CONTENTS

SESSION 1: OPENING PLENARY

SESSION 2: IN SITU PROCESSING

SESSION 3: MODELING

SESSION 4: SURFACE PROCESSING

SESSION 5: CHEMISTRY & GEOCHEMISTRY

SESSION 6: INTERNATIONAL UPDATES 1

SESSION 7: GEOLOGY & STRATIGRAPHY

SESSION 8: RESOURCE & ENVIRONMENT EVALUATION

SESSION 9: PYROLYSIS

35th Oil Shale Symposium

Thanks to our sponors...

Symposium hosted by...

The Governor's Office of Energy Development, the Utah Geological Survey, & the Center for Oil Shale Technology & Research, Colorado School of Mines

35th Oil Shale Symposium *October 5-9, 2015*

at the Hilton Salt Lake City Center, Salt Lake City, Utah

Welcome to the 35th Oil Shale Symposium

With a Recollection of James H. Gary

Jeremy Boak, Symposium Co-Chair

MINES

GOVERNOR'S OFFICE OF ENERGY DEVELOPMENT
Advancing Utah's Energy Future

35th Oil Shale Symposium Sponsors

energy API

Enefit

HATCH
Canada's 50 Best Managed Companies

INL Idaho National Laboratory

irati

QER
Creating new fuels for Australia

REDLEAF RESOURCES, INC.

SHALE TECH INTERNATIONAL

TOTAL

UMATAC Industrial Processes
A company of ThyssenKrupp Industrial Solutions

MINES

GOVERNOR'S OFFICE OF ENERGY DEVELOPMENT
Advancing Utah's Energy Future

James H. Gary (1922 - 2015)

- BS & MS degrees in Chemical Engineering from Virginia Polytechnical Institute
- PhD from University of Florida
- Worked for Standard Oil of Ohio
- At Colorado School of Mines:
  - Head of Department of Chemical Engineering and Petroleum Refining (12 years)
  - Dean of Faculty, and Vice President for Academic Affairs (7 years)
  - Chair of 1st through 25th Oil Shale Symposia (1964-1992)

MINES

GOVERNOR'S OFFICE OF ENERGY DEVELOPMENT

Jim Gary and the Oil Shale Symposia

- At the 26th Oil Shale Symposium, Dr. Gary was asked to make some comments from the perspective of the Chair of the previous 25 Symposia
- He dwelt at length on the first CSM sponsored Symposia - from 1918 to 1924
- Declared the 26th should be the 33rd Oil Shale Symposium

MINES

GOVERNOR'S OFFICE OF ENERGY DEVELOPMENT

Oil Prices and Symposium Pages

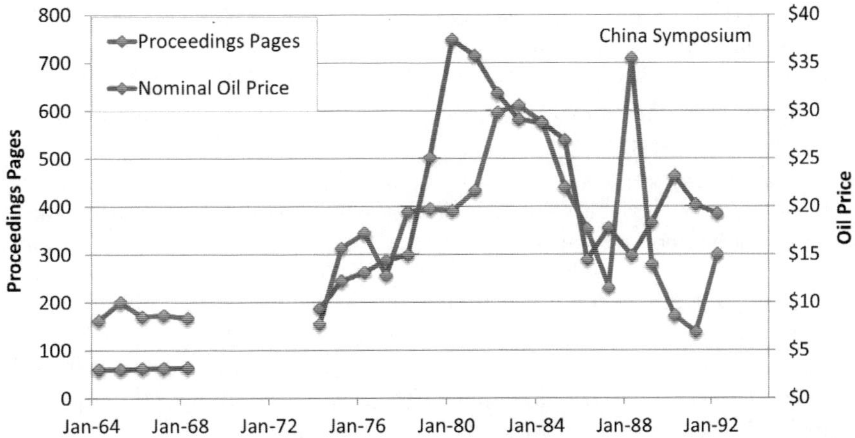

MINES

GOVERNOR'S OFFICE OF ENERGY DEVELOPMENT

The Gary Viewpoint - Optimistic

- ... there are many reasons to believe that now the circumstances are in favor of the development of a shale oil industry to supplement the liquid and gaseous fuel needs of the country. (1964)
- Perhaps the 13th Oil Shale Symposium will signify the approach of the turn because everything is now favorable for proceeding except for the resistance of the no-growth groups. (1980)
- Unless there are major unforeseen impacts on the construction and development phases now underway, there should be a production of 50,000 to 90,000 barrels per day of shale oil by the mid-1980s. (1981)
- It is my hope this program will continue under new leadership to support and encourage research and development to produce oil from shale economically in an environmentally acceptably manner. In my opinion, this is essential for the future strength and security of the United States. (1992)

MINES

GOVERNOR'S OFFICE OF ENERGY DEVELOPMENT

The Gary Viewpoint - Frustrated

- The distortion of the industry position by news media has created a crisis with respect to industry credibility and a chasm between those dependent upon petroleum fuels and those most able to supply their needs. (1979)
- .... the political atmosphere changed from one of encouragement to one of laws and regulations designed to prevent development. (1981)
- Although there is still a more than adequate supply of government indecision ... When the industry does develop, it will be one of the most studied industries in the history of the world. If as much thought was placed on political issues, there is a very good chance that Thomas Jefferson and his cohorts would still be wrestling with the Declaration of Independence and the Constitution would be several more centuries down the road. (1984)
- Unfortunately, both government and industry have very short memories and resort to short-term planning. This has resulted in the rollercoaster history of the shale oil industry. (1986)

MINES

GOVERNOR'S OFFICE OF ENERGY DEVELOPMENT
Advancing Utah's Energy Future

The Gary Viewpoint – Forward-Looking

- The keynote session of the Eighth Oil Shale Symposium was devoted to reporting "Input-Output Energy Studies for Development of Oil Shale Resources' and guideline discussions were encouraged. This provided a series of papers on net energy relationships as well as the limitations and restrictions of each of the studies. (1974)
- Whenever there is a shortage of petroleum, crash research and development programs are initiated and everyone gets into the act. Later, when crude oil becomes more available, these programs are phased out and the expertise developed is allowed to dissipate. In the long run this is very costly to industry and the government because much of the same work is repeated and new people have to be educated to the needs of the industry. (1986)
- In the future, our energy, from necessity, will come from many sources (solar, so-called alternate fuels, etc.) rather than the few of the past (oil, gas, coal, and nuclear). (1992)

MINES

GOVERNOR'S OFFICE OF ENERGY DEVELOPMENT
Advancing Utah's Energy Future

Recollections from Tom Sladek

- Jim was immensely competent in his field, but he also was a true southern gentleman - kind, courteous, patient, and tolerant of the opinions of others, even when they conflicted with his own.
- He would not long suffer fools, whether they be self-promoting executives, elected officials, bureaucrats, academicians and their administrators, or graduate students.
- He was fully retired .... in 2007, but ... still keenly interested in the oil shale situation in Jordan and generous with his time. He recalled his visits to the Kingdom with complete clarity and especially how kindly he and Jane had been treated by the Jordanian people.

MINES

GOVERNOR'S OFFICE OF ENERGY DEVELOPMENT

Recollections from Tom Sladek

- Jim spent his professional career trying to make America a safer place through the development of fuels and chemicals from unconventional domestic resources....
- Although he believed in the promises of technology, Jim was also aware of the potential for great harm if technology is not applied with understanding and caution.
- Jim was a friend, a colleague, a mentor, and a model of what an engineer can become.
- His passing has made our world a smaller and less inspiring place.

MINES

GOVERNOR'S OFFICE OF ENERGY DEVELOPMENT

We Stand on the Shoulders of Giants

James Gary

Victor Alderson

MINES

GOVERNOR'S OFFICE OF ENERGY DEVELOPMENT

UTAH SCHOOL AND INSTITUTIONAL TRUST LANDS ADMINISTRATION (SITLA)

STATE TRUST LANDS

- At statehood, Congress granted most western states land to support public schools and universities.
- SITLA manages 3.5 million acres (1.4mm ha), plus 1mm acres (400,000 ha) severed minerals.
- Land income funds Permanent School fund, a perpetual endowment supporting each K-12 public/charter school in Utah.

SITLA

- Historic state management of lands was conservative and bureaucratic.
- SITLA was created in 1994 to manage state trust lands independently from state government, "run like a business"
- Revenue from trust lands includes conventional oil and gas, other minerals (OIL SHALE!), real estate development, commercial investments.

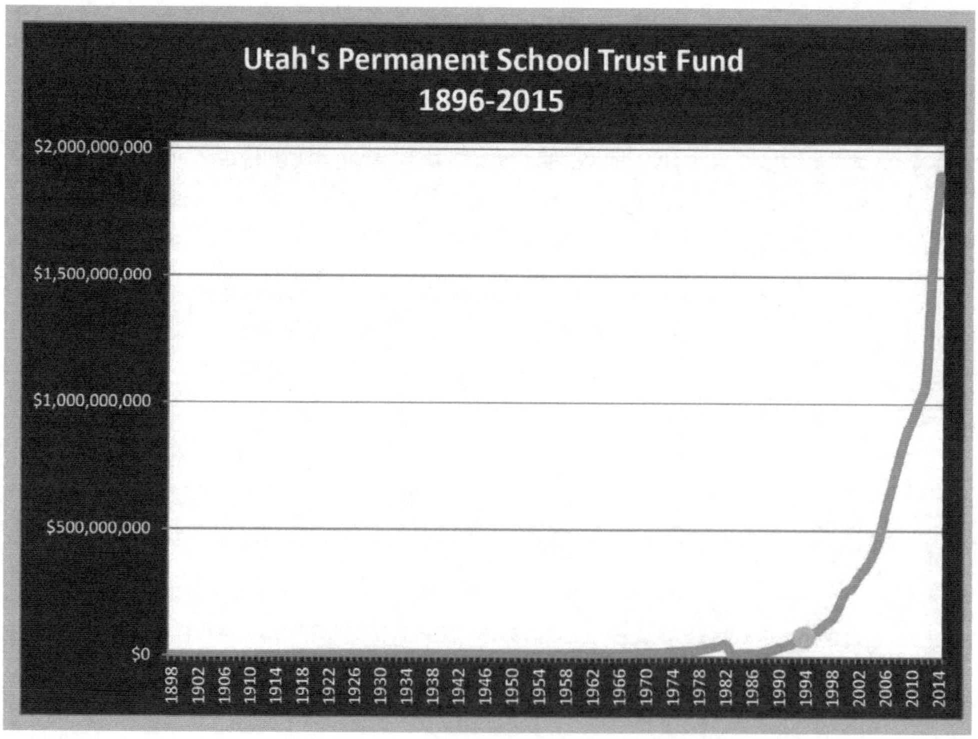

OIL SHALE RESOURCES ON SITLA LANDS

- 118,656 acres (48,018 ha) of oil shale resources, in both larger blocks and scattered 1 square mile sections
- In-place resource estimated @12 billion barrels of 25+/gpt
- Larger blocks of land acquired through 1980s land exchanges with federal government

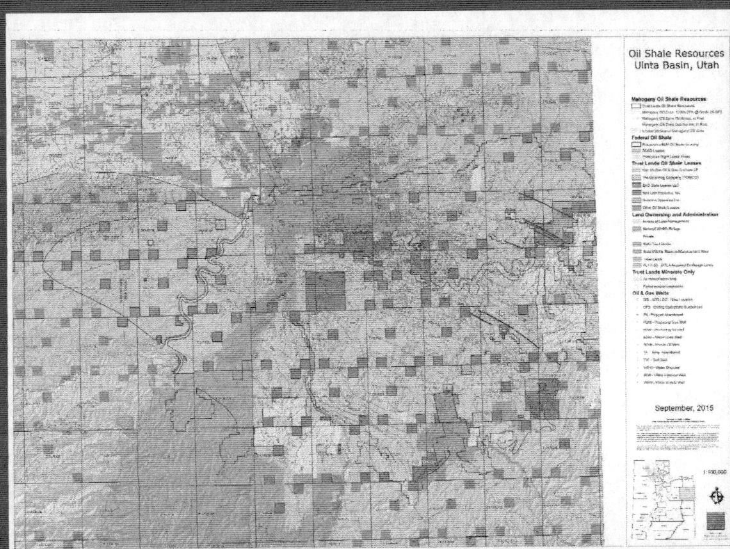

SITLA OIL SHALE LEASING PROGRAM

- All oil shale leases issued by SITLA director with consent of Board
- 10 year primary term, renewable for 10 additional years, then can be held by production
- Production royalty begins at 5%, escalates 0.5% annually up to 12.5%
- Contacts: Tom Faddies, John Blake

THOUGHTS ON OIL SHALE

- Unlike the federal government, State of Utah actively supports continued development of the oil shale industry.
- Almost all near term development of oil shale in Utah will be on SITLA and private lands.
- SITLA continues a long-term program of consolidating state lands into contiguous blocks where federal role is limited.
- Up to industry to demonstrate economics, ability to operate in carbon-constrained environment.

Oil Shale:

Treasure Trove or Pandora's Box?

A 10 Week Study and eBook

by

Ron Stites, Lee Yuan Ming and Matthew James

NANYANG TECHNOLOGICAL UNIVERSITY

Stites & Associates, LLC
The Technology Development Network

Introduction

- **Grew Out of Two Projects**
    - Extensive Development Project by SALLC
    - Joint Internship Between Nanyang and SALLC
        - Introduced by O. J. Schneider of Ettanet
- **Included Extensive Research**
    - Library – History, Resources & Technology
    - Interviews – Industry, Regulators & Academicians
    - Physical Visits to Sites
    - Regulatory, Technical and Financial Evaluations
- **Resulted in an eBook**
    - Tremendous Amount of Information
- **Resulted in Conclusions & Recommendations**
    - Rather Unexpected Compared to Initial Reactions
    - Great Learning and Cultural Experience

NANYANG TECHNOLOGICAL UNIVERSITY

Stites & Associates, LLC
The Technology Development Network

The Road Trip

- **Visits Physical Site**
  - Shale Tech Facility near Rifle, CO
  - Enefit Box Cut near Dragon, UT
  - CUOS Site near Dragon, UT
- **Interviewed Industry Experts**
  - Mr. Justin Bilyeu, Shale Tech
  - Mr. Kevin Biehle, Shale Tech
  - Mr. Jeff Hartley, RedLeaf
- **Interviewed Academic Experts**
  - Dr. Jennifer Spinti and others at University of Utah
  - Dr. Jeremy Boak at Colorado School of Mines

Not ALL Work

NANYANG TECHNOLOGICAL UNIVERSITY

Stites & Associates, LLC
The Technology Development Network

eBook Audience?

- **Concerned US Citizen**
  - Owns Much of the Resource
  - Determines Much of If/How Developed
  - Poorly Informed
- **Oil Shale Industry**
  - Some Misconceptions
  - General Malaise & Defeatism
- **Regulators**
  - Often Intimidated by the Advocacy Industry

NANYANG TECHNOLOGICAL UNIVERSITY

Stites & Associates, LLC
The Technology Development Network

The Need for Shale Oil

- **eBook Deals with the Need for Reliable Oil Source**
  - Often Forgotten is the Amount that is in the West
  - Often Forgotten is the Correlation Between Standard of Living and Abundance of Energy
  - Often Forgotten is the Danger of NOT having Reliable Oil Sources
- **eBook Deals with Relationships Between Climate Change & Energy**
  - This is a Very Difficult Topic
  - Mostly Rhetoric – Few Facts

A Surprise – The Advocacy Industry Wouldn't Engage

NANYANG TECHNOLOGICAL UNIVERSITY

Stites & Associates, LLC
The Technology Development Network

What is Oil Shale/Shale Oil?

- **Interns Came with Only the "Public View"**
- **Spent Many Weeks Learning**
  - What is Oil Shale
  - Geology/Chemistry
  - History of Development
  - Technologies
  - Development Outside of the US
- **Surprising Difference between Reality & "Public View"**
  - The Claim "Can't be Done" Absolutely Wrong
  - The Claim "Environmental Disaster" Way Overstated
  - The Claim "Too Resource Intensive" an Issue that Can be Solved

A Goal of the eBook – Provide Objective Evidence About Oil Shale

NANYANG TECHNOLOGICAL UNIVERSITY

Stites & Associates, LLC
The Technology Development Network

Basic Oil Shale Concepts

- **How is Oil Shale Graded?**
  - What are Some of the Differences within Grades?
- **"Standard" Grading Test – Modified Fisher Assay (MFA)**
  - How Done?
  - What is "Says"
  - What are Some Known Limitations
- **What are the Primary Technologies?**
  - Ex-Situ Retorting
  - In-Situ Retorting
  - Ex-Situ Extractive/Reactive Methods
  - Hybrid Methods

The eBook Digs Deeper Not Enough to Speak in Generalities

NANYANG TECHNOLOGICAL UNIVERSITY

Stites & Associates, LLC
The Technology Development Network

What are the Key Issues?

- **What are the Key Impediments to Commercialization?**
  - High Cost – Especially Capital Intensive
  - Significant Technical Risks
  - Very Significant Business Risks
- **What are the Key Positive and Negative Issues?**
  - **Positives:**
    - Energy Security
    - Employment
    - Payout to Owners – Especially US Citizens
  - **Negatives**
    - Water Use – Pollution & Water Rights
    - Air Use -- Pollution
    - Land Use
    - Green House Gas – Carbon Emissions

eBook Recognizes the Issues and Deals with Each in an Objective Fashion

NANYANG TECHNOLOGICAL UNIVERSITY

Stites & Associates, LLC
The Technology Development Network

Focusing on Business Risk

- **Assessing Risk versus Reward – The Concept of ROI**
  - Presents Return on Investment (ROI) Concepts
  - Discusses the Relationships between Risk and ROI
  - Presents General Rules of Thumb
  - Discusses the Kiss of Death – Technical Risk
- **Digs Deeper into Financial Modeling**
  - Fixed Costs
  - Variable Cost
  - Time Value of Money
- **Presents a Generic Model for Three Types of Technologies**
  - Ex-Situ Classic Retorting
  - In-Situ Heating
  - Ex-Situ Advanced Treatment

Generic Models Available at a Link Given in the eBook: Example

eBook Recognizes There is No Free Lunch

NANYANG TECHNOLOGICAL UNIVERSITY

Stites & Associates, LLC
The Technology Development Network

A New Concept Return on Resources

- **The Mineral Owners Dilemma**
  - The Concept of Mineral Rights and Return on Resources (ROR)
  - Oil Shale Rights Mostly Owned by US Citizens
  - Many "Rights" Balanced Against Mineral Development
  - Never Easy – Never Simple
  - Generally Leads to Partnerships Between Mineral Owner and Developer
- **Partnership A Model for Oil Shale Development**
  - The US Citizens have Needs – Oil Shale Can Meet Some of Them
  - Development Has Issues – Developers Can Meet Most of Them
- **Specific Western Issues**
  - Water Rights and Needs – Mechanisms for Balancing Exist
  - Air Pollution Issues – Mechanisms for Solving Well Developed
- **Partners Need to Recognize Competing Technologies**
  - Tight Formation Gas & Oil
  - Nuclear
  - Renewables

A Partnership Between US Citizens and Oil Shale Developers Needs to be the Model

NANYANG TECHNOLOGICAL UNIVERSITY

Stites & Associates, LLC
The Technology Development Network

Conclusions and Opinions

- **A Surprising Number of Conclusions**
  - Things Were Actually Learned
  - Many Were Contrary To "Public View"
    - The Development Of Oil Shale (That Is Oil From Oil Shale) Is Technologically And Financially Feasible Without Creating An Environmental Disaster.
    - Oil Shale Will Not Be Developed In The Us At The Scale Anticipated In The 1970's For Many Decades.
    - There Are Several Different Technologies That Can Be Used For Developing Oil Shale.
    - There Are Several Organizations Dedicated To Environmentally Responsible Development Of Oil Shale.

NANYANG TECHNOLOGICAL UNIVERSITY

Stites & Associates, LLC
The Technology Development Network

Conclusions and Opinions

Tried to Answer Three Questions

- Should Oil Shale be Developed?
- If So What Would It Look Like?
- What are the Chances Oil Shale Will Be Developed in Next 10 Years

Opinions Slightly Different but More Positive than Original

- Matthew James
  - "Hugging Trees Won't Save Us"
  - "The Danger of the 'Single Story'"
- Lee Yuan Ming
  - "Possible"
  - "Helpful"
  - "Technology Improvements Needed"
- Ron Stites
  - "Should be Done"
  - "At a Reasonable Scale"
  - "On Private or SITLA Land"
  - "No Government Money for Commercialization"

Available on Amazon Kindle at:
http://www.amazon.com/dp/B013PSCOVU

NANYANG TECHNOLOGICAL UNIVERSITY

Stites & Associates, LLC
The Technology Development Network

Acknowledgments

We must recognize the indispensable help we received from four other Nanyang Interns who did much original research on specific issues:

- Lee Sey Chee
- Shao Ying Lam
- Hui Jie Tey
- Grayson Wen Guo.

We also received help from a number of member of the Oil Shale Industry:

- Mr. Justin Bilyeu of Shale Tech
- Mr. Kevin Biehle of Shale Tech
- Mr. Jeff Hartley and others at Red Leaf Resources
- Michelle Faulk and others at Enefit American Oil

Within the Regulatory Community we were helped by:

- Mr. Mark Schlagenhauf of the Colorado Department of Natural Resources

In Academia we received help from:

- Dr. Jennifer Spinti and many others from the University of Utah Institute for Clean & Secure Energy,
- Dr. Jeremy Boak of the Colorado School of Mines
- Mr. Loh Pui Wah and Mr. Stephen Lee Siang Guan both of Nanyang Technological University

NANYANG TECHNOLOGICAL UNIVERSITY

Stites & Associates, LLC
The Technology Development Network

The Authors

Lee Yuan Ming, B.Eng.Sci. and M.S. (both expected 2016)

Lee Yuan Ming is an undergraduate student in his senior year at Nanyang Technological University in Singapore, where he is majoring in Chemical Engineering under the Renaissance Engineering Programme. At the current phase of his life, he has his eyes set on a career in the energy industry as an engineering consultant, where he hopes he can play his part in helping the world meet its energy challenges and transiting to renewable energy. His future plans include building sustainable businesses and pro bono social enterprise consulting. An avid amateur basketball player, he regularly strikes fear into his opponents' hearts on the court as all 5'5" of him towers over them. He also enjoys chemically altering the properties of raw ingredients and turning them into delectable delicacies in the kitchen. He occasionally entertains by engaging in the production of aurally satisfying tones with his vocal chords on stage as well. Holler at lee@tek-dev.net

NANYANG TECHNOLOGICAL UNIVERSITY

Stites & Associates, LLC
The Technology Development Network

The Authors

Matthew James: B.Eng.Sci. and M.S. (both expected 2016)

Matthew holds a Diploma in Chemical Engineering and currently pursuing a dual-degree in Bachelor of Engineering Science (Chemical Engineering) and Master of Science (Technology Management). He is very eager to join the petrochemical industry upon graduation. Aside from his academic life, Matthew is also very keen on product design and development. He started off as a freelance graphic designer, which eventually led him to a product development stint under Kraft Foods Singapore. He has also designed and built a portable oil-spill skimmer using biodegradable adsorbent that is capable of performing 6 times better than synthetic adsorbent. The project was featured in the Young Water Talents Symposium 2012 held in Singapore. Most recently, Matthew ideated an urban donation platform for the San Francisco Bay Area. The project was applauded by potential partners and investors, and was featured in the City Innovate Summit 2015. He may be reached at matthew@tek-dev.net

NANYANG TECHNOLOGICAL UNIVERSITY

Stites & Associates, LLC
The Technology Development Network

The Authors

Ron Stites, B.S. Chemistry, M.B.A. Finance and Accounting

Ron Stites has served as the Director of Research and similar positions in a number of technology driven companies – including several start-ups. He holds several patents in the areas of gasification and catalysis. He has several other patents pending in the area of alternative fuels, carbon capture and unique fuel applications. He holds a BS Degree in Chemistry and an MBA in Finance and Accounting. He began his career in environmental testing as an analytical chemist. He has built and managed several laboratories doing such diverse testing as hazardous waste testing, radiological testing, petroleum testing, catalyst testing and engineering studies. More recently his focus has been on industrial chemistry and chemical applications – including oil shale. He has written extensively and given seminars on R&D Management, Analytical Techniques and the Design of Experiments. He has been operating his own independent consulting company, Stites & Associates, LLC, since 1996. He may be reached at ron@tek-dev.net

NANYANG TECHNOLOGICAL UNIVERSITY

Stites & Associates, LLC
The Technology Development Network

1.5 **Lt. Governor's Welcome to Utah**

Spencer Cox

Lieutenant Governor of the State of Utah, Salt Lake City, UT, USA

Lieutenant Governor Spencer Cox will discuss energy development in Utah. Oil Shale has a long history in the state and Utah is its way to being the only state in the US with an operating oil shale industry. Lieutenant Governor Cox will discuss oil shale's importance to Utah and how its development fits into the Governor's larger energy development plan.

THE INSTITUTE FOR CLEAN AND SECURE ENERGY
AT THE UNIVERSITY OF UTAH

Economic Analysis of In Situ Oil Shale Development in the Uinta Basin

J. Wilkey, M. Hradisky, J. Spinti, P.J. Smith, T. Ring, K. Kelly

The Institute for Clean and Secure Energy
University of Utah

October 5, 2015

THE INSTITUTE FOR CLEAN AND SECURE ENERGY
AT THE UNIVERSITY OF UTAH

Oil Shale / Oil Sands Program

Mission: Pursue research that improves industry's ability to utilize the vast energy stored in oil shale and oil sands resources with minimal environmental impact

Three main research areas:

1. Basin scale simulation of environmental and economic impacts of oil shale and oil sands development
2. Secure liquid fuel production by in situ thermal processing of oil shale
3. Environmental, legal, economic and policy issues

THE INSTITUTE FOR CLEAN AND SECURE ENERGY
AT THE UNIVERSITY OF UTAH

Introduction

- Richest oil shale deposits in Utah are located at depths that are best accessed in situ
- Number of technologically feasible in situ processing methods
- No proven costs
- Questions
  - What oil supply price (OSP) is required for in situ oil shale processing to be economically viable?
  - How does OSP vary as f(retorting, economic) input parameters?

THE INSTITUTE FOR CLEAN AND SECURE ENERGY
AT THE UNIVERSITY OF UTAH

Approach Summary

- Use design of experiments (DoE) to find oil supply price (OSP)
  - Pick input parameters and ranges
    - CFD: 6 parameters
    - Economic: 7 parameters
  - Latin hypercube sampling to sample design space
  - CFD simulations of in situ retort to find mass and energy balance
  - Discounted cash flow to find OSP
  - Regression analysis to quantify impact of input parameters on price

THE INSTITUTE FOR CLEAN AND SECURE ENERGY
AT THE UNIVERSITY OF UTAH

Location

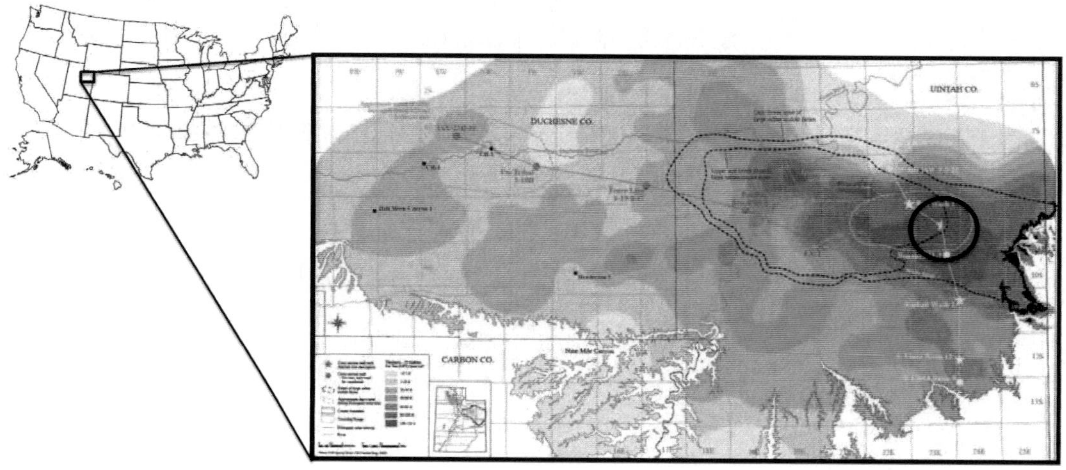

THE INSTITUTE FOR CLEAN AND SECURE ENERGY
AT THE UNIVERSITY OF UTAH

In Situ Process Diagram

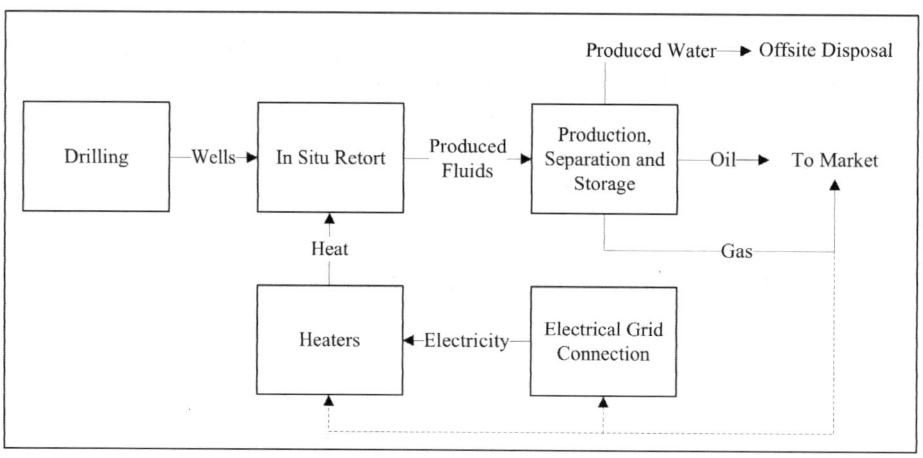

THE INSTITUTE FOR CLEAN AND SECURE ENERGY
AT THE UNIVERSITY OF UTAH

CFD Geometry

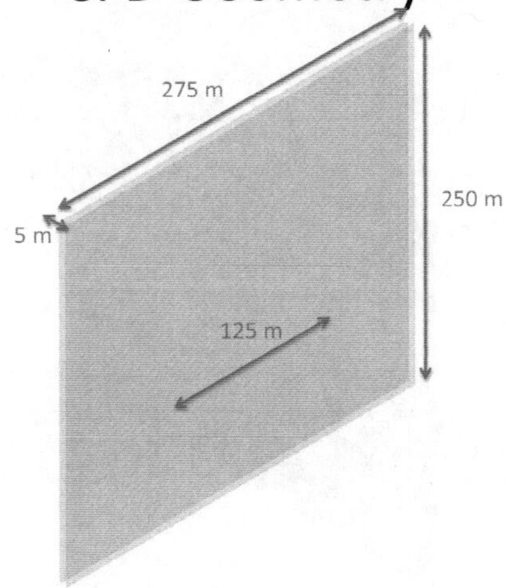

Periodic Boundary Conditions

THE INSTITUTE FOR CLEAN AND SECURE ENERGY
AT THE UNIVERSITY OF UTAH

CFD Geometry

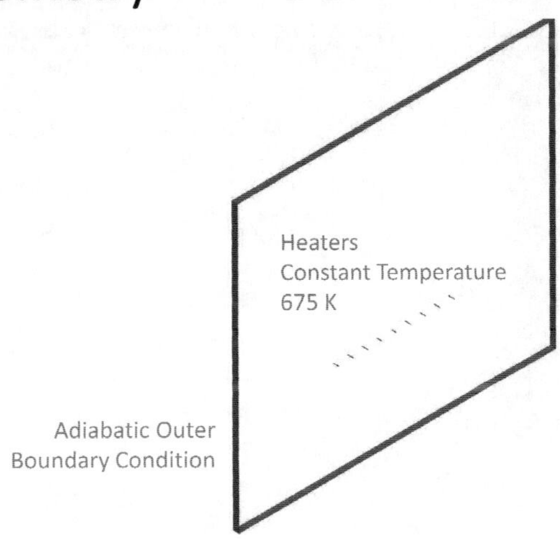

THE INSTITUTE FOR CLEAN AND SECURE ENERGY
AT THE UNIVERSITY OF UTAH

CFD Geometry

Detailed Stratigraphy from UGS cores

Mesh resolution: 5 cm

- Heat transfer mode
  - Conduction
- Kinetic Parameters
  - Determined from Uinta Basin samples
- Production
  - Gas/oil - variable
- Simulation time
  - 7 years of heating
- Computational time
  - Approx. 24 hrs per simulation
  - 360 cores

THE INSTITUTE FOR CLEAN AND SECURE ENERGY
AT THE UNIVERSITY OF UTAH

Parameters

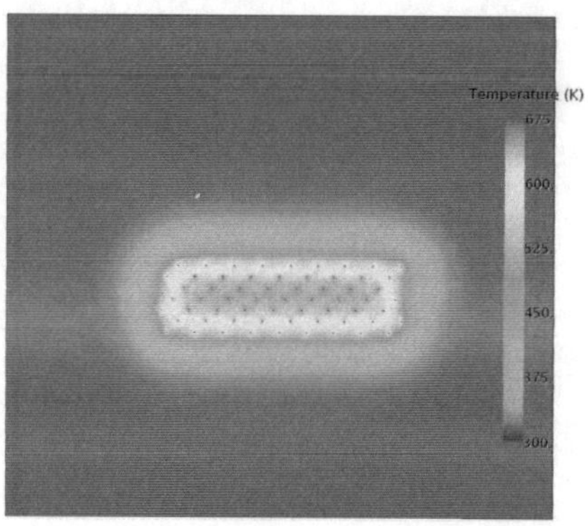

CFD PARAMETER	RANGE
Horizontal well spacing (ft)	17 – 82
Vertical row spacing (ft)	8 – 81
Vertical offset angle (degrees)	0.1 – 63
Vertical location of target region	Spanning Domain
Well radius (in)	[4.0, 4.5, 5.0, 5.5, 6.0]
Number of well rows	1 – 10
Number of wells*	5 – 232

- Scenarios
  - 242 CFD simulations

THE INSTITUTE FOR CLEAN AND SECURE ENERGY
AT THE UNIVERSITY OF UTAH

Annual Cash Flow Equation

- $CF=S-C_V-C_F-T-R-C_{TDC}-C_L-C_S-C_{RIP}-C_P-C_{DC}\pm C_{WC}$
- Sales = (Volume Product)(Product Price)
- Operating Costs
  - Variable: Utilities
  - Fixed: Labor, maintenance, insurance, property taxes
- State and Federal Taxes

$$TI=S-C_V-d-C_F-D-R-ST$$

- Total Depreciable Capital

$$C_{TDC}=F_C(C_{TBM}+C_{site}+C_{serv}+C_{alloc})$$

THE INSTITUTE FOR CLEAN AND SECURE ENERGY
AT THE UNIVERSITY OF UTAH

Discounted Cash Flow and Oil Supply Price

- Discount annual cash flows by interest rate i to find net present value (NPV):

$$f_n = 1/(1+i)^n$$

$$NPV = ©f_n CF_n$$

- Find OSP by
  - Specify $i=IRR$
  - Solve for OSP that gives NPV = 0

THE INSTITUTE FOR CLEAN AND SECURE ENERGY
AT THE UNIVERSITY OF UTAH

Parameters

- Scenarios
  - 242 CFD simulations
  - 512 economic scenarios

CFD PARAMETER	RANGE
Horizontal well spacing (ft)	17 – 82
Vertical row spacing (ft)	8 – 81
Vertical offset angle (degrees)	0.1 – 63
Vertical location of target region	Spanning Domain
Well radius (in)	[4.0, 4.5, 5.0, 5.5, 6.0]
Number of well rows	1 – 10
Number of wells*	5 – 232

ECONOMIC PARAMETER	RANGE
Drilling and completion time (days)	38 – 657
Well capital cost (million 2014 USD)	$1.9 – $6.6
Well lateral length (ft)	3,920 – 17,080
Recovery mass fraction (x_r)	0.4 – 0.8
Gas mass fraction (x_g)	0.0 – 0.5
Wellhead gas price ($/MCF)	$0 - $5.87
Internal rate of return	10% - 20%

THE INSTITUTE FOR CLEAN AND SECURE ENERGY
AT THE UNIVERSITY OF UTAH

Parameters

- Scenarios
  - 242 CFD simulations
  - 512 economic scenarios
  - Total: (242)(512) = 123,904 scenarios

CFD PARAMETER	RANGE
Horizontal well spacing (ft)	17 – 82
Vertical row spacing (ft)	8 – 81
Vertical offset angle (degrees)	0.1 – 63
Vertical location of target region	Spanning Domain
Well radius (in)	[4.0, 4.5, 5.0, 5.5, 6.0]
Number of well rows	1 – 10
Number of wells*	5 – 232

ECONOMIC PARAMETER	RANGE
Drilling and completion time (days)	38 – 657
Well capital cost (million 2014 USD)	$1.9 – $6.6
Well lateral length (ft)	3,920 – 17,080
Recovery mass fraction (x_r)	0.4 – 0.8
Gas mass fraction (x_g)	0.0 – 0.5
Wellhead gas price ($/MCF)	$0 - $5.87
Internal rate of return	10% - 20%

THE INSTITUTE FOR CLEAN AND SECURE ENERGY
AT THE UNIVERSITY OF UTAH

Capital and Supply Costs

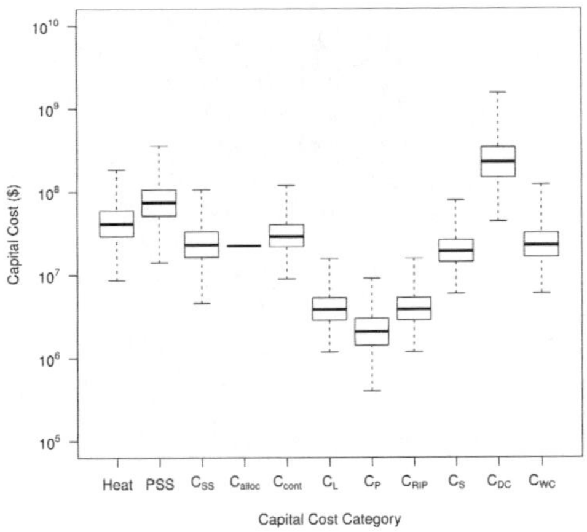

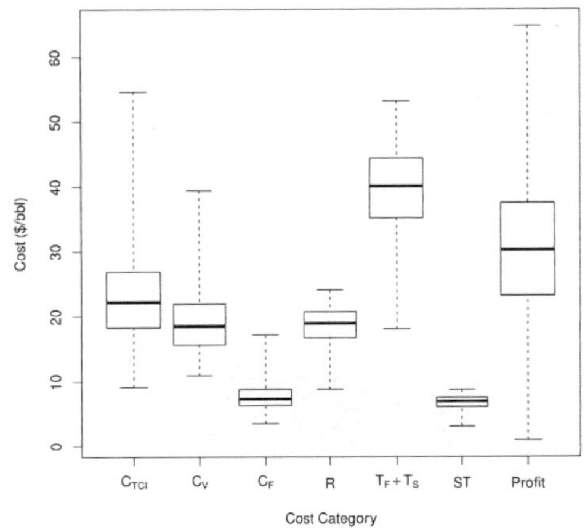

THE INSTITUTE FOR CLEAN AND SECURE ENERGY
AT THE UNIVERSITY OF UTAH

OSP Results

- Range of OSP results
  - Min: $70/bbl
  - Max: $85,000/bbl
- Small fraction of design points are potentially viable
  - Economic viability cutoff: $175/bbl
    - Assuming twice EIA forecasted average oil price (2014-2040)
  - 6.3% of scenarios ≤ $175/bbl

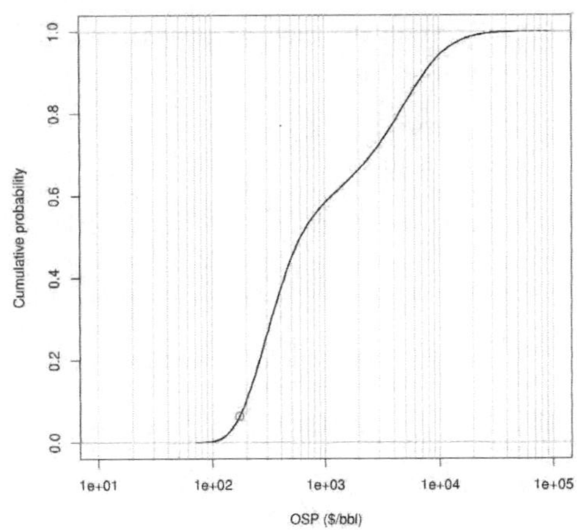

THE INSTITUTE FOR CLEAN AND SECURE ENERGY
AT THE UNIVERSITY OF UTAH

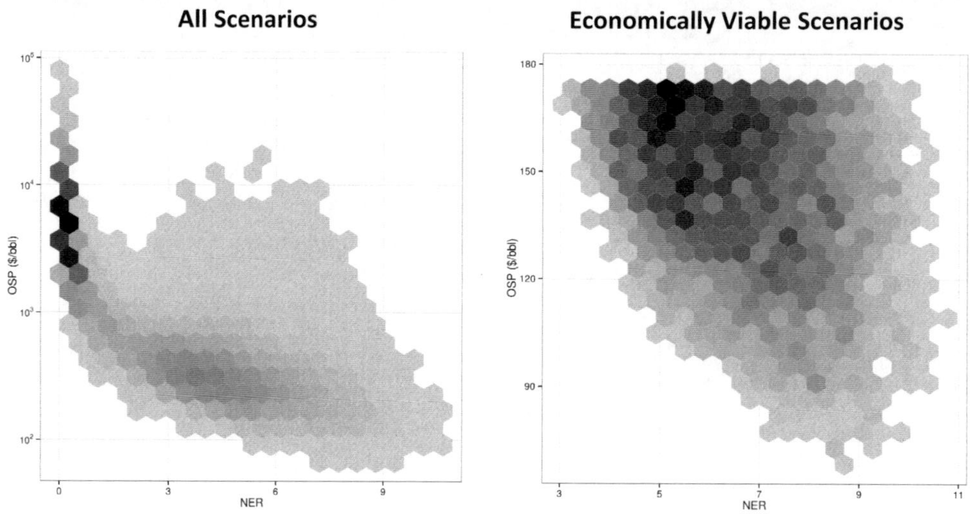

THE INSTITUTE FOR CLEAN AND SECURE ENERGY
AT THE UNIVERSITY OF UTAH

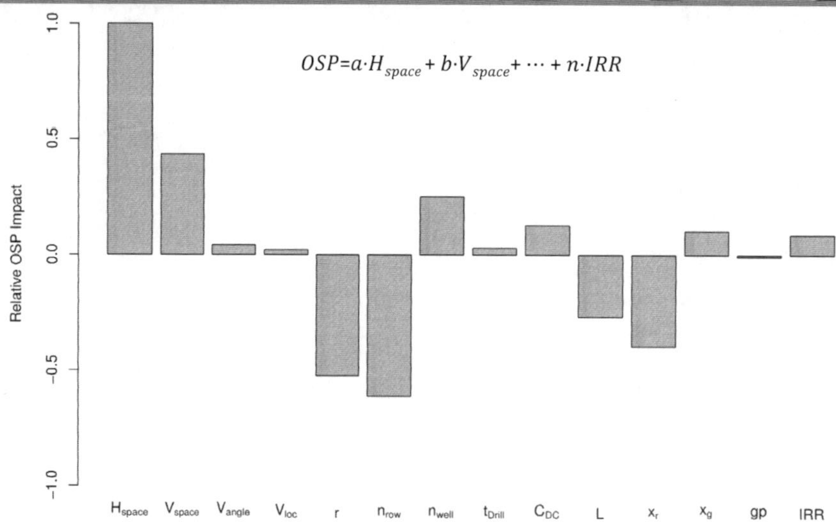

THE INSTITUTE FOR CLEAN AND SECURE ENERGY
AT THE UNIVERSITY OF UTAH

Optimal Input Parameters

Parameter	Optimal Range
H_{space} (ft)	40
V_{space} (ft)	20 – 40
V_{angle}(°)	---
V_{loc} (ft)	200
r (inch)	4.5
n_{row} (#)	≥ 3
n_{well} (#)	50 – 60

Parameter	Optimal Range
t_{drill} (days)	< 400
C_{dc} (10^6 \$)	↓
L (10^3 ft)	↑
x_r (---)	↑
x_g (---)	↓
gp (\$/MCF)	---
IRR	< 25%

THE INSTITUTE FOR CLEAN AND SECURE ENERGY
AT THE UNIVERSITY OF UTAH

Conclusions

- OSP results highly dependent on NER
  - NER > 3 required to get OSP ≤ \$175/bbl
  - Highest NER ≠ Lowest OSP
    - Time to first production and quantity of oil produced also very important
- Wells are most significant cost
  - Capital costs > operating costs
- Large uncertainty range for many parameters could be reduced with more laboratory and pilot-scale testing

QUALIFICATION OF HIGH-POWER-DENSITY MINERAL-INSULATED ELECTRIC HEATER CABLES

A. Burnham, C. Sandberg, K. Thomas, V. Reiling, J. Faull, L. Bourgeois, J. Huffman, L. Switzer

35th Oil Shale Symposium

Salt Lake City, UT, October 5-6, 2015

AMSO's CCR Process requires a high-heat-flux heater (10-15 kW/m)

2

- Initial non-MC AAA electric heaters failed in 2012-2013
- AMSO embarked on a heater qualification program for both electric cables and hot gases
- MC AAA developed a new MI cable with internal Cu-A180 conductor transitions
- An extensive set of tests were conducted to develop a lifetime model for the MC AAA cables

Failed splice in 2013 deployment

MC AAA internal splice

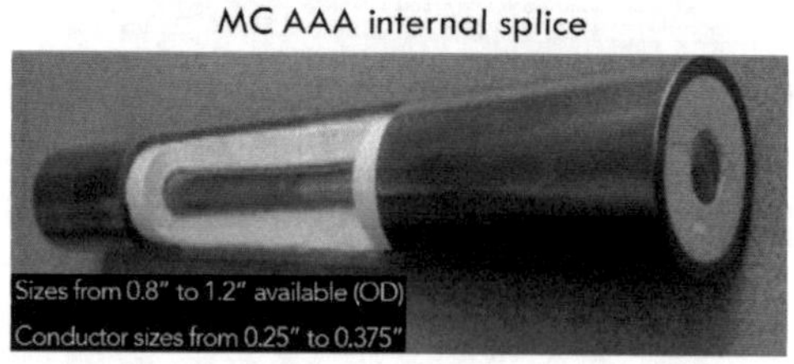

MC AAA commissioned a new manufacturing facility in the U.K. in 2013

3

- Conductor, MgO half-tubes, and sheath ribbon are fed into a continuous welding and rolling line
- Single cables up to 2000 m long can be made
- Diameter reduction of >50% during rolling and annealing

AMSO qualification test program

4

- High-temperature high-voltage tests
  - Measure MgO leakage current and break-down voltages for thermal exposures up to a week
- Multi-week high-temperature high-current endurance tests
  - Measure reduction in breakdown voltage
  - Test robustness of VFD power control system
- Transient and steady-state thermal conductivity tests
  - Gives low-temperature MgO thermal conductivity
- High-temperature conductor melt test
  - Gives high-temperature MgO thermal conductivity

High-voltage breakdown test details

5

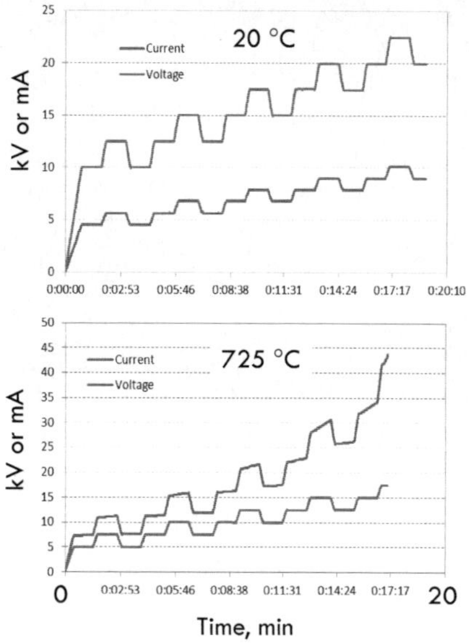

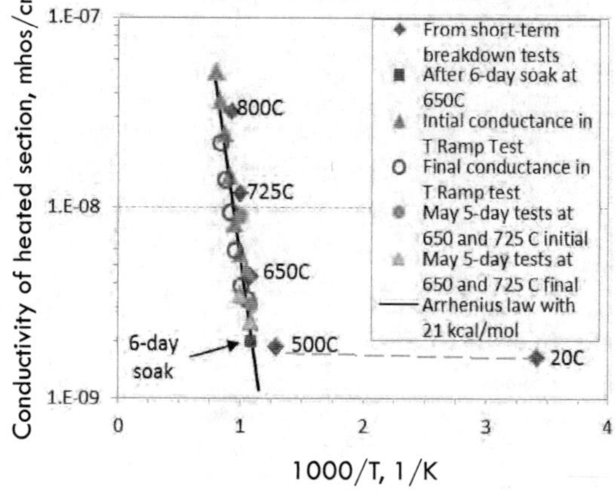

The conductivity mechanism changes with temperature (Freund, et al., 1993)

6

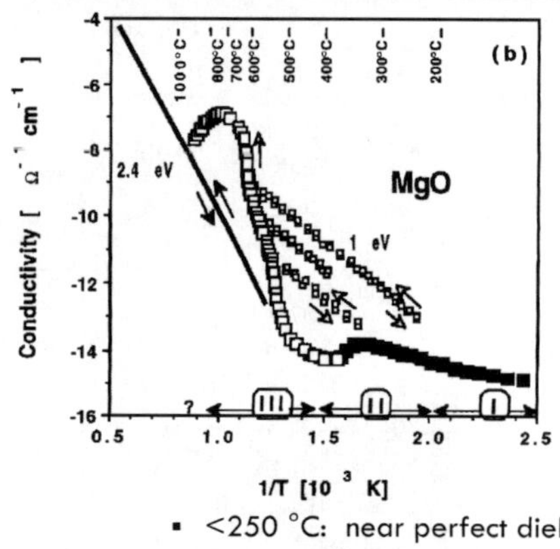

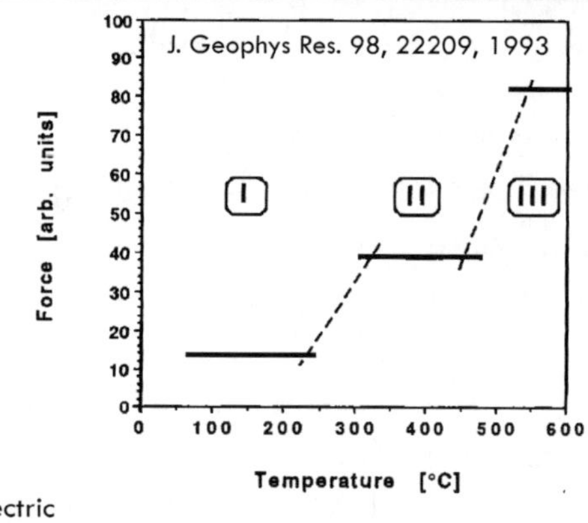

- <250 °C: near perfect dielectric
- 250-500 °C: delocalization of positive holes at point defects
- >500 °C: generation of mobile positive hole charge carriers

Changes in conduction mechanism could change the MgO aging mechanism at ~500 °C

Summary of MC AAA breakdown tests

7

Breakdown voltage decreases with MgO thermal exposure while under bias voltage of a few kV

Test Temperature (°C)	Failure (kV)
Ambient	18.1
500	17.5
650	17.5 (4 hrs)
725	17.4
800	14.9 7.5 (4 hrs)
725 for 6 days	7.8
675-975 over 1 wk	5.9

No breakdowns occurred at internal splices

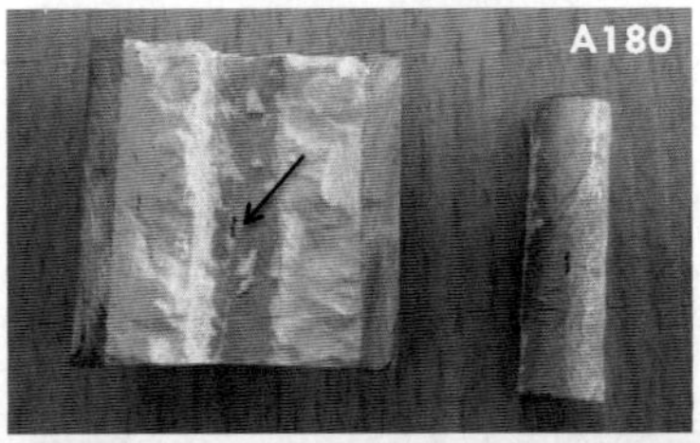

Multi-week tests for qualification at high temperature and current

8

- Assume heater lifetime depends on MgO temperature in Arrhenius fashion
  - Average MgO temperature is between sheath and conductor temperature
- Conductor temperature depends on sheath temperature, heat flux, and MgO thermal conductivity
- MgO thermal conductivity depends on porosity and grain size

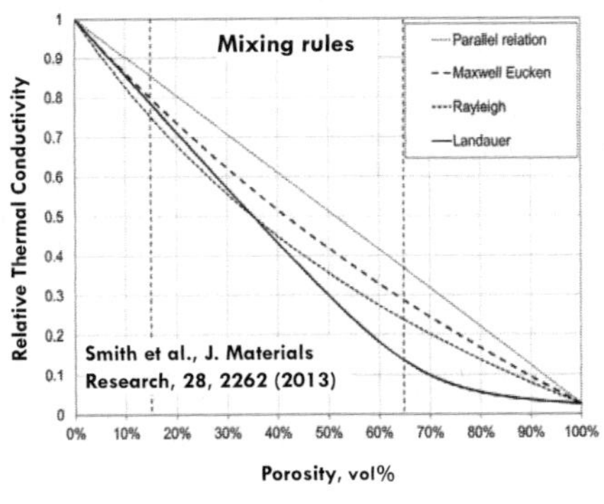

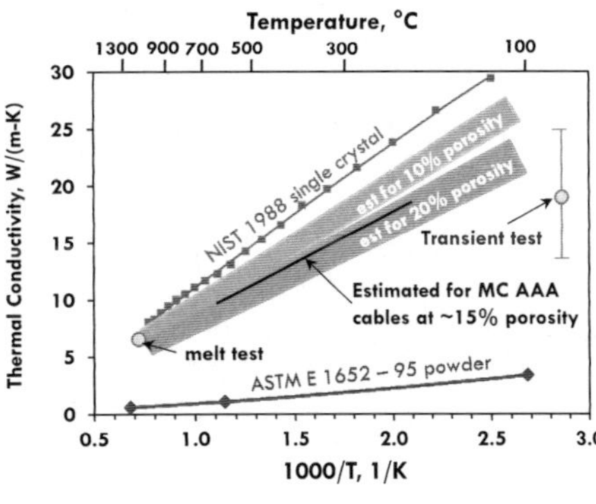

Conductor melt test to determine k_{MgO}

9

» Procedure:

- Self-heat insulated cables until melting is detected (change in resistance)
- Cool to solidify and repeat – hold for as long as possible

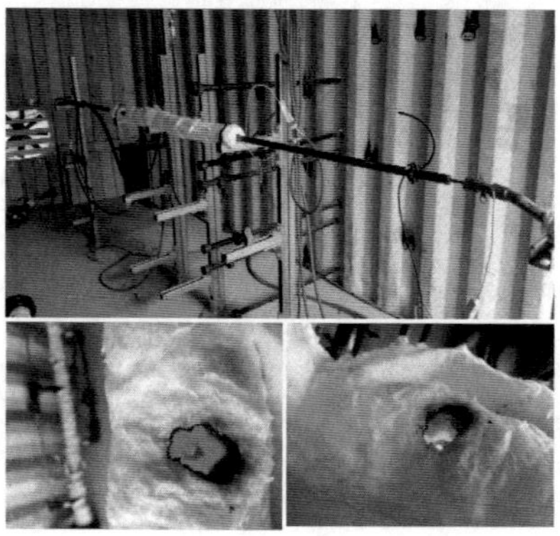

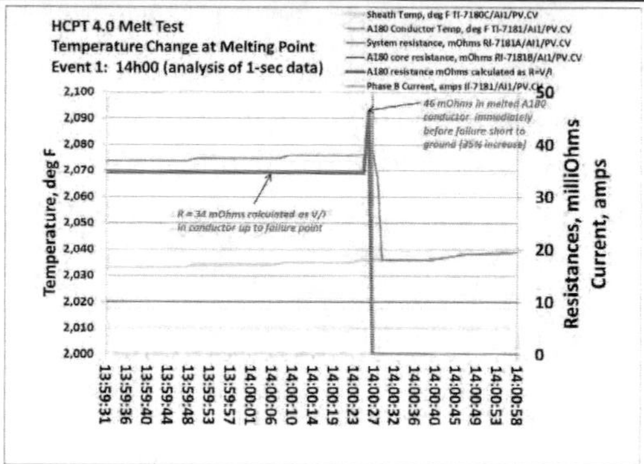

» Results:

- Cable showed a small resistance change upon melting, but failed later by shorting to sheath
- ΔT was only about 30 °C ($T_{surf} \approx 1115$ °C)
- k_{MgO} ~ 6-7 W/m-K

Multi-week high-current test results

10

» Operated at sheath temperatures of ~600 °C

- Considered a worst case in the pilot for the conductor with coke buildup

» Initial tests at ~1.5 kW/m for 28 days had no failures

» Increased power to limit of test equipment: ~3 kW/m

- Enables a single 3-phase heater to achieve AMSO Pilot requirements
- No failures after ~42 days

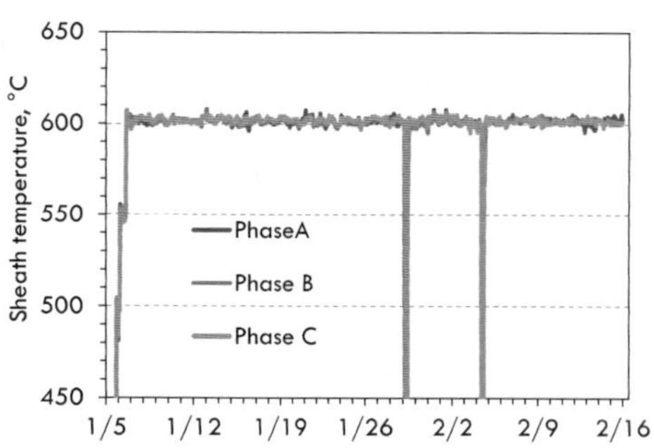

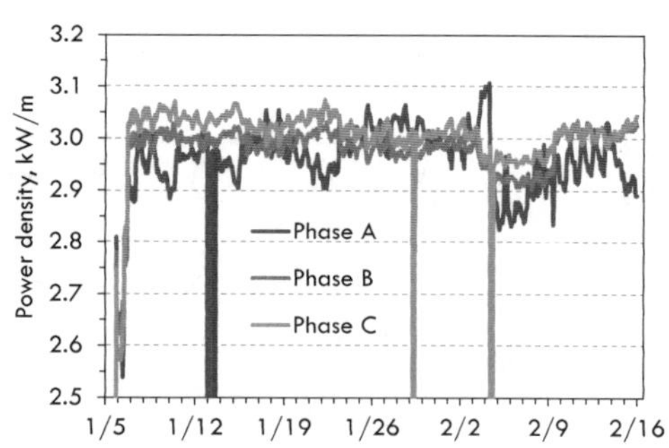

Breakdown voltage results for high-current test cables

11

Power Density (W/ft)	Current Frequency During Aging (Hz)	Sheath Temp (°C)	Est. Core Temp (°C)	Time Held (days)	Phenix Breakdown Test T (°C)	Phenix Breakdown @ 60 Hz (kV)
500	60	625	640	~28	720	**12**
500	60	560	575	~28	730	**12**
500	6	625	640	~28	720	**13**
500	6	560	575	~28	730	**11**
950	60	600	630	~42	725	**8 avg**

- Earlier tests gave ~17 kV at 725 °C (~same as for no thermal aging)
- More severe conditions yield lower breakdown voltages
- No apparent effect of frequency (capacitive reactance current) on aging itself

Arrhenius analysis of all breakdown data

12

- Use average MgO temperature
  - The MgO is aging, not the conductor
- Apparent E_a of ~210 kJ/mol
  - Assumed first-order decay to zero (highly conservative)
  - A more rigorous kinetic analysis is warranted
- Estimated lifetimes are favorable
  - Cable temperatures up to ~550 °C are acceptable for our Pilot test
  - Estimated lifetimes are ~10 years for 5 kV heaters operated at ~500 °C
  - Lifetimes for 300–500 °C might be higher or lower than predicted due the change in conductivity mechanism
- Caveat: 4σ point defects were not tested here due to short lengths

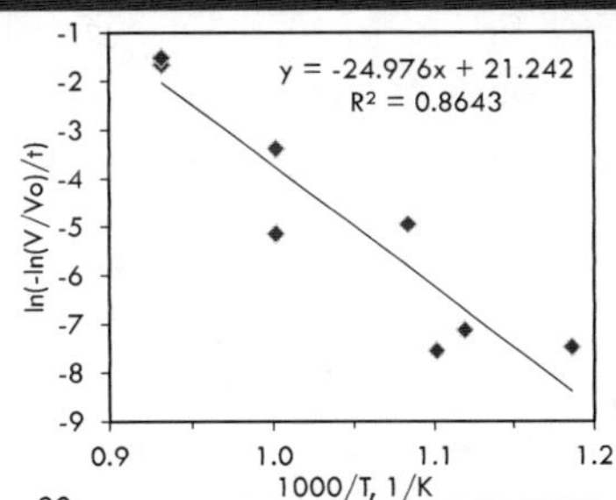

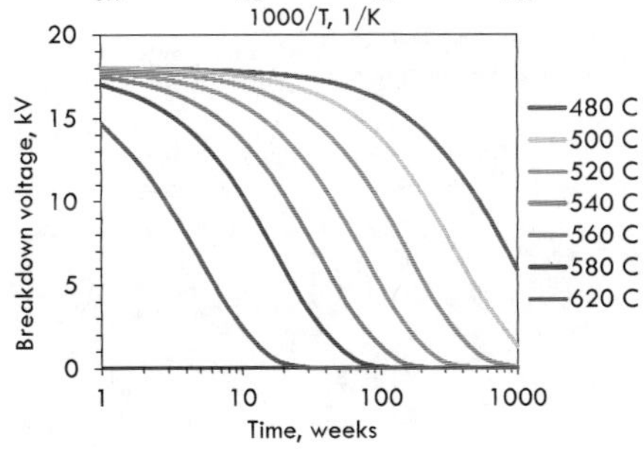

Initial thermal conditions for AMSO Pilot

13

- Initial operating temperatures will be closer to 400 °C for boiling oil in the CCR process
- From the estimated thermal conductivities, ΔT between sheath and conductor will be <50 °C for our power densities
- Estimated lifetime is years even with significant coke buildup

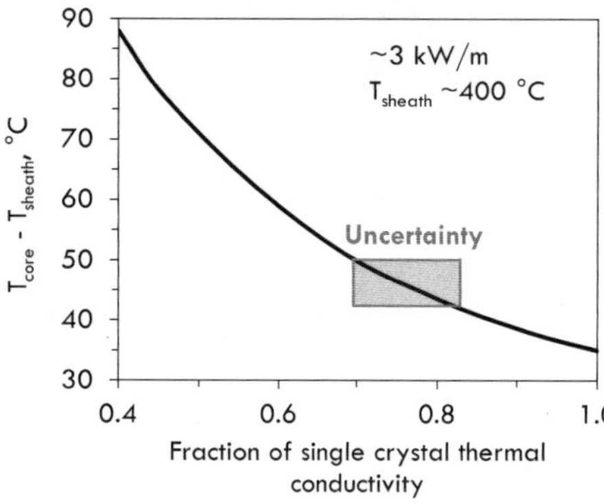

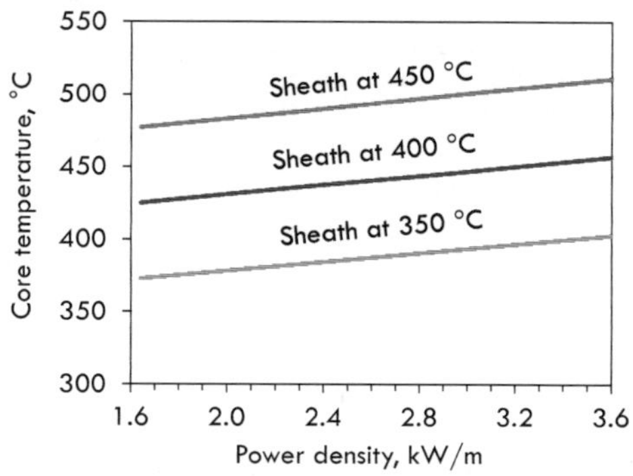

Cables do not degrade significantly during spooling and unspooling

14

- MC AAA designed a device to simulate the bending and unbending that occurs while loading and unloading on a spool
- Breakdown voltage at 500 °C after 10 cycles: 16.0 kV (σ =1.2)

Conclusions

15

- MC AAA MI cables with internal splices are robust to failure at AMSO Pilot operating conditions (the goal of this study)
- Attainment of low porosity (~15%) during manufacturing enables high thermal conductivities for the compacted MgO
- Breakdown voltages of cables from high-voltage and high-current tests at high temperatures can be analyzed in terms of an Arrhenius law to estimate cable lifetimes
- Estimated lifetimes are ~10 years for 5 kV heaters operated at ~500 °C or less
  - Additional long-term testing at lower temperatures is needed to validate these predictions
  - Additional testing at lower frequencies is also needed to adequately test any possible advantages

SURFACE QUALIFICATION SYSTEM FOR DOWNHOLE HEATERS

Authored by:
J. Faull, R LaFollette, A. Papineau,
V. Reiling, L. Switzer, V. Saubestre

35th Oil Shale Symposium
Salt Lake City, Utah, October 5, 2015

AMSO's Unique Heating Process – Conduction, Convection, Reflux (CCR™)

2

Conductive and convective heat transfer provided by refluxing oil (CCR) and *potentially* facilitated by thermo-mechanical fracture permeability

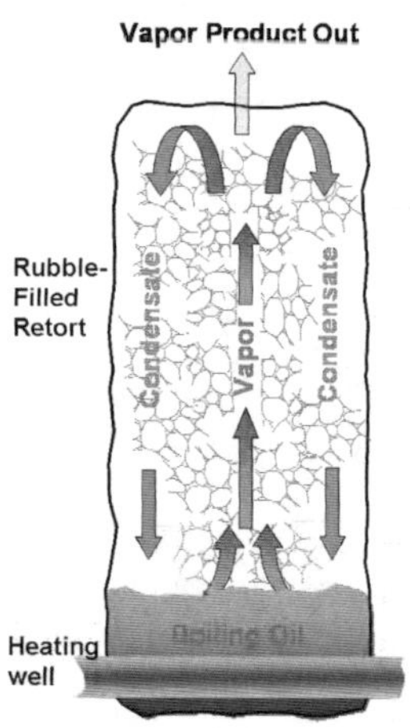

Confined on all but one side

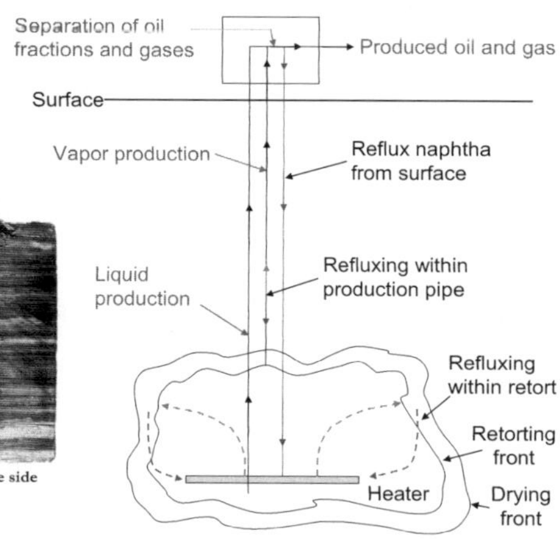

AMSO's Current Pilot Status

3

- Attempted electric heaters in 2012-2013
- AMSO has undertaken a comprehensive heater qualification program (ProQual™)
  - New, more robust electric cables (MC AAA)
  - Hot fluid circulation
- Developed several experiments to qualify different heaters
  1. Insulation test stand
  2. Propane flow loop
  3. Coking apparatus
  4. Heater cable power testing unit
  5. Water-HC thermodynamics
  6. ***Heater Test Stand***

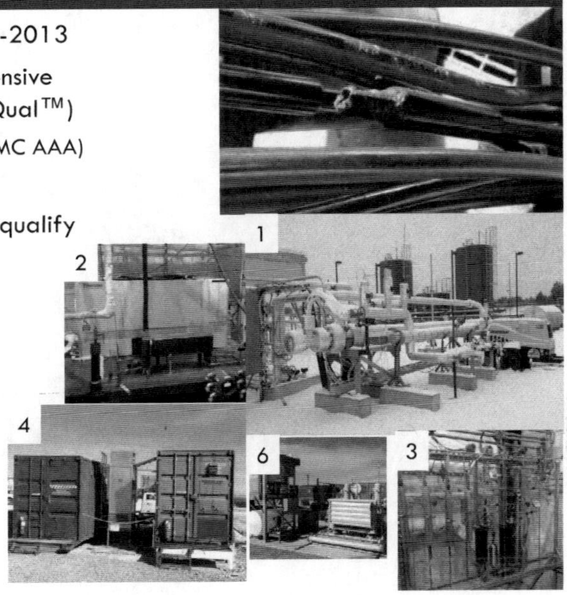

Goals and Features of Heater Test Stand

4

- Goals:
  - Build a "tool" to qualify heater designs in a surrogate wellbore environment
  - Understand the VLE of hydrocarbon mixtures with water
  - Develop scale-up relationships for the Pilot heater
- Features:
  - 40' long by 8" dia. pipe (surrogate wellbore)
  - Adjustable angle (~0-90°)
  - Design Limits: 750 °F; 750 psig
  - Highly instrumented (~100 TC's in main barrel)
  - Liquid level, P, and T control
  - Heat Flux Wall (HFW) to control heat losses
  - Water injection

Process Flow Diagram

5

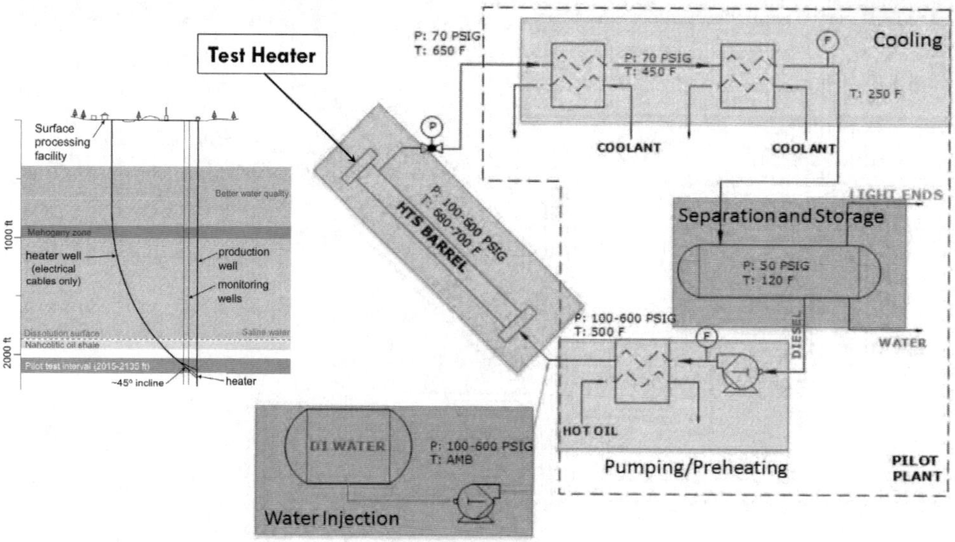

Temperature Control of Emulated Well-Bore

6

- 6 temperature zones, measuring:
  - Cable/heater
  - Bulk and wall
  - Top vs bottom
- Heat loss is controlled through the Heat Flux Wall (HFW)
  - 2 layers of refractory with MI heat tracing in between
  - 10 thermocouples calculate and control the heat lost in each top/bottom section

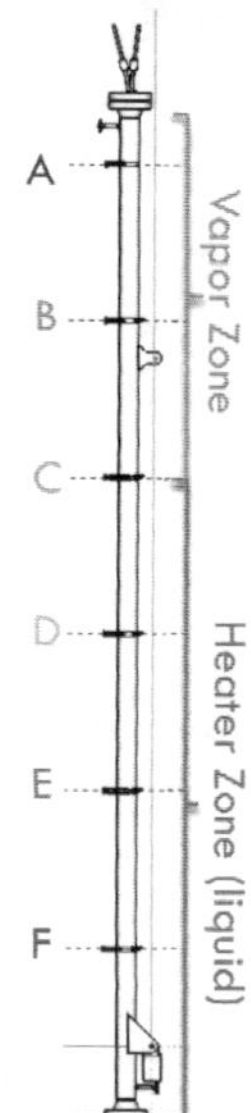

Construction & Deployment Early 2015

7

90° (VERTICAL) orientation – **6+ weeks**

45° (ANGLED) orientation – **6+ weeks**

» Operations
- Power delivered: 22~60 kW
- 3 electric MI cables; external wye
- ~100 stable operating conditions
- No safety incidents

VLE: Consistent with Thermodynamic Models

8

» P-T values in agreement with Thermodynamic Models & Bench Data

» VLE independent of vessel orientation, heat losses

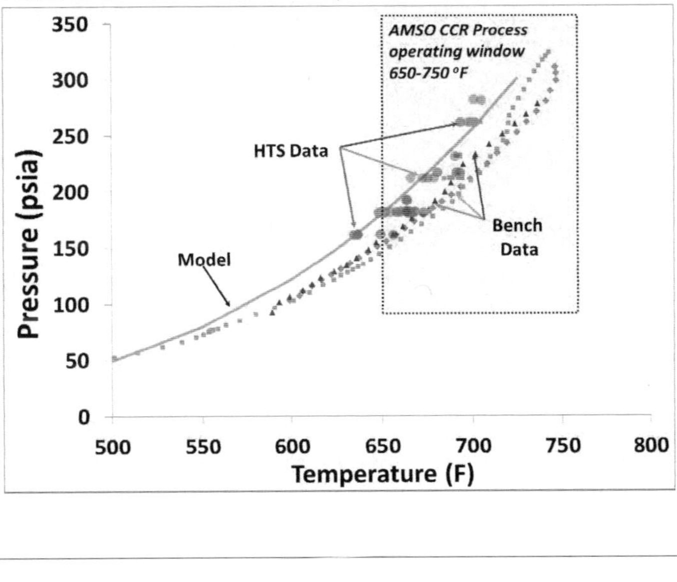

Operating Lines : Coupled Heat & Mass Balance

9

- Energy balances closed to within ±5%
- Operating lines change with H_2O injection & simulated heat uptake to shale formation, consistent with energy balance, independent of vessel orientation

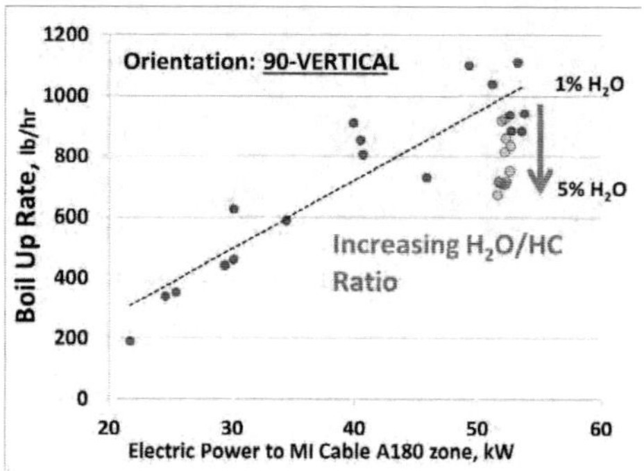

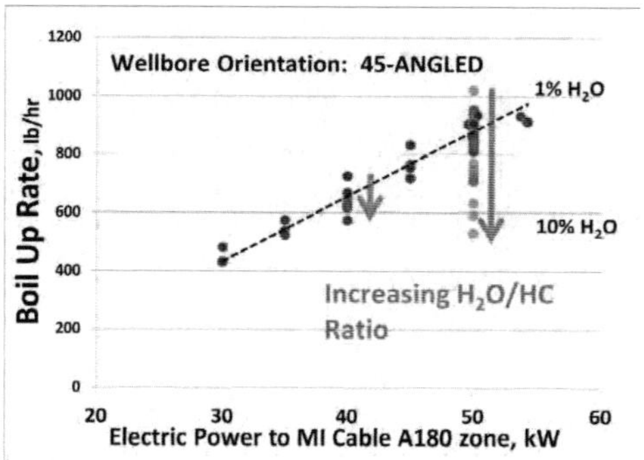

Overall Heat Transfer Coefficient (U)

10

- Data suggest high HTC regardless of <u>wellbore orientation</u>

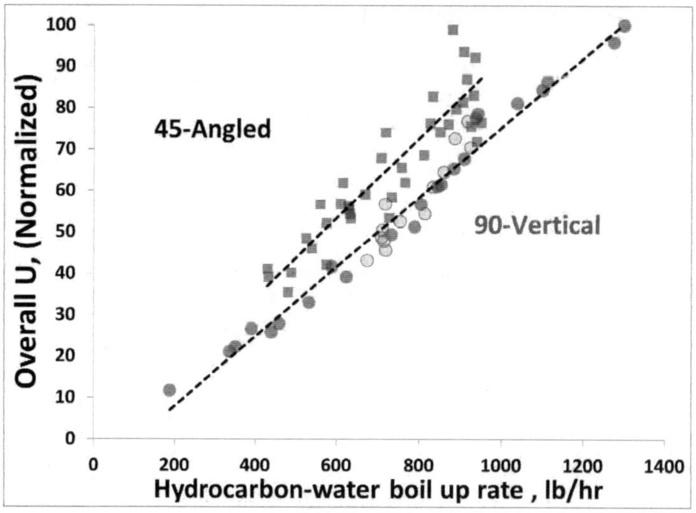

Local Heat Transfer Coefficients

11

- Local ***h*** rise with increasing nucleation/convective boiling
- Local ***h***'s proportional to q (Heat Flux)

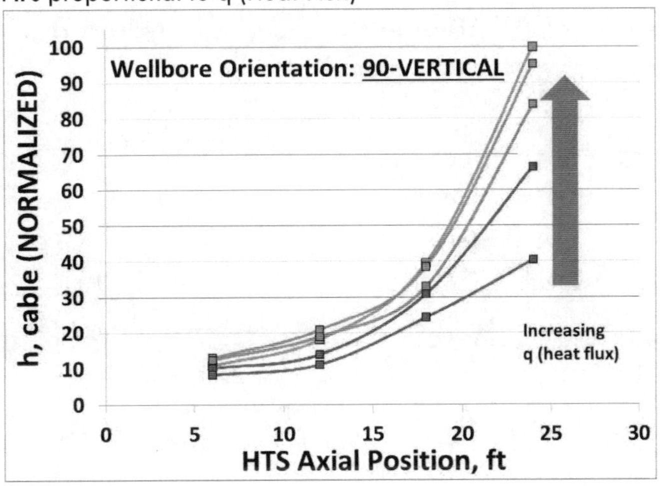

Evolution of Diesel

12

- Some change in color of diesel samples under "normal" operation
- Heat transfer coefficients did not change during "fouling" events mid-way through 2nd campaign

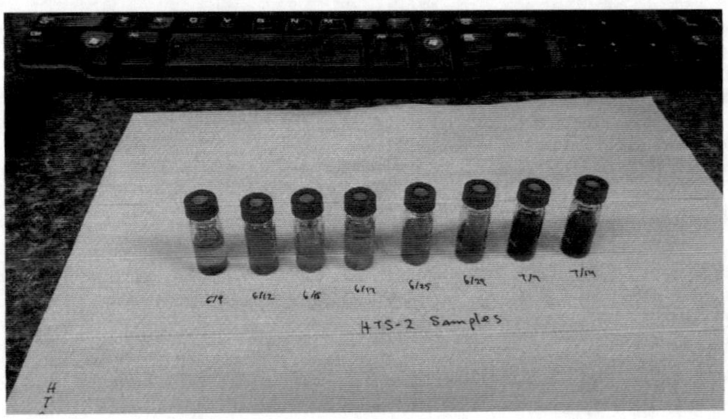

Evolution of Hydrocarbons in HTS

13

- Observed mild cracking of longer chain hydrocarbons into shorter chain hydrocarbons (expected)
- No change in Heat Transfer despite changes in hydrocarbon compositions during >3 months operation

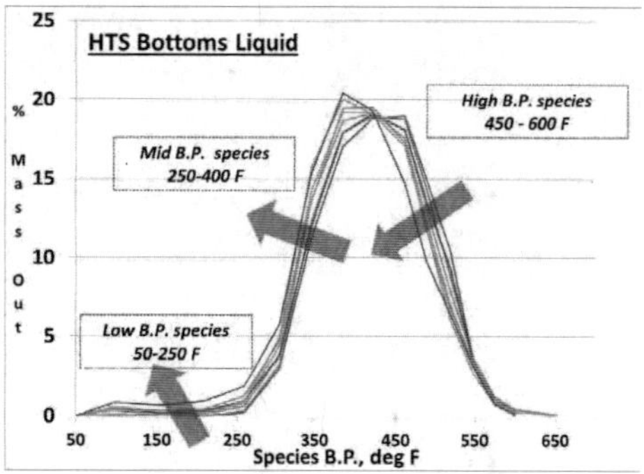

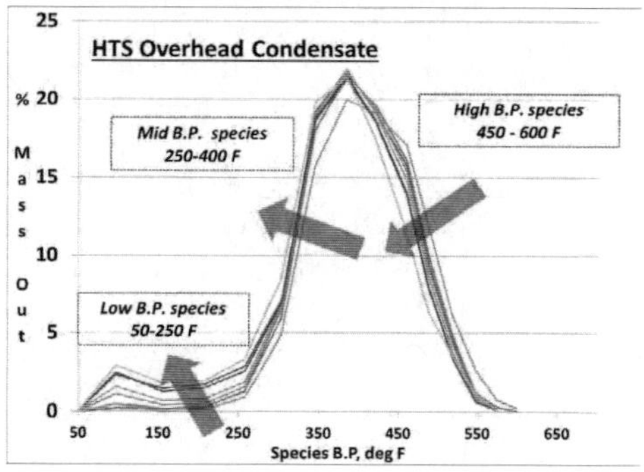

Scale-Up Implications

14

- With realistic design assumptions, scale up appears feasible

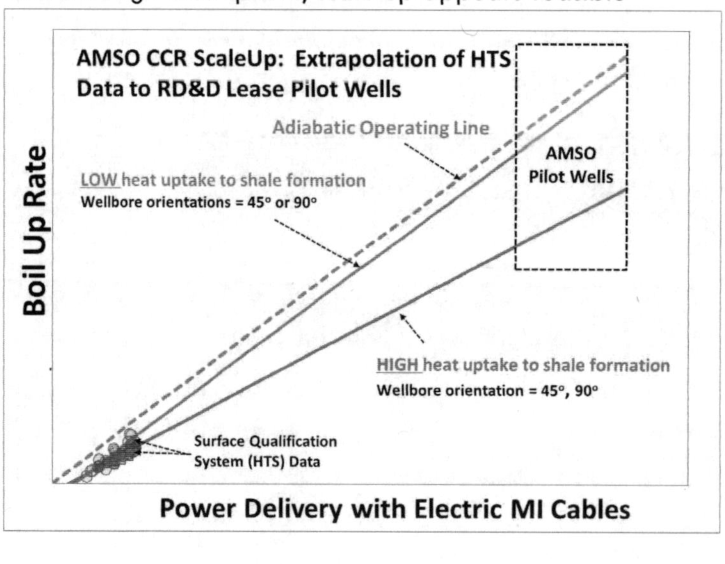

HTS Future Plans

15

- Expand the electric heater operating envelope
  - Varying hydrocarbon composition to simulate downhole fluids
  - Higher power, alternative cable geometries and internals
- Alternate heater designs
  - Hot Fluid Circulation (**HFC**)
    - → Hot pipe with circulating fluid to indirectly heat the oil pool
  - Hot Fluid Direct Injection (**HFDI**)
    - → Inject high temperature fluids directly into HTS barrel (well bore)

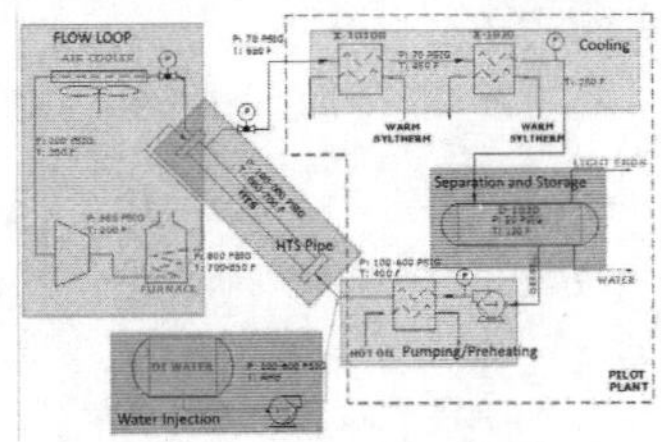

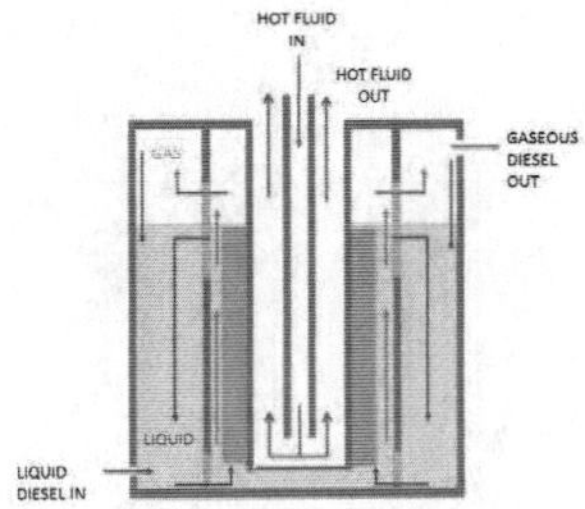

Summary & Conclusions

16

- Operated the HTS with electric cables > 2200 hours
  - Demonstrated heating in a surrogate well bore consistent with design basis
    - → Data are repeatable and valid
    - → Operated at conditions necessary (650-750 °F) to pyrolyze kerogen
    - → Simulated both low and high heat uptake to shale oil formation
    - → Simulated both angled (45°) and vertical (90°) wellbore configurations
  - VLE behavior in agreement with best-in-class thermodynamic models and Lab data
  - No significant fouling on electric cables issues; mild thermal cracking of long chain hydrocarbons
- Collectively, these HTS results provide AMSO with confidence in its scale-up and forward build-out strategy

Questions ?

AMSO Forward Plan

18

- AMSO has made significant progress in the technical evaluation of heating systems and interpretive tools for the Pilot
  - Multiple experimental systems have been commissioned and operated by ***Internal*** resources
- Experimental and design work are yielding excellent results which will help support a ***Pilot heater*** decision

- Goal: **AMSO will choose a heater system that gives the best chance of success in heating the Pilot retort**

Post Mortem
(+2200 hours in operation)

19

- Insulation Resistance and cable resistance testing show the cables are fully in-tact with no obvious signs of degradation after >2200 hours in boiling diesel service
- There is a thin film coke covering the cables. Over 1,000 film thickness readings were taken to understand the film build-up
  - Rate of film buildup ~1 micron /week vapor phase, 3~6 micron per week liquid phase
  - Samples of residue taken from HTS cables and shroud to help assess fouling of cables
- *No apparent change in heat transfer coefficient during 12+ weeks operation*

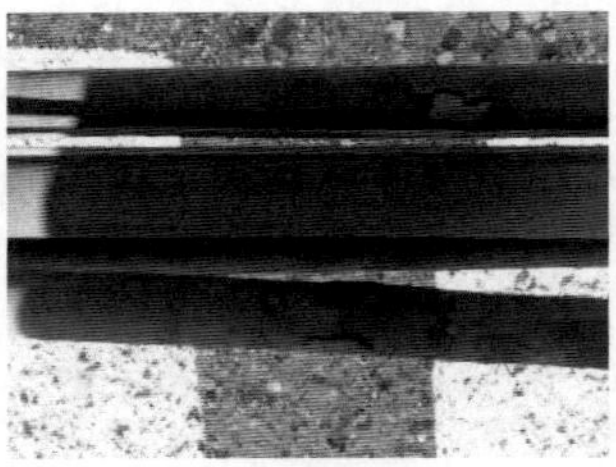
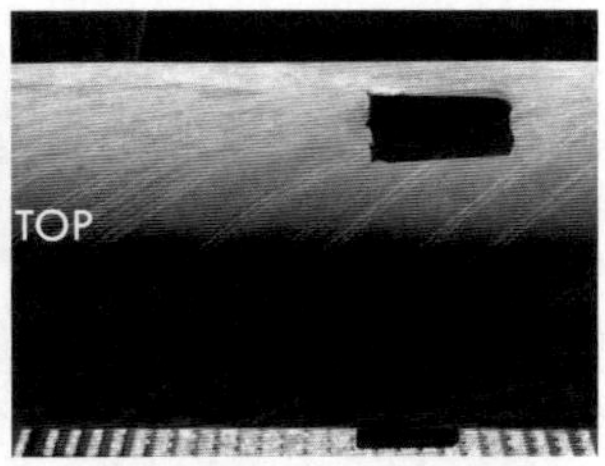

Geothermic Fuel Cells for in-situ oil shale processing

Colorado Fuel Cell Center

Neal P. Sullivan[1], Buddy Haun[1], Mark Daubenspeck[1], Gladys Anyenya[1], Robert J. Braun[1],

[1]Colorado Fuel Cell Center, Mechanical Engineering Department

Colorado School of Mines, Golden, Colorado, USA

Joseph V. Bonadies[2], Bernhard Fischer[2], Adam Wright[2], Gerald Jones[2], Robert Li[2]

[2]Delphi Powertrain Systems, Fenton, Michigan, USA

Mark Wall[3], Alan Forbes[3], Marshall Savage[3]

[3]IEP Technology, Parker, Colorado, USA

Presented to:

35th Oil Shale Symposium, Salt Lake City, Utah, USA

Monday, October 5, 2015

Colorado Fuel Cell Center

DELPHI

1

Geothermic Fuel Cells (GFCs) present an ambitious opportunity for unconventional oil and gas recovery

Colorado Fuel Cell Center

- **Utilize heat from solid-oxide fuel cells to convert oil shale to oil**
- **Application to other unconventional energy resources**
  - **Tar sands: down-hole steam generation**

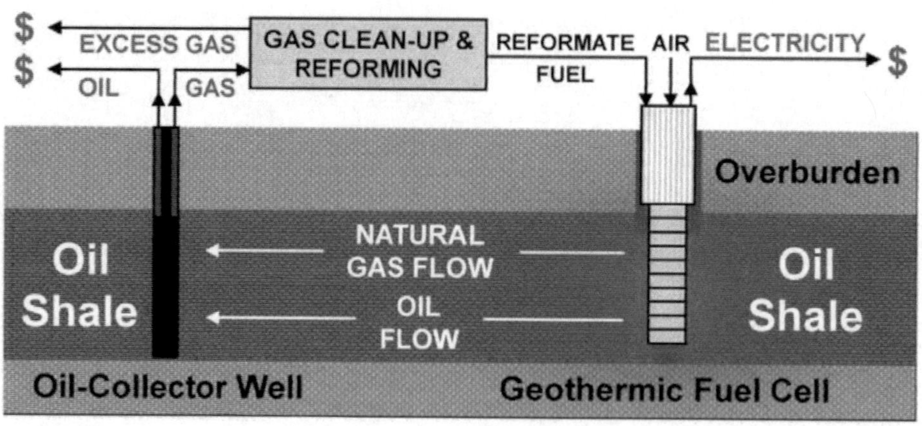

2

When fully implemented, Geothermic Fuel Cells will liberate oil trapped within oil shale

Colorado Fuel Cell Center

- Fuel cells for oil-shale conversion
  - Fuel cell operates at high temperature (750 °C)
  - Fuel cell liberates heat (~ 500 kW)
  - Heat converts kerogen into oil
- Shale buried below ~ 300-m overburden
  - Supply pipeline acts as heat exchanger
- Module specifications
  - 6-ft tall; three fuel cell stacks per module
  - 160 modules in parallel
  - 1000-ft total GFC length
  - 350 kW electrical / 500 kW thermal power
- 7 MW of gross electricity generation per acre

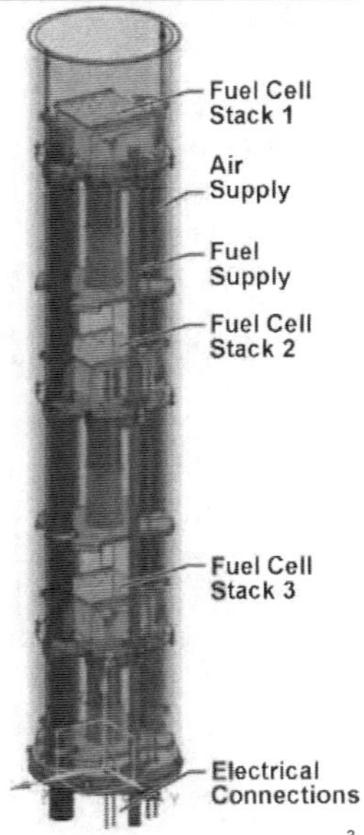

3

Project team mates industrial leaders in solid-oxide fuel cell technology with "balance-of-plant" suppliers

Colorado Fuel Cell Center

- IEP Technology: Project sponsor
- Delphi: Design and construction of GFCs
- Precision Combustion, Inc.
  - 35 kW_{th} auto-thermal natural-gas reformer
- Saridea, Inc.
  - Gas pre-heater / heat exchanger design
- Colorado School of Mines: Performance testing
  - Laboratory testing (First year)
  - In-ground testing (Second year)
- Results to date
  - Laboratory testing of world's first two GFCs
    - Fueled by City of Golden natural gas
    - Thermally self-sustained operation
  - First in-ground GFC installed and operated
    - 1 month continuous in-ground operation

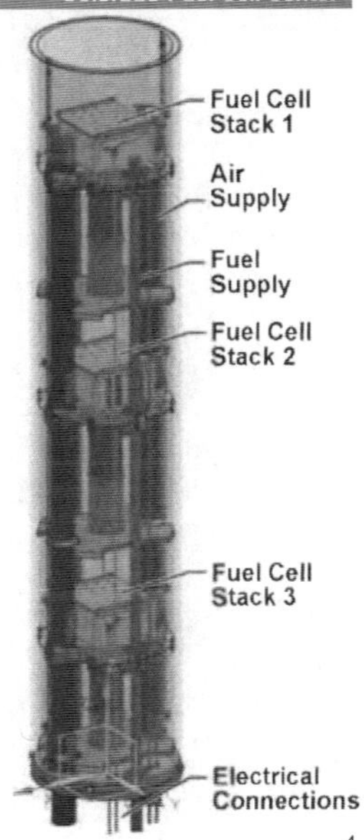

4

A thirty-foot-tall, three-module Geothermic Fuel Cell was operated in-ground at CSM for one month

Colorado Fuel Cell Center

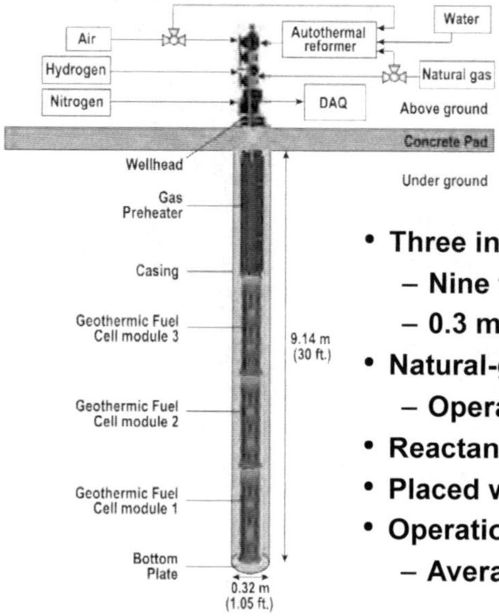

- Three integrated GFC modules
  - Nine fuel-cell stacks
  - 0.3 m diameter; 6 m total length
- Natural-gas fuel processor
  - Operation on municipal natural gas
- Reactant-gas preheater
- Placed within in 9-m-deep bore hole
- Operation during December 2014
  - Average outdoor air temperature: 1 °C

5

In-Ground, Outdoor Geothermic Fuel Cell Test Site at Colorado School of Mines

Colorado Fuel Cell Center

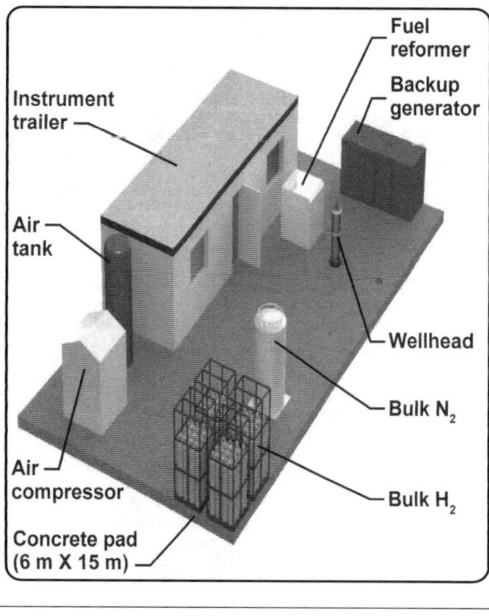

6

Continuous operation over 600 hours; initial data set collected, with lessons learned

Colorado Fuel Cell Center

Total chemical power into system (kW)	60.8	(Power in NG to reformer) + (Power in NG into preheat combustor) + (Power in supplemental H_2 to GFC)
Total heat into system (kW)	12.5	(Preheating of reactants at ATR) + (GFC internal electric heaters)
Stack electric power (kW)	4.4	(Electric current) * (GFC terminal voltage)
System electrical efficiency (%)	7.2	(Stack electric power) / (Total chemical power into system)
Fuel utilization (%)	25.8	(Moles of H_2 consumed to produce electric current) / (Moles of fuel [H_2 + CO + 3*CH_4] fed to GFC)
GFC heat loss to geology (kW)	18.8	[Total chemical and sensible energy in GFC inlet gases - Total energy in GFC exhaust gases] - GFC electric power
Preheater heat loss to geology (kW)	10.4	[Total (chemical and sensible) energy in the preheater inlet gas streams] - [Total energy in preheater outlet gas streams]
Total heat loss to geology (kW)	29.1	(Heat loss from GFC) + (Heat loss from pre-heat unit)
System heating efficiency (%)	47.8	(Total heat loss to surroundings) / (Total chemical power into system)
Combined heat and power efficiency (%)	55.0	(System electrical efficiency) + (System heating efficiency)
Heat loss per unit length of GFC (kW/m)	3.2	(Heat loss to geology) / (6 m GFC length + 3 m of Pre-Heater length)
Reformer efficiency (%)	75.1	(Chemical power in reformate) / (Chemical power in NG fuel)
Heat exchanger efficiency (%)	65.1	(Heat gained by GFC inlet gases) / [(Heat from NG combustion) + (Heat loss from GFC exhaust streams)]

7

Significant heat transfer to the geology demonstrated; far-higher heat transfer possible

Colorado Fuel Cell Center

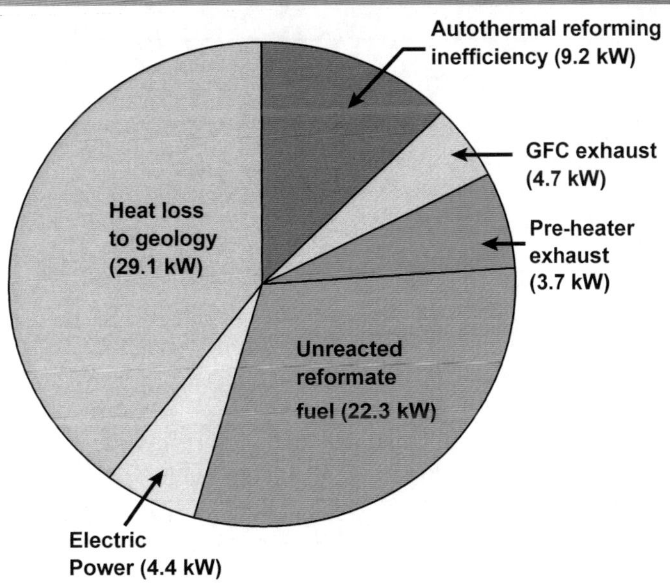

8

Geology heating demonstrated; much-higher temperatures possible

Colorado Fuel Cell Center

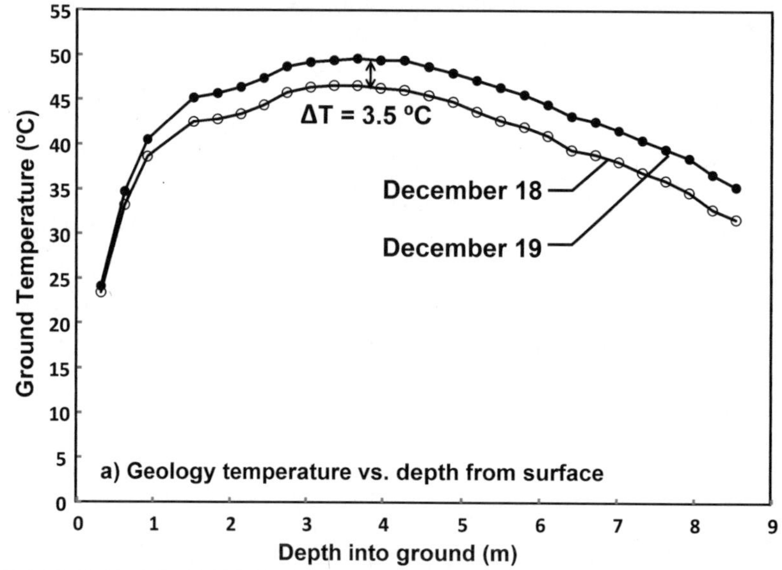

9

Geology-temperature decrease follows logarithmic decay; helps to define spacing of GFCs in the field

Colorado Fuel Cell Center

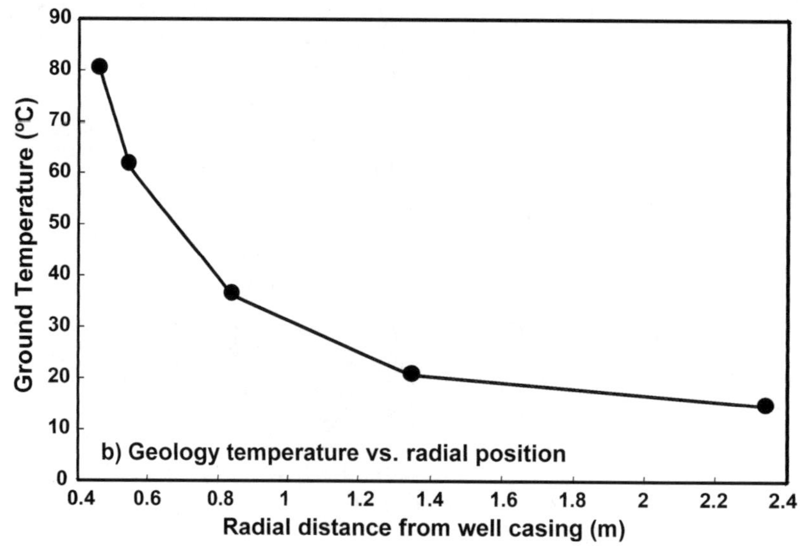

10

Future plans for Geothermic Fuel Cells involve oil-shale field demonstration and extended applications

Colorado Fuel Cell Center

- Field-test design and planning are underway
  - Three (3x) 4-m-long Geothermic Fuel Cells installed in oil-shale field
    - Depth of 60 meters; elongated pre-heat section
  - Orient in triangular configuration with central collector well
  - Operate over period of years to demonstrate oil conversion
- Techno-economic analysis
  - Optimize field layout and GFC operation for lowest lifting cost
  - Add fidelity to the current $38 / barrel production-cost estimate
- Other applications for Geothermic Fuel Cells are being explored
  - Down-hole steam generation
  - Heavy-oil extraction
  - Combined heat, electricity, and clean water
    - Cascata Energy, LLC

11

In summary, the Geothermic Fuel Cell concept shows promise for in-situ oil-shale processing

Colorado Fuel Cell Center

- Delphi-built GFCs are being tested at the Colorado School of Mines
- One month of outdoor, in-ground testing completed
  - Three integrated Geothermic Fuel Cell modules
  - 0.3 m diameter; 6 m length
  - Natural-gas operation
- Important data collected and lessons learned
  - Combined heat-and-power efficiency reached 56%
  - Over 3.3 kW/m heat generation
  - Far-from-optimal operation
  - Further increase to 67% possible with modest plumbing changes

12

Acknowledgements

Colorado Fuel Cell Center

- CSM thanks Delphi for development of outdoor test facility
- US Department of Energy
  - Office of Fossil Energy
  - Solid-State Energy Conversion Alliance
  - Supports Delphi fuel-cell stack development
  - This material is based upon work supported by the Department of Energy under Award Number DE-FE00011769
- Primary investigator thanks CSM supporting departments
  - Capital Planning and Construction
  - Facilities Management
  - Colorado School of Mines central administration

13

A Thermoplasticty Model for Oil Shale

35TH OIL SHALE SYMPOSIUM, OCTOBER 5TH 2015

SHABNAM SEMNANI[1], JOSHUA WHITE[2], DAVID CAMP[2], AND ALAN BURNHAM[1,3]

[1] STANFORD UNIVERSITY
[2] LAWRENCE LIVERMORE NATIONAL LABORATORY
[3] AMERICAN SHALE OIL

Outline

- Goal
- Introduction
- Isotropic Model
- Isotropic Model: Comparison with Experimental Data
- Anisotropic Model
- Anisotropic Model: Comparison with Experimental Data
- Ongoing and Future Work
- Conclusion

Oil Shale Symposium 2015 2

Goal

- Our goal is to understand (and model) oil shale behavior during in situ processing.
- A coupled **thermo-hydro-mechanical-chemical** process which has very practical consequences:

  ✓ Permeability alteration
  ✓ Wellbore closure and stability issues
  ✓ Subsidence

Terminology

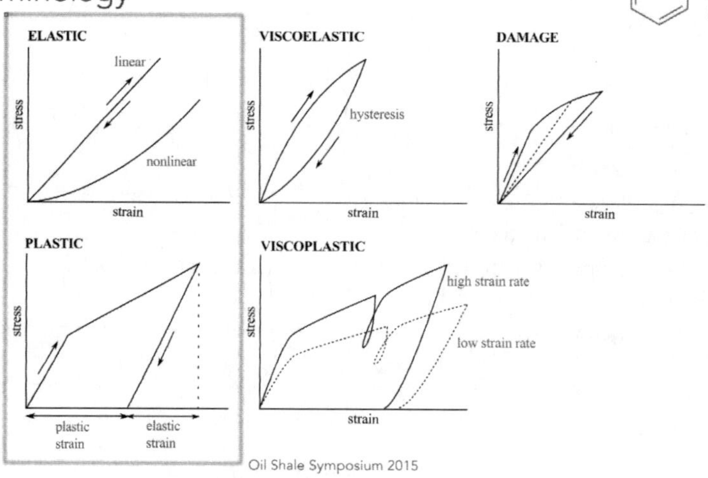

Introduction

- Oil shale exhibits complex thermo-mechanical behavior.

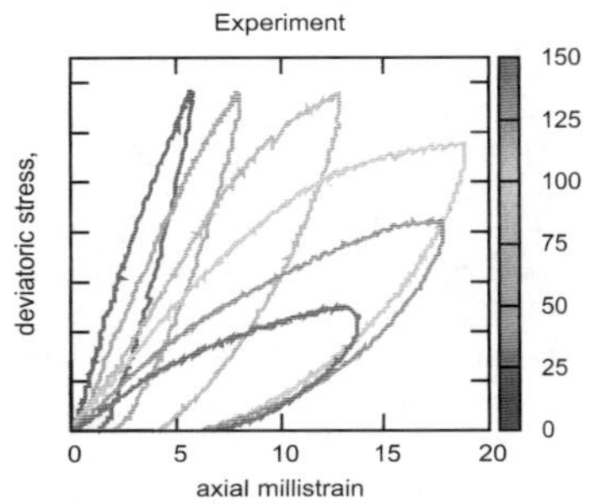

Figure: Triaxial stress-strain response of an oil shale sample progressively heated to 175 C (below pyrolysis temperatures).

Introduction

- At higher temperatures, pyrolysis and subsequent reactions fundamentally alter the material microstructure.

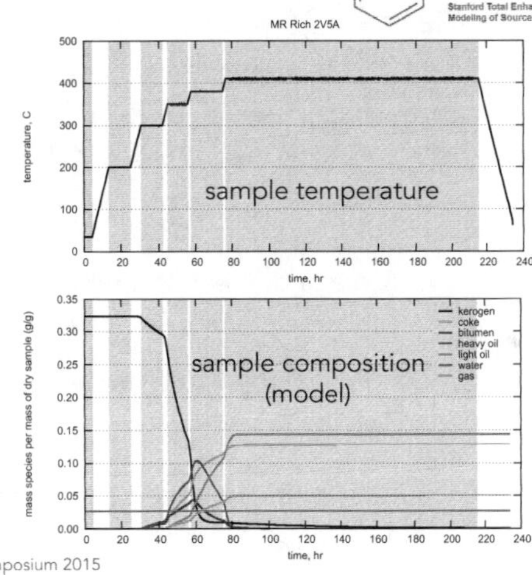

STEMS
Stanford Total Enhanced Modeling of Source rock

Laboratory testing program

Laboratory	Test Types	Temperature
Agapito Associates (AA)	Triaxial	Room temperature
New England Research (NER)	Triaxial	Multi-temperature up to 250 C
MetaRock Laboratories (MR)	Uniaxial Triaxial	Multi-temperature up to 410 C
Advance Terra Testing (ATT)	Uniaxial	Previously retorted

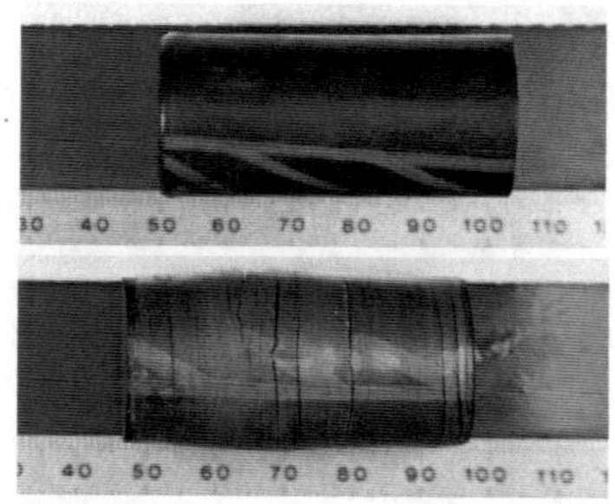

Pre- and post-experiment, 300C, 45 gallons per ton (GPT)

STEMS
Stanford Total Enhanced Modeling of Source rock

Material Model Development

Main mechanisms to consider

1. Elasticity – recoverable deformation
2. Plasticity – irrecoverable deformation
3. Thermal deformation
4. Softening effect of temperature
5. Microstructural changes – conversion of constituent phases
6. Anisotropy – bedding parallel vs. perpendicular properties

Elastic multi-temperature response

- Elastic response in multi-temperature experiments: Examine secant unloading stiffness.

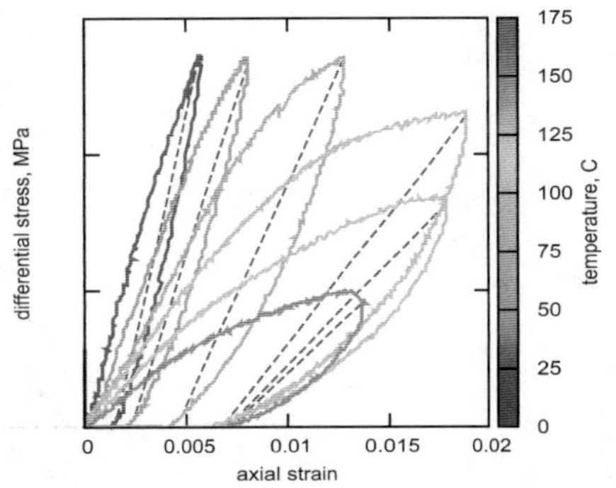

Figure: Triaxial stress-strain response of an oil shale sample progressively heated to 175 C (below pyrolysis temperatures).

Elastic multi-temperature response

Empirical Model: $E(\Delta T) = (E_0 - E_\infty)\exp(-\gamma \Delta T) + E_\infty$

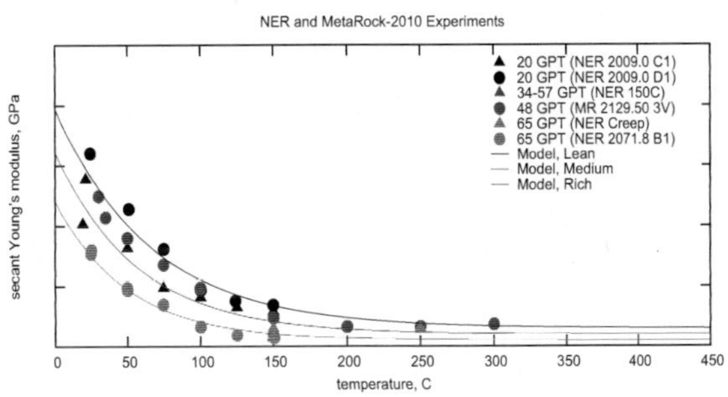

Material Model: Plastic Behavior

- Samples exhibit both brittle and ductile behavior
- Specific response depends on grade, confining pressure, and thermal history

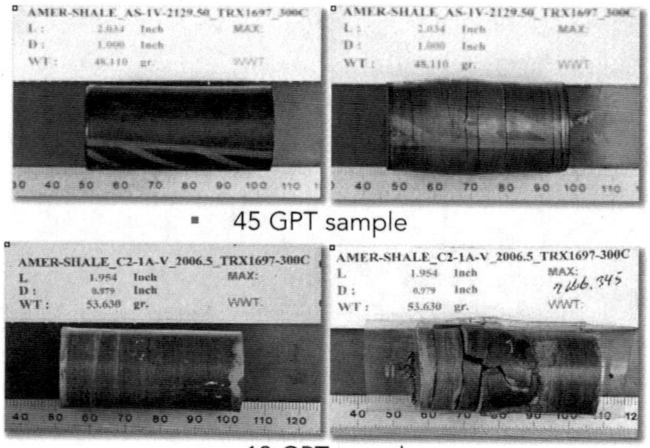

- 45 GPT sample
- 13 GPT sample

Isotropic Model: Yield Surface

- Brittle cone similar to Drucker-Prager surface (Drucker and Prager 1952)
- Ellipsoidal cap similar to Modified Cam-Clay (MCC) (Roscoe and Burland 1968)

p : mean pressure
q : deviatoric stress
p_c : pre-consolidation pressure

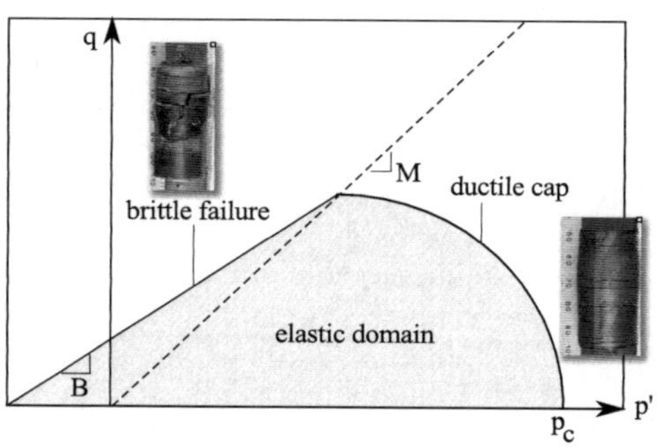

Isotropic Model: Hardening law

STEMS
Stanford Total Enhanced Modeling of Source rock

Change in strength = Mechanical hardening + Thermal softening + Softening due to phase change

$$\dot{p}_c = H_v \dot{\epsilon}_v^p - H_t \langle \dot{T} \rangle + H_k \dot{k}$$

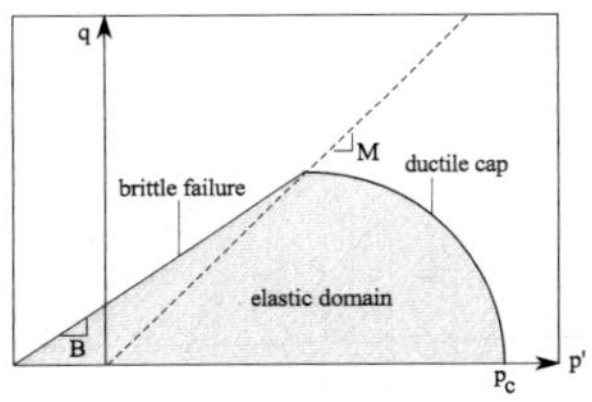

Isotropic Model: Comparison with Experiments

STEMS
Stanford Total Enhanced Modeling of Source rock

- Uniaxial Experiments at 45 C

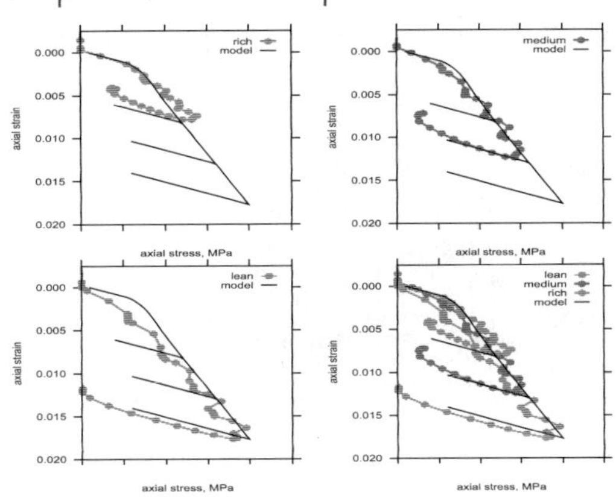

Isotropic Model: Comparison with Experiments

STEMS
Stanford Total Enhanced Modeling of Source rock

- Multi-temperature triaxial experimental data, 59 GPT.

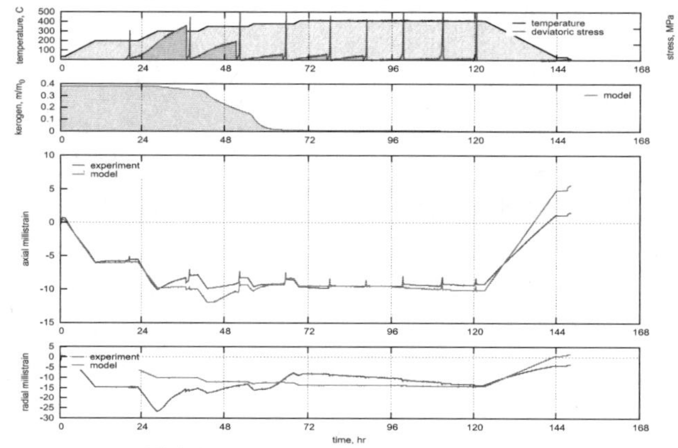

Oil Shale Symposium 2015

15

Anisotropic Extension: Motivation

STEMS
Stanford Total Enhanced Modeling of Source rock

- Strong evidence of anisotropy in the structure of shale and laboratory experiments.

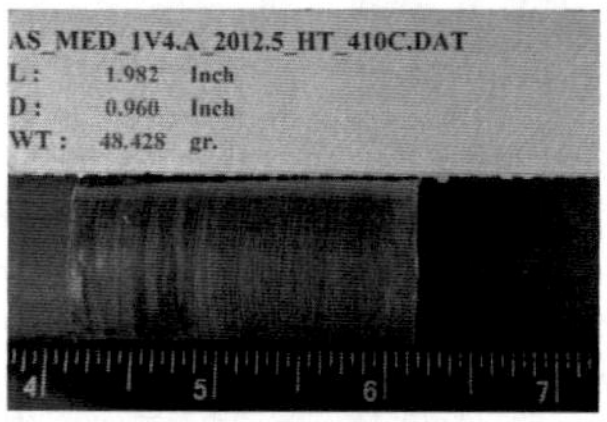

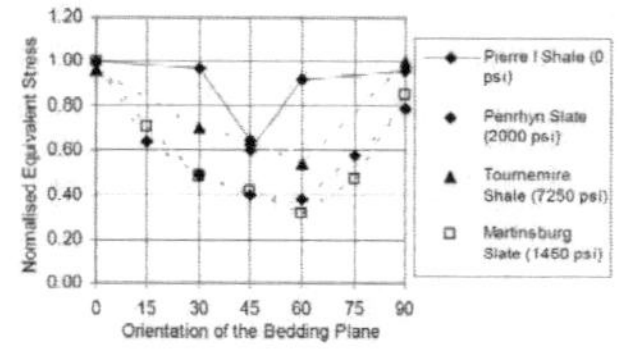

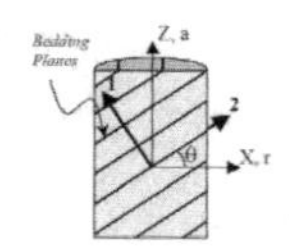

Variation of the compressive strength with orientation of bedding plane (After Crook et al., 2002)

Reference: Crook et al. SPE/ISRM Rock Mechanics Conference. Society of Petroleum Engineers, 2002.

Oil Shale Symposium 2015

16

STEMS
Stanford Total Enhanced Modeling of Source rock

Anisotropic Extension: Motivation

- Various modes of failure observed.
- Depends on bedding plane orientation, confining pressure, grade, etc.

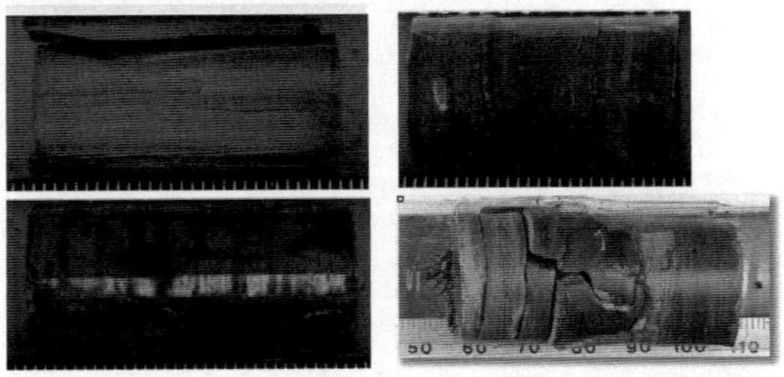

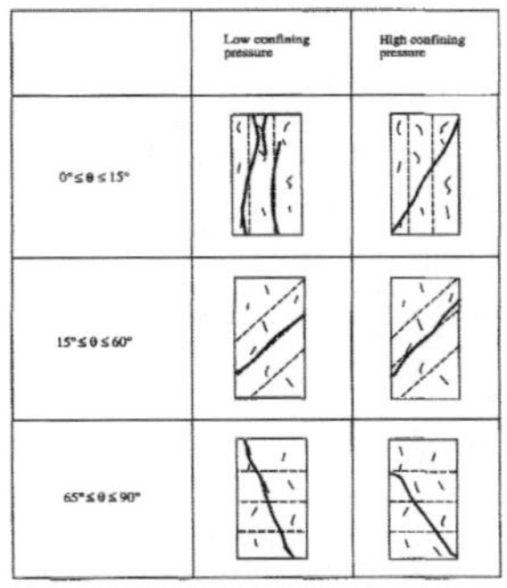

Various modes of failure, from Niandou et al. (1997)

Reference: Niandou, H., Shao, J. F., Henry, J. P., Fourmaintraux, D. (1997). International Journal of Rock Mechanics and Mining Sciences, 34(1), 3-16.

STEMS
Stanford Total Enhanced Modeling of Source rock

Anisotropic model

	Elasticity	Plasticity
Isotropic	➢ 2 independent elasticity parameters: E and ν	➢ 2 parameters to determine plastic deformations: M and p_c
Anisotropic	➢ 5 independent elasticity parameters: $E_1, E_2, \nu_{12}, \nu_{23}$, and G_{12}.	➢ 3 Additional independent scaling parameter: α, β and γ

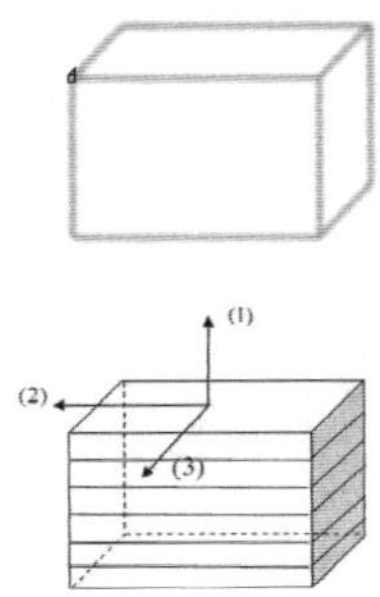

➢ Reduces to isotropic MCC model for $\alpha = \beta = \gamma = 1$

Anisotropic model: Comparison with Experiments

STEMS
Stanford Total Enhanced Modeling of Source rock

- The proposed model was calibrated using triaxial test data for Tournemire shale presented by Niandou et al. (1997)
- for two confining pressures (P=30 MPa and P=50 MPa) :

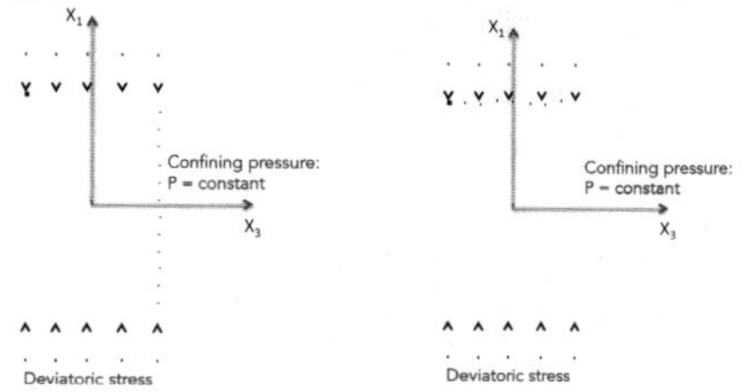

Reference: Niandou, H., Shao, J. F., Henry, J. P., Fourmaintraux, D. (1997). International Journal of Rock Mechanics and Mining Sciences, 34(1), 3-16.

Anisotropic model: Comparison with Experiments

STEMS
Stanford Total Enhanced Modeling of Source rock

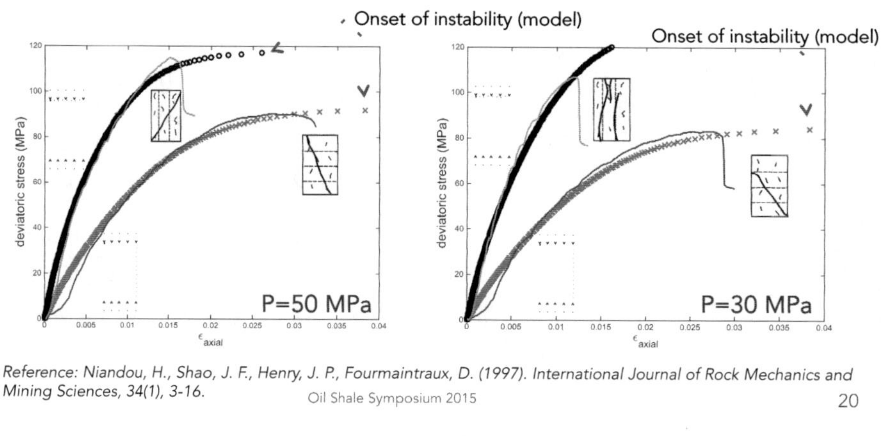

Reference: Niandou, H., Shao, J. F., Henry, J. P., Fourmaintraux, D. (1997). International Journal of Rock Mechanics and Mining Sciences, 34(1), 3-16.

Ongoing and future work:

STEMS
Stanford Total Enhanced Modeling of Source rock

- Comparison with oil shale experimental data using anisotropic model
- Further laboratory experiments
- Analysis of mode of failure and instability
- Including rate dependence
- Including finite deformations

Oil Shale Symposium 2015 21

Conclusion:

STEMS
Stanford Total Enhanced Modeling of Source rock

- Behavior of oil shale strongly depends on the kerogen content and temperature.
- Temperature dependent material model was proposed and calibrated to laboratory data.
- The model was extended to take into account anisotropy.
- The model was able to capture the difference in the behavior of material due to loading in different directions.

Oil Shale Symposium 2015 22

Acknowledgments

STEMS
Stanford Total Enhanced Modeling of Source rock

- This work was supported by the Total through the STEMS project.
- Portions of this work were completed by Lawrence Livermore National Laboratory for the Department of Energy under contract number DE-AC52-07NA27344.

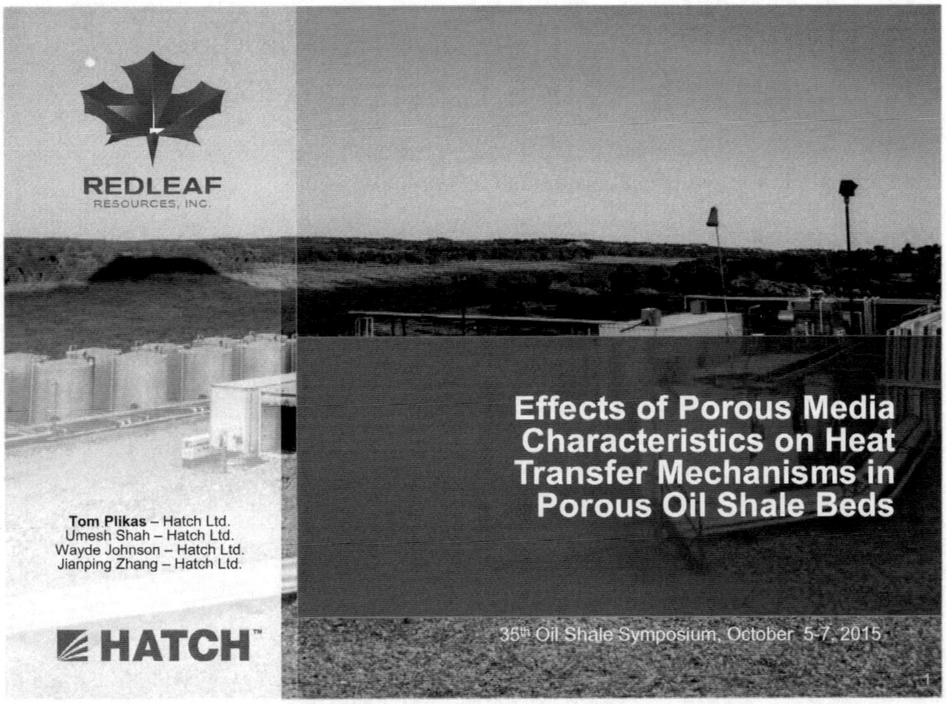

Introduction

- Slow heating applications (<10°C per day), such as EcoShale™ and in-situ processes, have considerably different heat transfer dynamics than fast heating applications, such as surface retorts.
- In slow heating applications, whether the shale is heated directly or indirectly, heating can span the range from conduction dominated heat transfer to steady, convective heating, and on through to highly transient, chaotic Rayleigh-Benard convection.
- The physical characteristics of the porous shale bed in terms of porosity, particle size distribution (PSD), and permeability can dictate which fluid dynamic and heat transfer regime the system is in.
- Numerical experiments have been performed on porous shale beds with different porous media characteristics, with dimensionless numbers (Rayleigh, Grashof, and Reynolds) providing a good indication of which fluid dynamic and heat transfer regime the system is in.

REDLEAF RESOURCES, INC.

2

Heating Rate for Different Technologies, Compiled from Various Literature Sources.

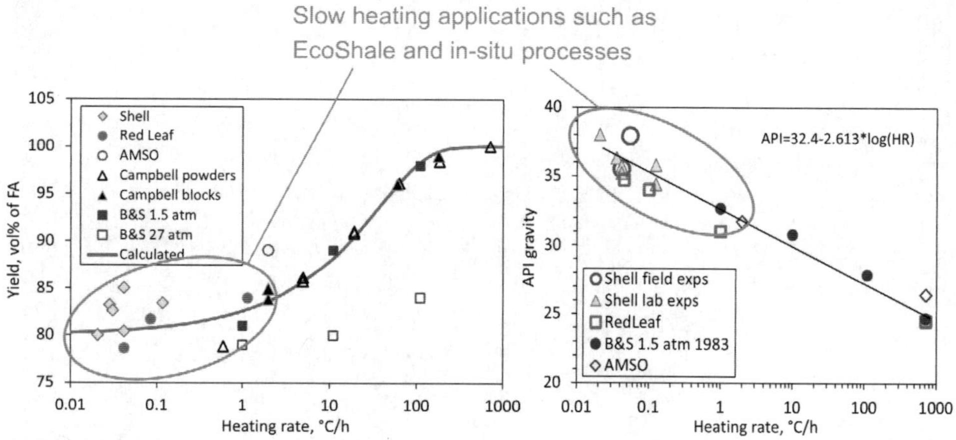

REDLEAF RESOURCES, INC.

3

Examine Dimensionless Numbers to Determine Fluid Dynamics and Heat Transfer Regimes

- Ratio of Grashof to Reynolds Number:

$$\frac{Gr}{\mathrm{Re}^2} = \frac{g\beta\Delta TH}{u^2}$$

- If $\frac{Gr}{\mathrm{Re}^2} < 1$, forced convection dominates over natural convection
- If $\frac{Gr}{\mathrm{Re}^2} > 1$, forced convection is negligible, and system is naturally driven
- Rayleigh number, which is Grashof multiplied by Prandtl number:

$$Ra = Gr\,\mathrm{Pr}$$

REDLEAF RESOURCES, INC.

4

mbient temperature
ansion coefficient
orous shale medium
dium
rence between hot bottom and cold top of porous

vity of porous shale medium
erage temperature between hot & cold

Rayleigh Number
For a fluid medium only:

$$Ra = \frac{\rho_o g \beta H^3 \Delta T}{\mu (k / \rho Cp)}$$

ρ_0 = *fluid density at ambient temperature*
β = *fluid thermal expansion coefficient*
H = *height of fluid medium*
ΔT = *temperature difference between hot bottom and cold top of fluid medium*
μ = *fluid viscosity*
k = *thermal conductivity of fluid medium*
ρ = *fluid density at average temperature between hot & cold*
C_p = *fluid specific heat*

REDLEAF RESOURCES, INC. 5

Rayleigh Number – Modified for Porous Media

$$Ra = \frac{\rho_o g \beta K H \Delta T}{\mu (k_m / \rho Cp)}$$

ρ_0 = *fluid density at ambient temperature*
β = *fluid thermal expansion coefficient*
K = *permeability of porous shale medium*
H = *height of fluid medium*
ΔT = *temperature difference between hot bottom and cold top of porous shale medium*
μ = *fluid viscosity*
k = *thermal conductivity of porous shale medium*
ρ = *fluid density at average temperature between hot & cold*
C_p = *fluid specific heat*

Ra	Flow condition
Ra < 40	Conduction only heating
40 < Ra < 400	Steady natural convection
400 < Ra < 1750	Unsteady Rayleigh-Benard convection
1750 < Ra	Chaotic, highly unsteady Rayleigh-Benard convection

1. Nield and Bejan, "Convection in Porous Media", 3rd Edition, 2006.
2. Otero et. al., "High-Rayleigh number convection in a fluid-saturated porous media", *J. Fluid Mech.* (2004), vol 500, pp. 263-281
3. Hewitt et. al., "High Rayleigh number convection in a three-dimensional porous medium, *J. Fluid Mech.* (2014), vol. 748, pp. 879-895
4. Martinet & Adrian, "Rayleigh-Benard convection: experimental study of time-dependant instabilities", *Experiments in Fluids* (1988), vol. 6, 316-322
5. Weber and Kimmich, "Rayleigh-Benard percolation transition of thermal convection in porous media: computational fluid dynamics, NMR velocity mapping, NMR temperature mapping", *Physical Review E.*, (2002), Vol. 66
6. Hidalgo et. al., "Scaling of Convective Mixing in Porous Media", *Physical Review Letters* (2012), Vol. 109

REDLEAF RESOURCES, INC. 6

Consider the Following 2 Dimensional Porous Shale Bed Heated Directly by a Gas

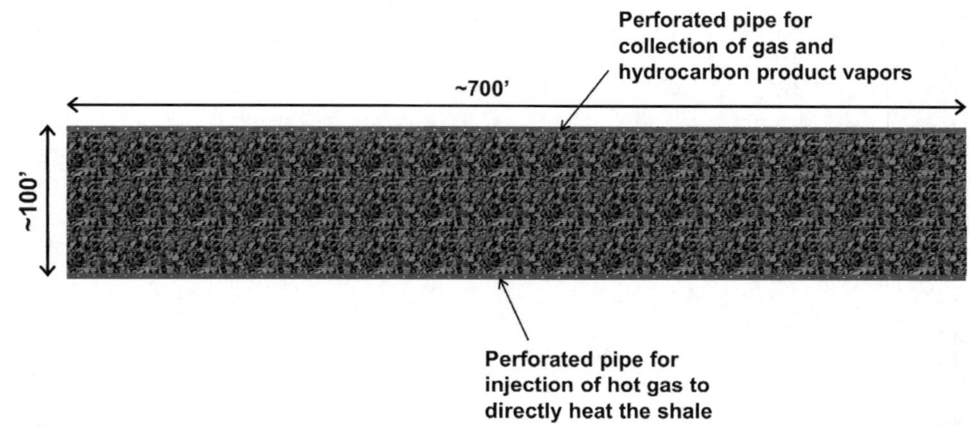

REDLEAF RESOURCES, INC.

7

Summary of Scenarios Considered

Permeability (m^2)	Permeability (darcy)	Porosity (%)	Ra	Hot gas flow rate per ton of shale	Gr/Re^2
6.0E-10	6.1E+02	17	7	0.30	17.0
4.1E-08	4.2E+04	28	340	0.30	8.0
3.0E-07	3.0E+05	36	2,000	0.30	5.8
6.0E-07	6.1E+05	39	4,000	0.30	5.7
3.0E-06	3.1E+06	48	20,000	0.30	5.7
6.0E-07	6.1E+05	39	4,000	456	3E-6

REDLEAF RESOURCES, INC.

8

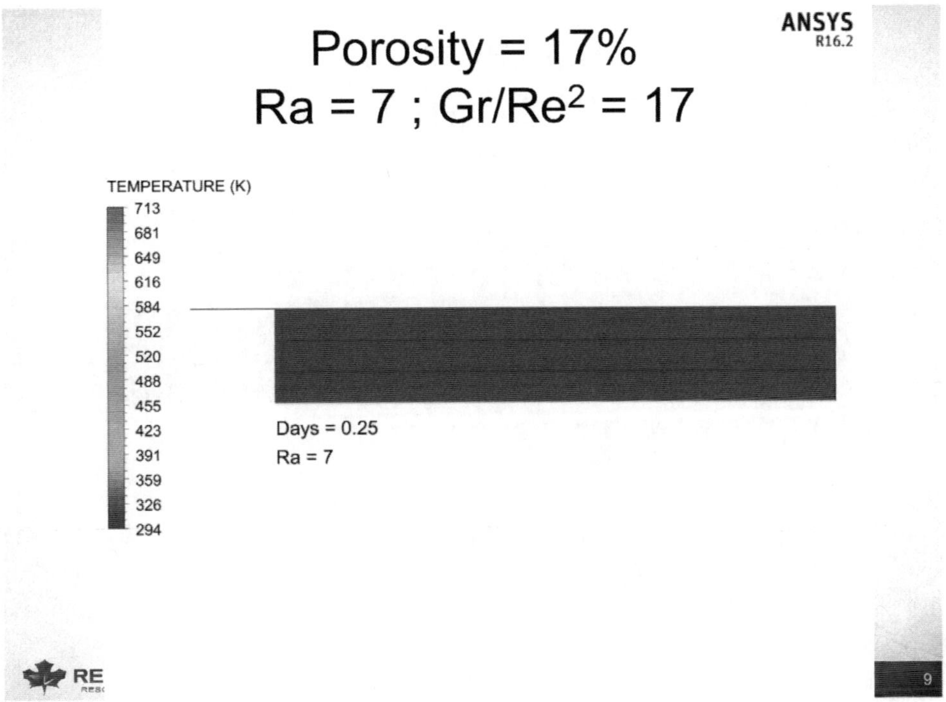
ANSYS
R16.2
Porosity = 17%
Ra = 7 ; Gr/Re2 = 17
TEMPERATURE (K)
713
681
649
616
584
552
520
488
455
423
391
359
326
294
Days = 0.25
Ra = 7
RE
9

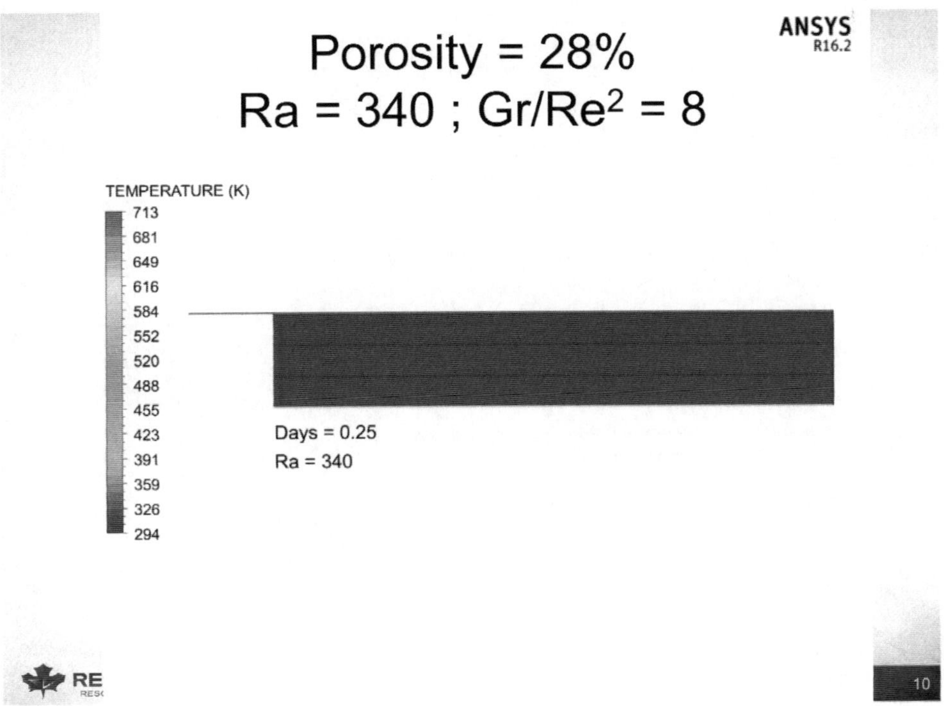
ANSYS
R16.2
Porosity = 28%
Ra = 340 ; Gr/Re2 = 8
TEMPERATURE (K)
713
681
649
616
584
552
520
488
455
423
391
359
326
294
Days = 0.25
Ra = 340
RE
10

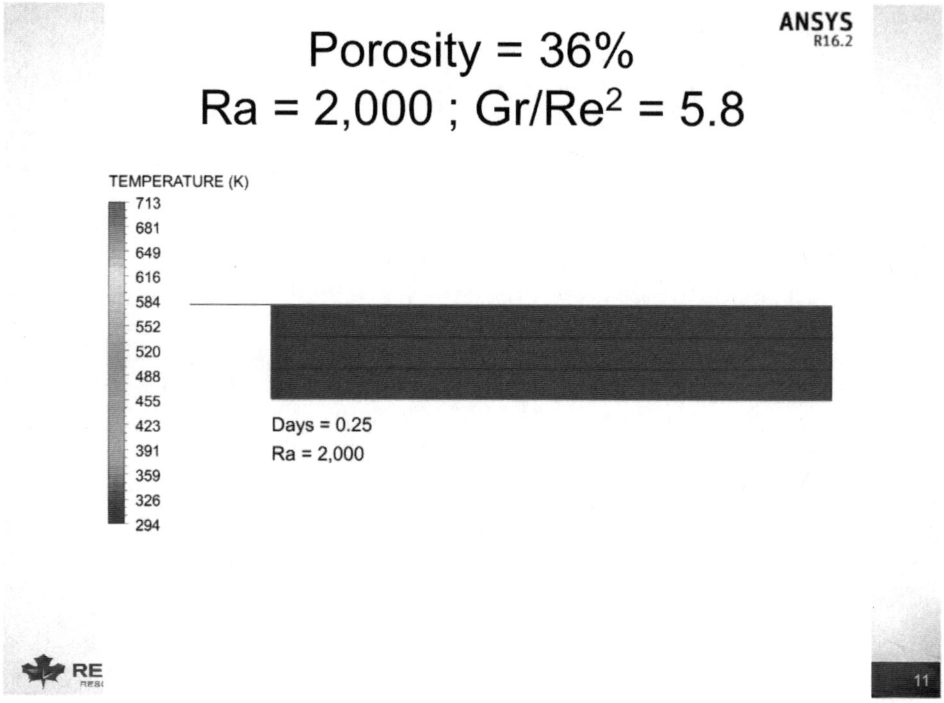
ANSYS
R16.2
Porosity = 36%
Ra = 2,000 ; Gr/Re2 = 5.8
TEMPERATURE (K)
713
681
649
616
584
552
520
488
455
423
391
359
326
294
Days = 0.25
Ra = 2,000
RE
11

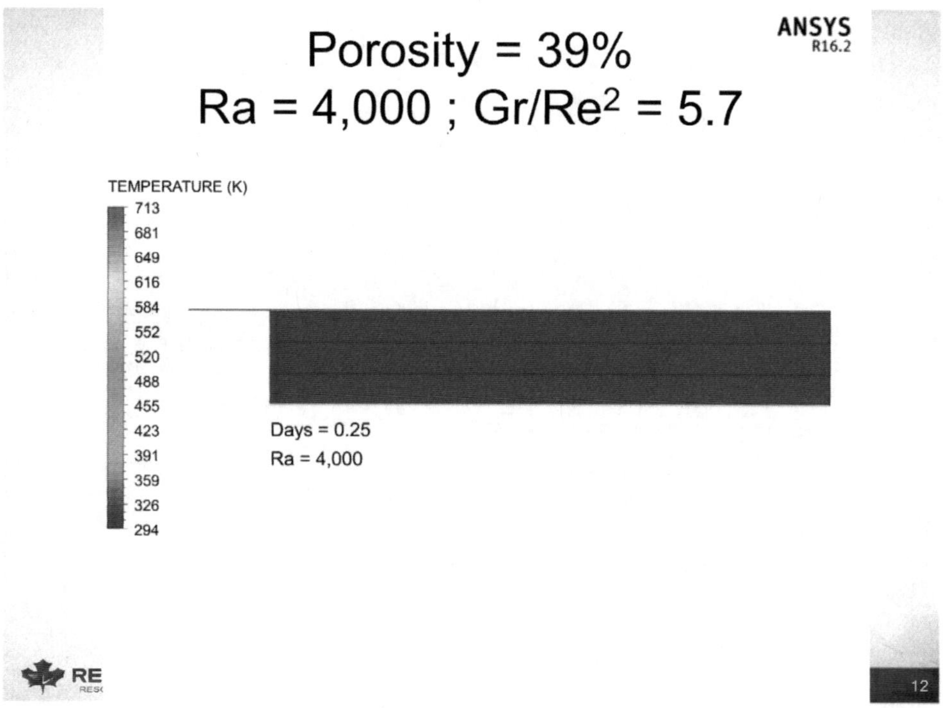
ANSYS
R16.2
Porosity = 39%
Ra = 4,000 ; Gr/Re2 = 5.7
TEMPERATURE (K)
713
681
649
616
584
552
520
488
455
423
391
359
326
294
Days = 0.25
Ra = 4,000
RE
12

ANSYS
R16.2

Porosity = 48%
Ra = 20,000 ; Gr/Re^2 = 5.7

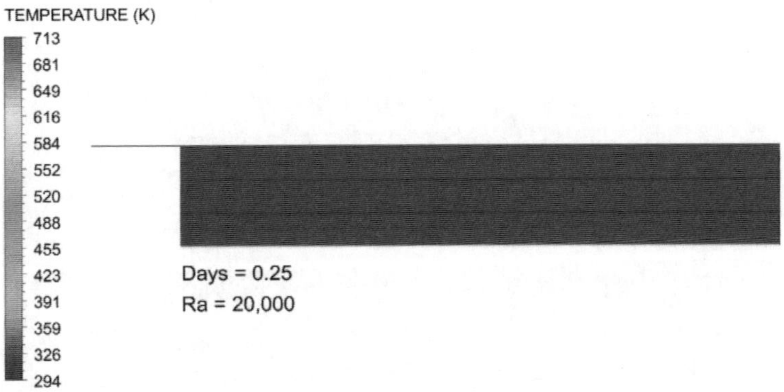

13

Ra = 4,000 ; Porosity = 39%
Gr/Re^2= 5.7 versus Gr/Re^2 = 3e-6

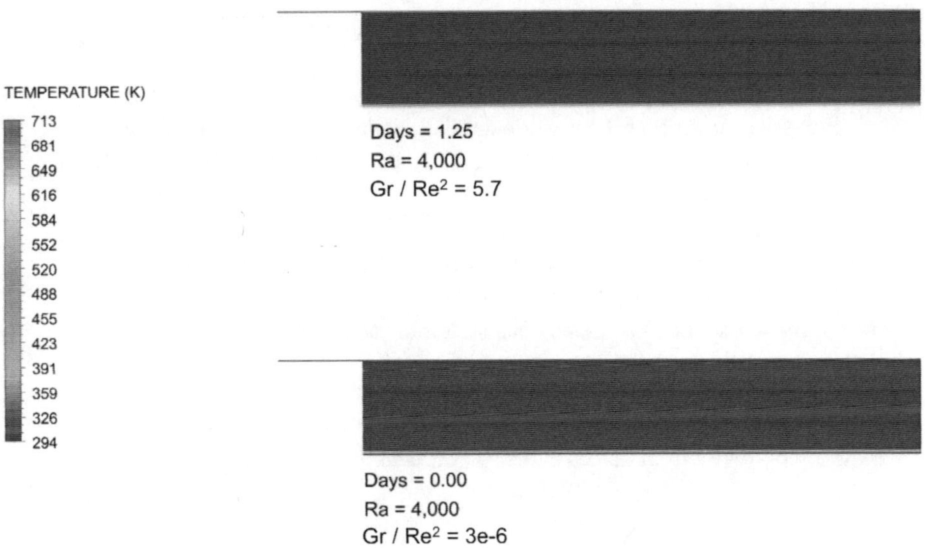

Key Engineering Design Parameters

Permeability of porous shale bed, which is determined by PSD and porosity

Height of the porous shale bed. Direct driver of free convection

ΔT between the heating source and porous shale bed.

$$Ra = \frac{\rho_o g \beta K H \Delta T}{\mu(k_m / \rho Cp)}$$

Ra	Flow condition
Ra < 40	Conduction only heating
40 < Ra < 400	Steady natural convection
400 < Ra < 1750	Unsteady Rayleigh-Benard convection
1750 < Ra	Chaotic, highly unsteady Rayleigh-Benard convection

REDLEAF RESOURCES, INC.

Implications

- Rate of heating of a porous oil shale bed is dependent on the heat transfer regime, which affects the overall process duration.
- Because pyrolysis kinetics are highly non-linear with temperature, rates of oil and gas production will vary substantially depending on the heat transfer regime the system is in. The API of the liquids will also be affected by the different temperature-time histories.
- Understanding and accounting for this fundamental transport phenomenon is necessary in any engineering assessment of slow heating applications such as EcoShale^{TM} and in-situ processes.
- Fortunately, most commercially available numerical codes that incorporate full implementation of the Navier-Stokes equations and handling of solid-gas interactions (direct multiphase or through porous media formulation) can simulate this process, and the operating parameters that influence this, and how to control them to achieve desired process performance.

REDLEAF RESOURCES, INC. 16

Thank You

Questions?

REDLEAF RESOURCES, INC.

17

RELATIONSHIP BETWEEN PARTICLE SIZE DISTRIBUTION (PSD) AND VOID FRACTION IN RUBBLIZED OIL SHALE BEDS

James W. Bunger, Ph.D..
JWBA, Inc
James W. Patten, Ph.D.
RLR, Inc

CSM Oil shale Symposium October 5-6, 2015

Purpose of PSD/Void fraction modeling

- Void fraction directly determines density of ore in capsule
- Void fraction directly controls capsule subsidence
- PSD controls void channel sizes, affecting heat and mass transfer

REDLEAF RESOURCES, INC.

Objectives

- To understand the relationship of particle size distribution to void fraction
- To identify the ideal blend mix (maximum density) for a given particle size range
- To understand the range of packing efficiencies
- To validate mathematical scaling and truncating methods
- To create a model that will estimate the void fraction from any PSD
- To apply teachings to control of blasting, crushing, blending and loading of capsule

REDLEAF RESOURCES, INC.

Measuring Particle Aspect Ratios

Define:

X = long axis
Y = short axis
Z = third dimension

REDLEAF RESOURCES, INC.

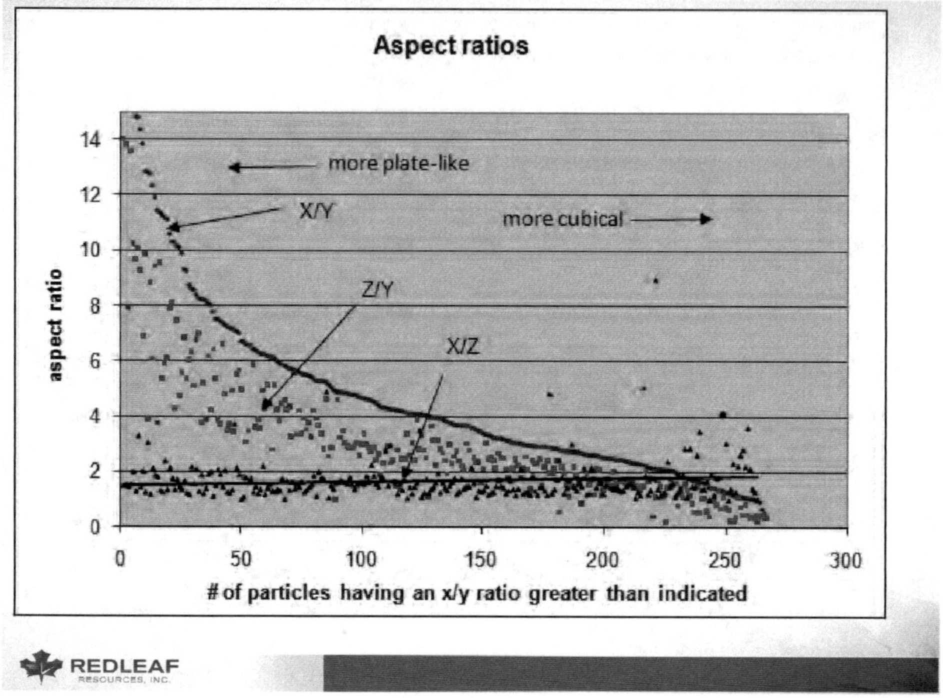

Volumes of geometric shapes	
Rectangular box	=1.0*X*Y*Z
Planer triangle	=0.5*X*Y*Z
Seep Ridge particle (average)	=0.44*X*Y*Z
4-Sided pyramid	=0.333*X*Y*Z

REDLEAF
RESOURCES, INC.

Average particle shapes

- The average particle, when placed in a rectangular box of dimensions x, y, and z will occupy 44% of the space. (56% voids)
- The average shape of a particle is midway between a planar triangle and a 4-sided pyramid (based on volume comparisons).
- Non-compacted beds of uniform-size particles typically exhibit 51-55% voids.
- Compacted/settled beds of the same screen size are typically 47% voids.
- Severe settling of uniform particles can result in someone lower voids (low 40s % voids).

REDLEAF RESOURCES, INC.

Void fraction of narrow particle size ranges

Bin	screen size	void fraction
1	0.625-.75"	0.472
2	0.5-.625"	0.464
8	#8-#4	0.474
10	#16-#10	0.465
12	#30-#20	0.481

The void fraction of narrow particle size ranges is not dependent on the particle size (for once-through crushing). This implies that the aspect ratio ranges of particles are generally similar, irrespective of particle size. Visual observation tends to confirm this conclusion.

REDLEAF RESOURCES, INC.

Typical behavior of two narrow particle sizes

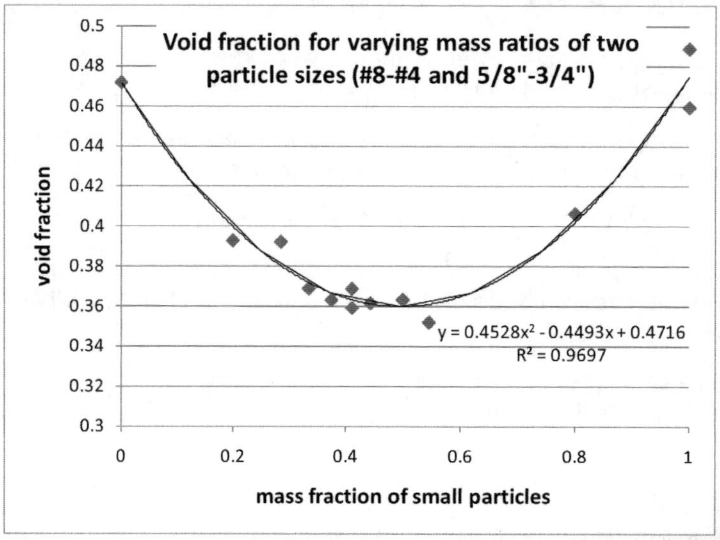

REDLEAF
RESOURCES, INC.

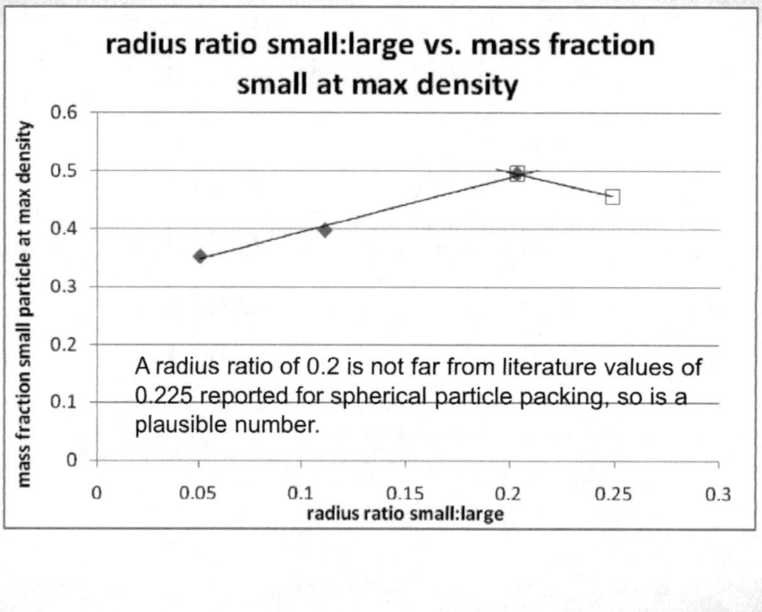

REDLEAF
RESOURCES, INC.

Mass recipe for a 4-component mixture, with each component having a successively smaller particle radius of 0.2.

Tetramodel PSD (wt%)	
1/2-5/8"	0.530
#8-#4	0.176
#30-#20	0.156
#140-#60	0.138
	1.000

This recipe is based on successive sizes being 47/53 (0.8868) of the previous size. It still may not be 'ideal'; so we are calling it 'standard'.

REDLEAF RESOURCES, INC.

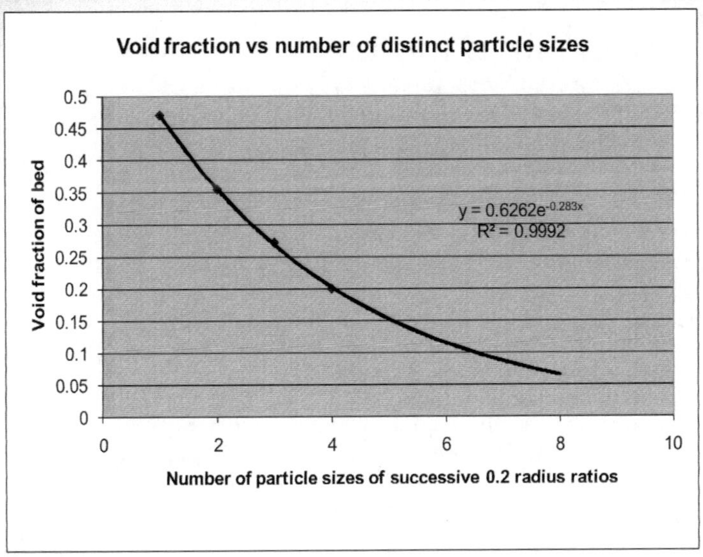

Void fraction is directly related to the range of particle sizes

REDLEAF RESOURCES, INC.

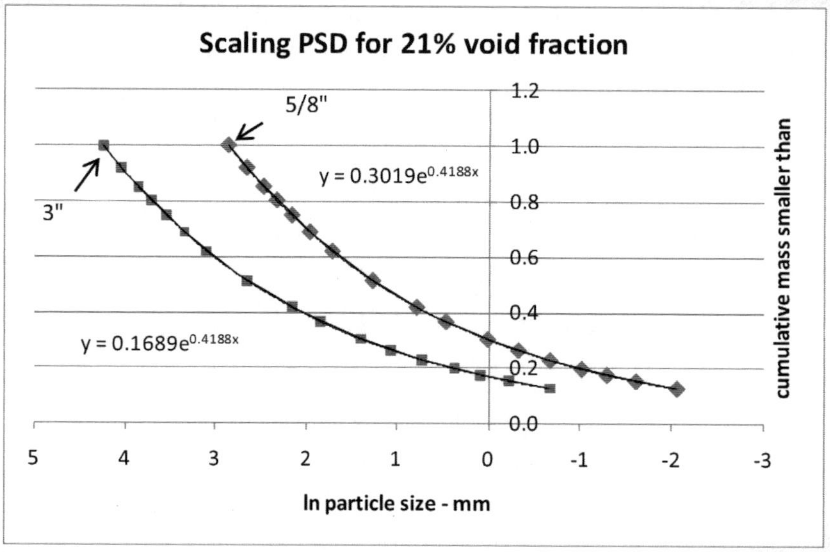

REDLEAF RESOURCES, INC.

Truncating the curve

(specifying a mix of a desired void)

1. Determine the PSD range needed for the desired void fraction.
2. Reallocate the bins (screen sizes) over that range.
3. Fit to a (cumulative passing) curve.
4. Use curve to establish the recipe for screen sizes available.

REDLEAF RESOURCES, INC.

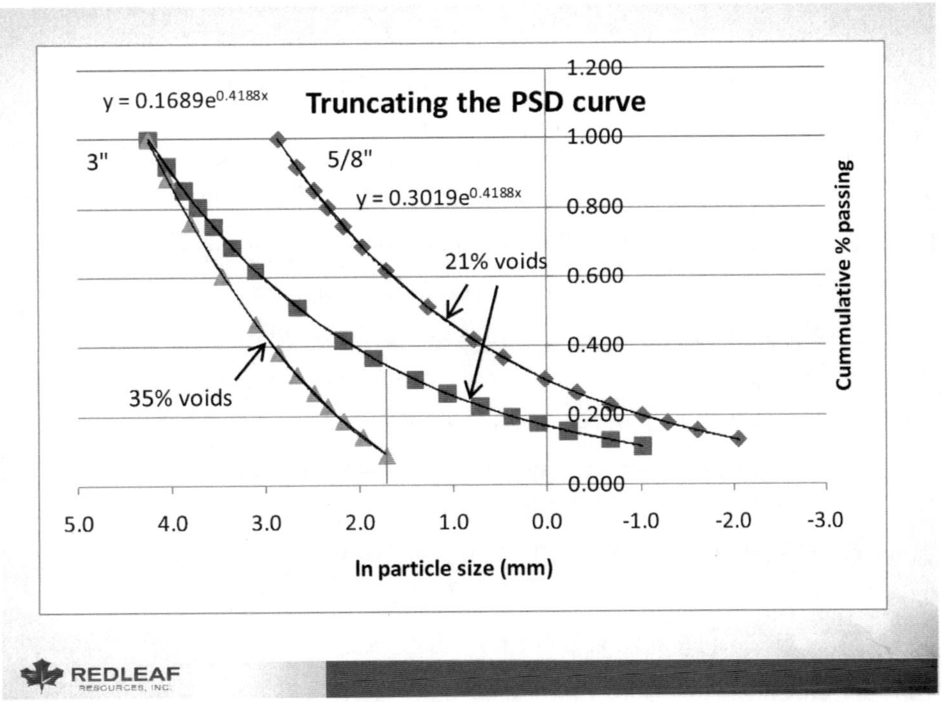
Truncating the PSD curve
y = 0.1689e^0.4188x
3"
5/8"
y = 0.3019e^0.4188x
21% voids
35% voids
1.200
1.000
0.800
0.600
0.400
0.200
0.000
5.0
4.0
3.0
2.0
1.0
0.0
-1.0
-2.0
-3.0
ln particle size (mm)
Cummulative % passing
REDLEAF
RESOURCES, INC.

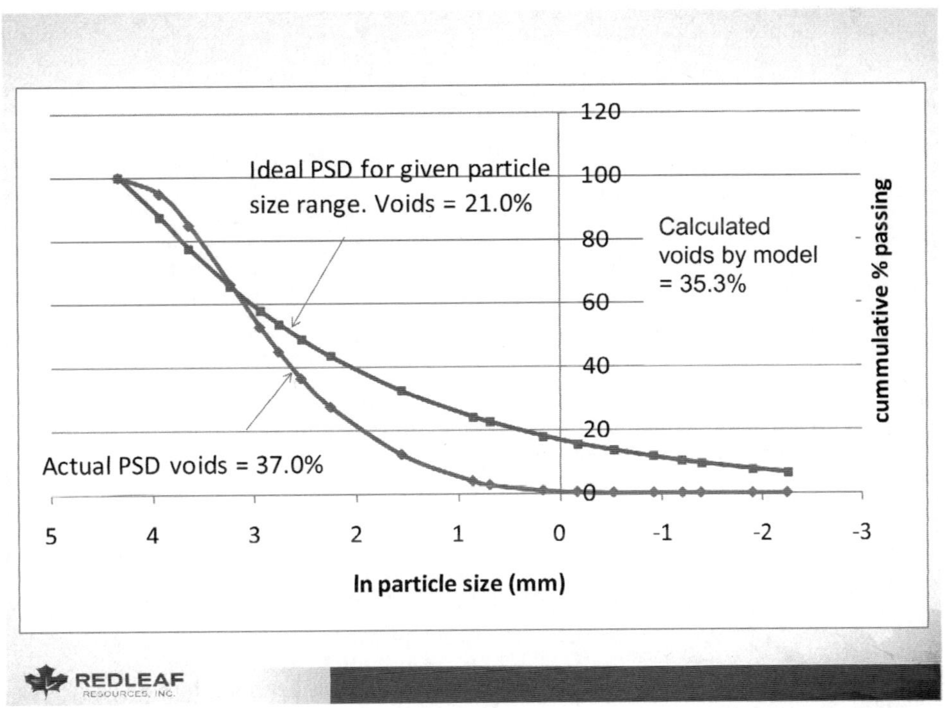
Ideal PSD for given particle
size range. Voids = 21.0%
Calculated
voids by model
= 35.3%
Actual PSD voids = 37.0%
120
100
80
60
40
20
0
5
4
3
2
1
0
-1
-2
-3
ln particle size (mm)
cummulative % passing
REDLEAF
RESOURCES, INC.

Methodology for estimating void fraction from a random PSD

1. Note the particle size range of the random PSD
2. Calculate a standard PSD for that range
3. Note the theoretical void fraction for that range
4. For each bin (screen size) take the absolute difference in fraction of mass
5. The sum of all differences is the fraction inefficiency
6. Calculate the void fraction from the prorated difference between 0% efficient and 100% efficient
7. Add offset for packing efficiency, if necessary

REDLEAF RESOURCES, INC.

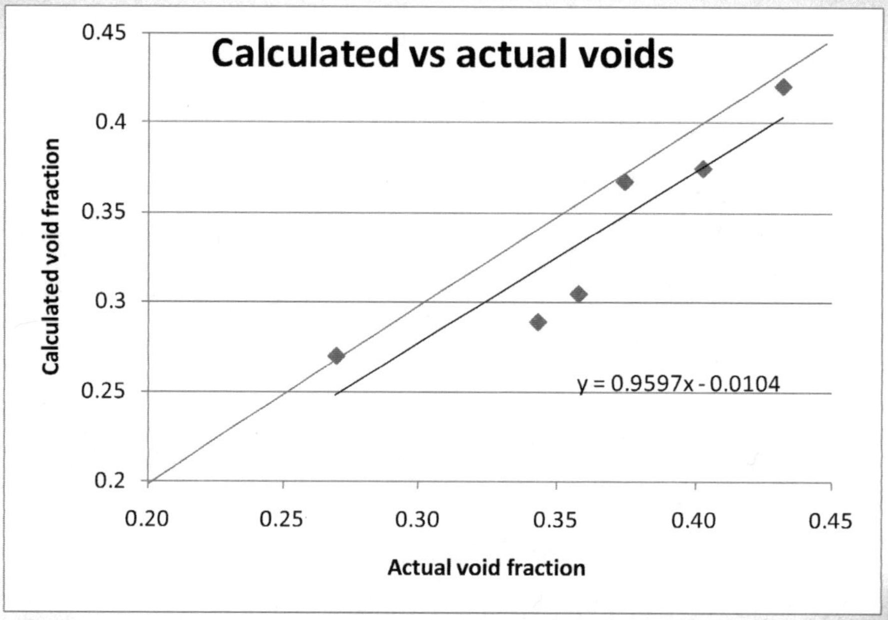

REDLEAF RESOURCES, INC.

Summary

- An ideal, or near ideal, PSD has been elucidated for particle shapes generated during mining and crushing of Seep Ridge oil shale.
- Methods for scaling and truncating the studied PSD to generate beds of predictable voids have been demonstrated.
- Progress has been made to construct a model for estimating void fractions for any random PSD.
- Along with correlations of blasting and crushing behavior, the model can be used to optimize ore preparation processes.

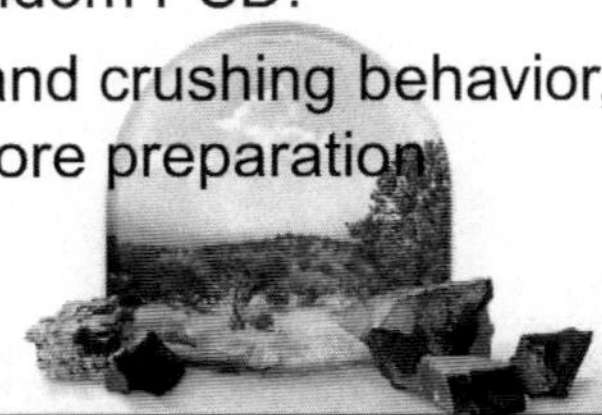

REDLEAF RESOURCES, INC.

Disclaimer

Some or all of the information contained in the presentation represents views, estimates and assumptions of Red Leaf Resources, Inc. and JWBA, Inc., Hatch Ltd. and Total E&P USA Oil Shale, LLC (together, the "Companies"), any of which may change without notice. Neither the Companies nor any of their respective shareholders, affiliates, partners, members, directors, officers, management, employees or representatives makes any representation or warranty, express or implied, as to the accuracy or completeness of any of the information or any other information (whether communicated in written or oral form) transmitted or made available to you.

REDLEAF RESOURCES, INC.

THE CHESS PROCESS: Environment Friendly, Low-cost Shale Oil Recovery

Thomas H. Fletcher
Brigham Young University, Provo, Utah
Carlos Adams
Energy Recovery Technology, American Fork, Utah

35th Oil Shale Symposium
Salt Lake City, UT
October 5-6, 2015

Need for Improved Oil Recovery

1. Size cost of current units

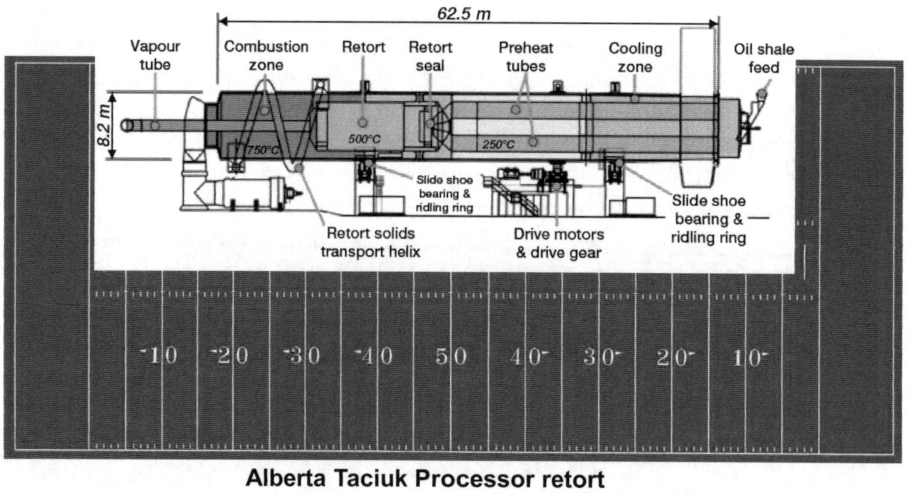

Alberta Taciuk Processor retort

Need for Improved Oil Recovery

2. Grade of ore that is economical

Comparative Oil Reserves

Grade (gpt)	% Org. Matter	Billion Barrels Oil
> 25	> 15	600
10 – 25	6 – 15	1,400
5 – 10	3 – 6	2,000
Total Green River Deposit *		**4,000**
World Crude Oil Reserves†		**1,000**

*Synthetic Fuels Data Handbook, Cameron Engineers
†Worldwide Pet. Ind. Outlook, 15th Edition, 1999-2003

Challenges with Green River Oil Shale Processing

- For Green River Oil Shale 10% organic (17 gpt), results in 8% liquid, 90% inorganic mineral
- 20% of organic material converts to char or coke
- Inorganic mineral is 33% dolomite, calcines above 1000°F
  - $CaMg(CO_3)_2 \rightarrow CaCO_3 + MgO + CO_2$
  - Energy cost of 380 Btu/lb (880 J/g)

Objective

Develop efficient process to

- Maximize oil yield
- Use residual char
- Keep ash temperature < 1000°F
- Minimize heat requirement & cost

Wanted:

MODERATE HEATUP

- Minimum heating rate should be > 10 K/min
- High heating rate (1000K/min), as in ash recycle systems:
  - NOT NECESSARY
  - UNDESIRABLE

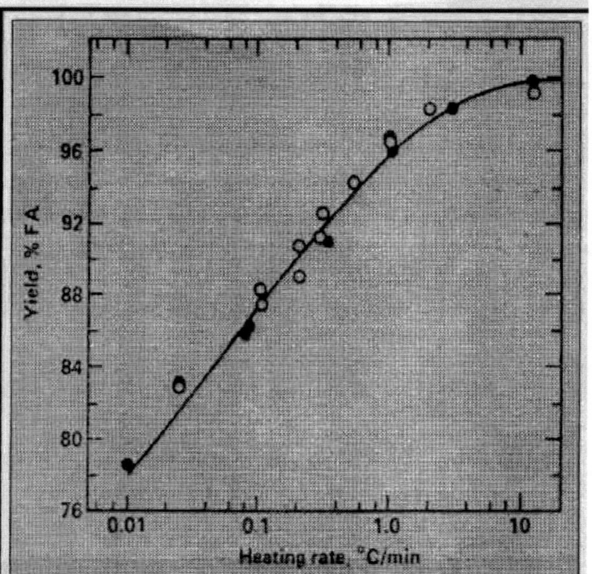

EFFECT OF HEATING RATE ON OIL YIELD
LOW HEATING RATES RESULT IN OIL LOSS DUE TO COKING. DATA POINTS ARE FROM TWO DIFFERENT SOURCES.
(After J. H. Campbell et al. IN Situ, 2(1), 1978).

MINIMIZE COKING REACTIONS

SURFACE COKING*

1200F, 0.2 SEC. = 14% LOSS
1200F, 1.2 SEC. = 40% LOSS
932F, 3 SEC. = 7% LOSS

METAL OXIDE SORBPTION† (OF OIL ON FINE ASH THEN THE SORBED OIL BURNS IN THE COMBUSTOR)

HYDROGEN ABSTRACTORS‡, S, O (present in recycle ash as Sulfates and as O2 in fine particle interstices)

*Bissel, et al., Shale Oil Cracking Kinetics, Ind. Eng. Chem. Process Des. Dev., 1985, 24, 383.
†Barney & Calson, US Patent #3,691,056
‡Lewis, LLNL Oil Shale Qtr. Report, UCID-16986-85-2

NO OXYGEN!

ROBINSON & HUBBARD:

- When oil shale retorted after preheat in gas with > 0.1% O_2, oil yield reduced by up to 16%
  - Equivalent increase in coke and water

(USBM RI-4787)

Oil shale pyrolysis and vapor formation occur at a relatively low temperature:

- 50% below 785°F
- 90% below 845°F
- 100% as T → 950°F

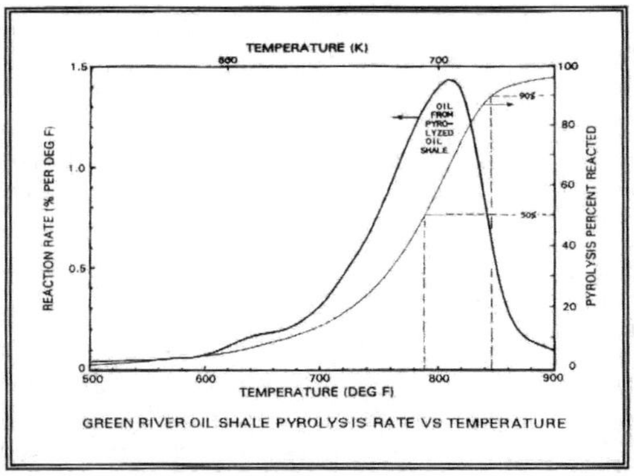

GREEN RIVER OIL SHALE PYROLYSIS RATE VS TEMPERATURE

Idea:

- Quickly withdraw at the formation temperature and cool the vapor immediately after it forms to avoid coking or cracking
- Heat residue to a relatively high T (950°F) thus providing the best of both effects
  - low temperature pyrolysis (minimal coking/cracking) and
  - high temperature (thorough) pyrolysis.

Challenge

- How do you efficiently heat the oil shale at ~ 10 K/min with no Sulfates or O_2 present?
  - Difficult when ash is recycled
- Most efficient heat exchange is counter-current (vs co-current).

EFFICIENT OIL RECOVERY

- Numerous stages for countercurrent heat exchange
- Low T vapor removal to reduce coking
- Optimal heating rate
- Optimal gas/solid fluid bed contacting
- Countercurrent oxidation of char to provide heat

- **Heat Transfer Problem!**

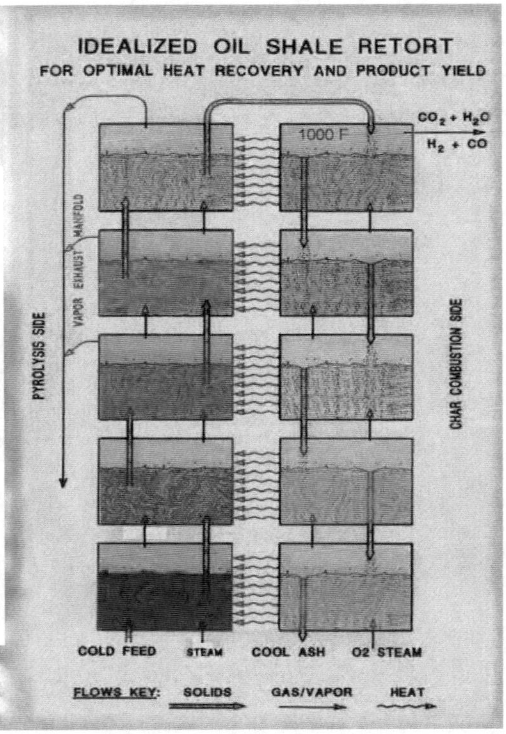

CHESS Process
(Counterflow Heat Exchange Retort)

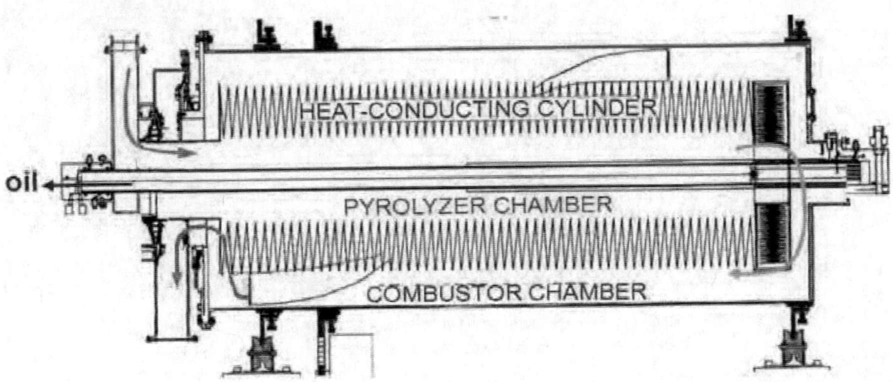

Counterflow Heat Exchange in Solid Streams

Process Diagram

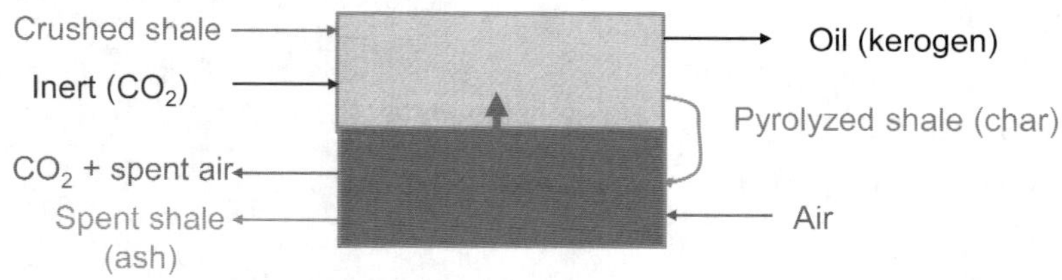

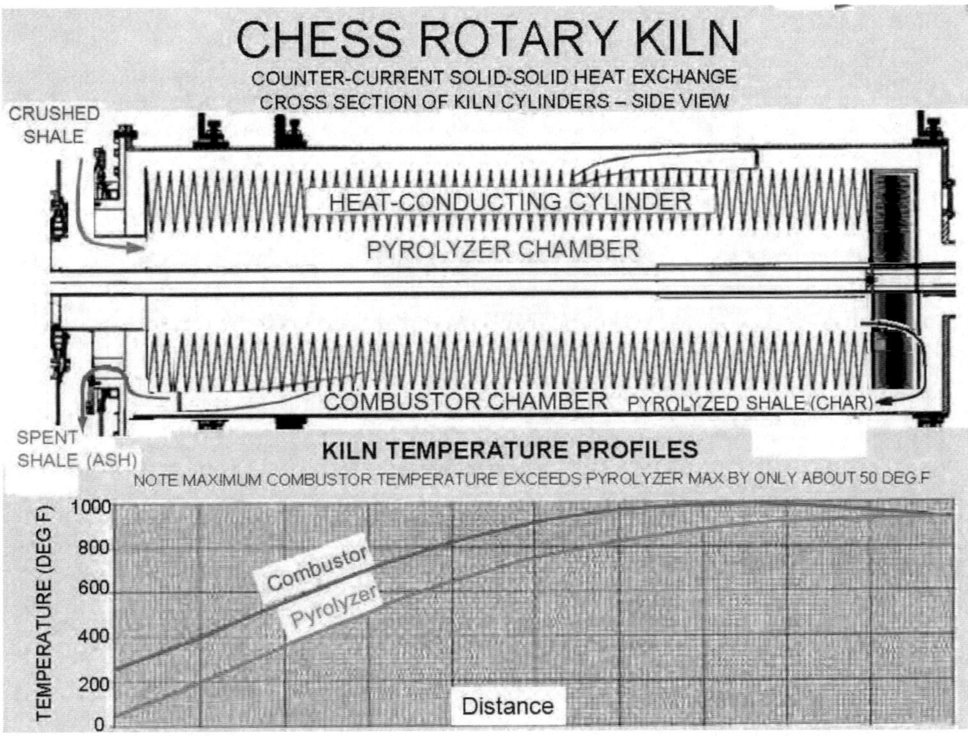

Demonstration: Batch Retort Data

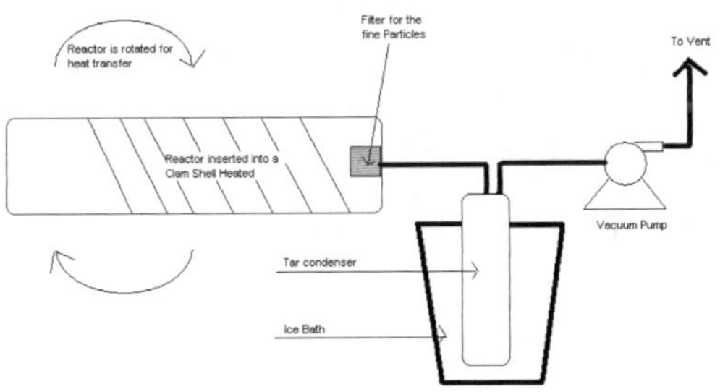

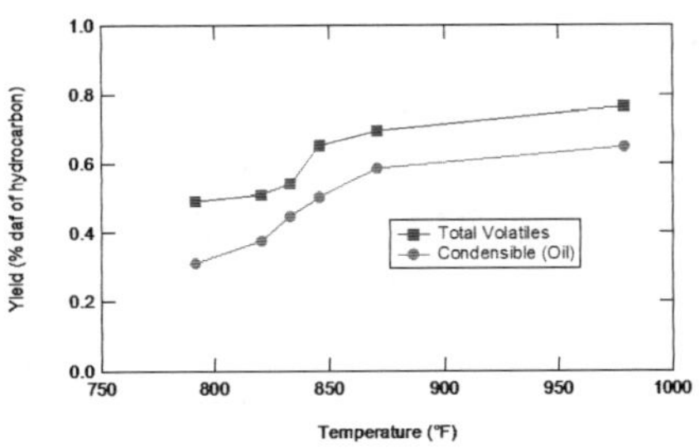

- Colorado Green River Oil Shale
- 10 K/min heating rate
  - Electric heaters
- Crushed to ~1/8 inch diameter
- Rotated in batch retort at rate similar to CHESS process
- 65% oil yield
  - (i.e., % of initial organic mass)
  - 103% of Fischer Assay

CHESS High Efficiency Features

- Recovers 80 to 90% of sensible heat
- Burns residual char for process heat
- Operates without recycling hot ash
- Burns coke at or below 1000°F to prevent calcination

Advantages to the CHESS Process

- **High oil yield**: 103% OF Fisher Assay – 15% above that anticipated (90% of FA) for other processes, plus 13.5% saved by burning coke rather than oil = 25% more oil
- **Reactive coke** (Low temperature retorting of shale provides for reactive coke and optimal coke use)
- **High Volume Specific Capacity** (VSC, ton/ft^3) achieved from the high heat exchange and high fill fraction of the CHESS system

Equipment Advantages

- Low vertical profile (saves $$)
- Compact single vessel multi-operation
  - 30% of kiln volume required by other processes
  - 30% of capital cost
- Simple design, easy scale-up
- 85% reduction in pollutant gases
- Minimal water use
- Ash is clean
  - Not leachable
  - Free of PAH
  - pH neutral

Comparative Size

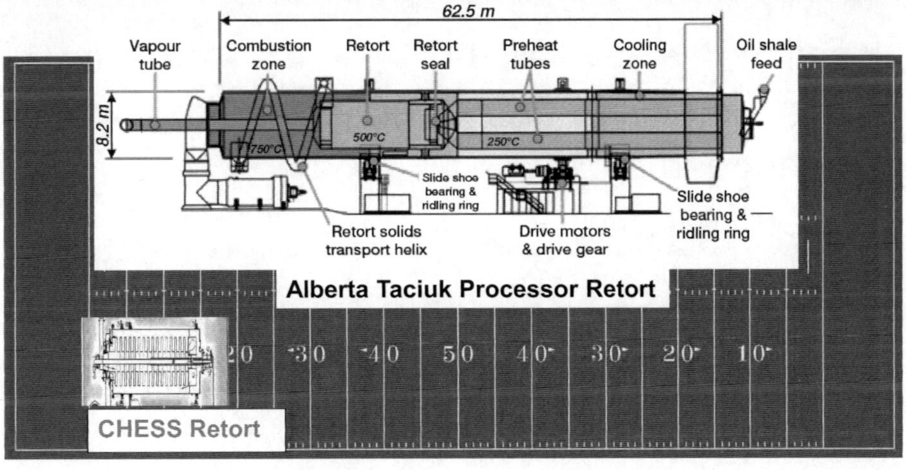

Cost Estimate for Oil Production

- Cost estimates include
  - Mining & handling
  - Retorting
  - Refining of oil product
  - Environmental remediation
- Ore grades > 5 gallons/ton
- $30/barrel
- Reviewed by DOE, NIST

Status

- Batch trials succesful
- 400 lb/hr continuous unit built and being debugged at BYU
  - Only 6 ft long

Thank you

REDLEAF
RESOURCES, INC.

Design, Construction, Process and Production Efficiency Implications of Shale Bed Void Fraction for the Red Leaf Resources EcoShale® Process

- James W. Patten (Red Leaf Resources, Inc.)
- James Bunger (JWBA, Inc.)
- Samuel Lethier (Red Leaf Resources, Inc. and TOTAL)
- Hamid Ghorbani (Hatch)
- Majid Maleki (Hatch)

Introduction

The Red Leaf Resources EcoShale® process for producing high quality liquid hydrocarbons from oil shale depends heavily on the interconnected porosity of a particulate (rubble) shale bed to support heat transfer by natural and forced convection, and to allow removal of produced hydrocarbons. This porosity can be engineered by managing both absolute particle size and relative particle size distribution. In this way void fraction can be engineered to vary from more than 50% to less than 20%

REDLEAF
RESOURCES, INC.

Figure 1

Particle Size Distribution Influence On Void Fraction

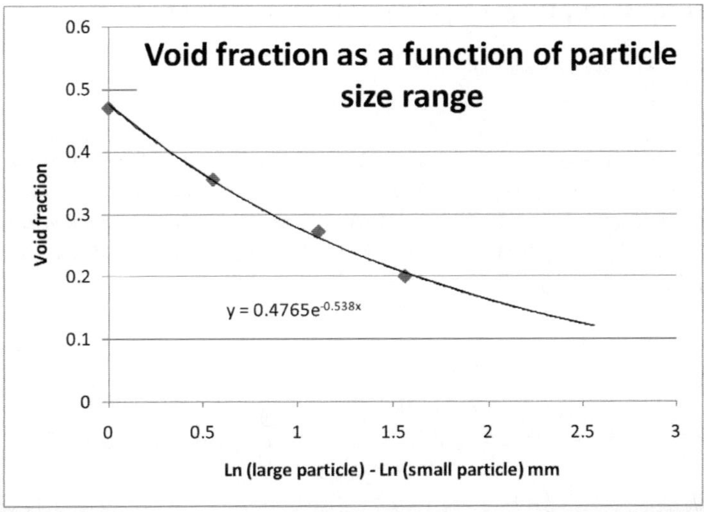

Void fraction is directly related to the range of particle sizes

REDLEAF RESOURCES, INC.

Particle Size Distribution Influence on Void Fraction

- A narrow particle size, 3/8-7/16”, was loose-filled in a cylindrical container. Void fraction was 0.511.
- This same material, when highly settled/compacted gave a void fraction of 0.406
- The degree of agitation/settling/compaction can influence the void fraction by up to 20%

REDLEAF RESOURCES, INC.

Void Fraction Directly Determines The Amount Of Shale Ore In A Given Capsule Volume

- 50% void fraction decreasing to 20% void fraction = 60% increase in mass/unit volume
- 50% void fraction decreasing to 40% void fraction = 20% increase in mass/unit volume
- 40% void fraction decreasing to 30% void fraction = 17% increase in mass/unit volume
- 35% void fraction decreasing to 30% void fraction = 8% increase in mass/unit volume
- 35% void fraction decreasing to 25% void fraction = 15% increase in mass/unit volume

REDLEAF RESOURCES, INC.

Direct measurement of void fraction and particle size distribution is important. Rubble pile surface analysis is deceiving, because small particles settle downwards. To illustrate this effect, a homogeneous shale rubble sample was placed in a 6 inch diameter cylinder in two lifts. The cylinder was then gently tapped. The top surface was photographed, and then 1 ½ inches was removed, and the newly exposed surface was photographed. This was repeated and surfaces were photographed at 3 inches from the top, and 4 ½ inches from the top. Material from the top 3 inches was also screened, with 23% passing through a #8 mesh. Material from the bottom 3 inches was screened in exactly the same way, and 60% passed through a #8 mesh.

REDLEAF RESOURCES, INC.

Figure 2

Top Surface of Rubble Pile/Column

REDLEAF RESOURCES, INC.

Figure 3

Section 1 ½ Inches From the Top of Rubble Pile/Column

REDLEAF RESOURCES, INC.

Figure 4

Section 3 Inches From Top of Rubble Pile/Column

REDLEAF RESOURCES, INC.

Figure 5

Section 4 ½ Inches Down From Top of Rubble Pile/Column

REDLEAF RESOURCES, INC.

- Void fraction directly determines rubble/pile column compressive behavior for a given shale richness/grade, and therefore directly influences load/stress on internal structures (pipe systems) and the design of the capsule roof, top edges, and corners to accommodate settling.
- Engineering to reduce void fraction can therefore be targeted to increase allowable capsule depth for a given settling limit
- Red Leaf conducts shale compressive tests under load temperature, and capsule retorting atmosphere representative of production capsules

REDLEAF
RESOURCES, INC.

Figure 6

Red Leaf Resources Hot Compression Test

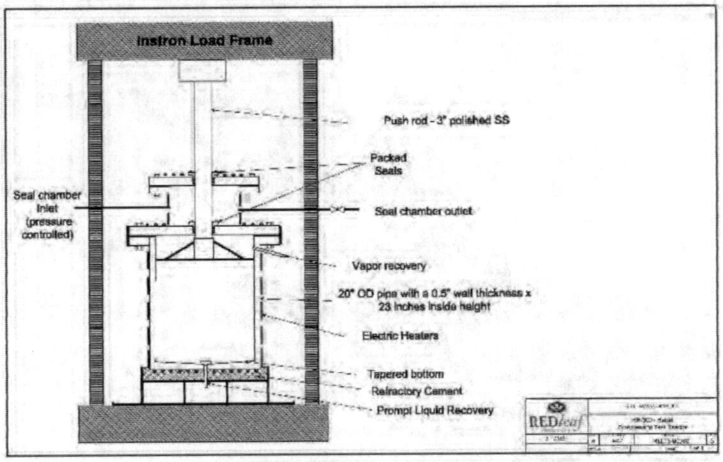

REDLEAF
RESOURCES, INC.

Figure 7

Hot Compression Testing Apparatus in Red Leaf Resources Laboratory

REDLEAF RESOURCES, INC.

Figure 8

Capsule Architecture Designed to Accommodate Settling on Retorting

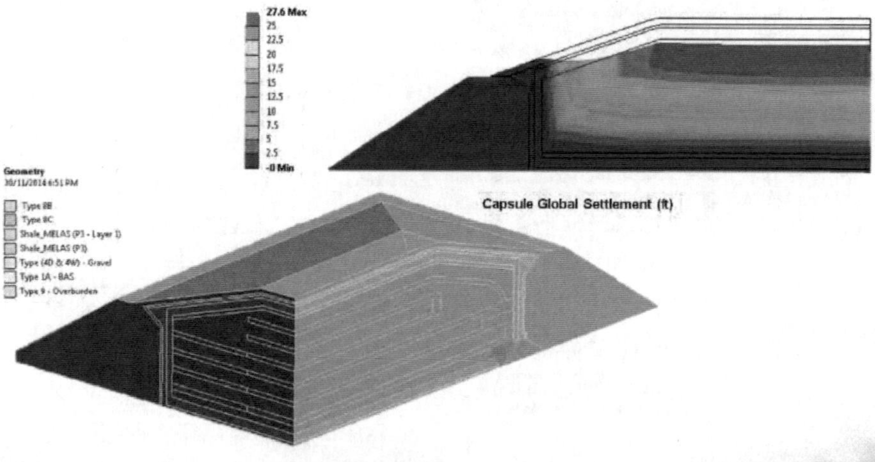

REDLEAF RESOURCES, INC.

Figure 9

Volumetric Strain vs. Void Fraction for 26 gpt Shale Ore

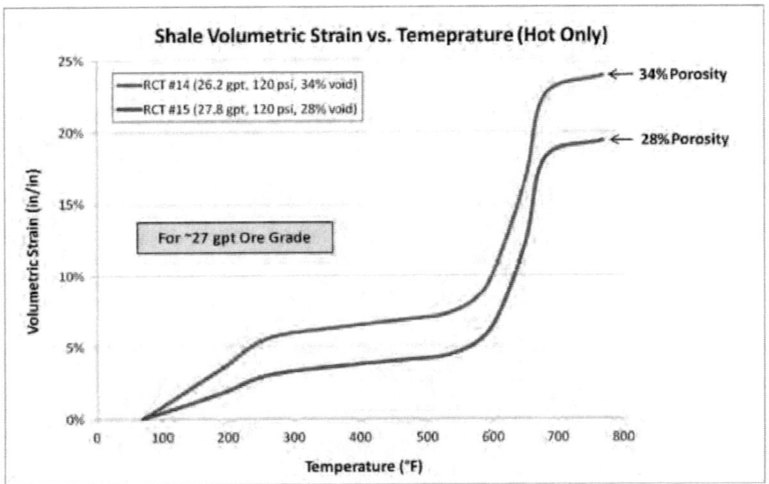

REDLEAF RESOURCES, INC.

Figure 10

Volumetric Strain vs. Void Fraction for 17 gpt Shale Ore

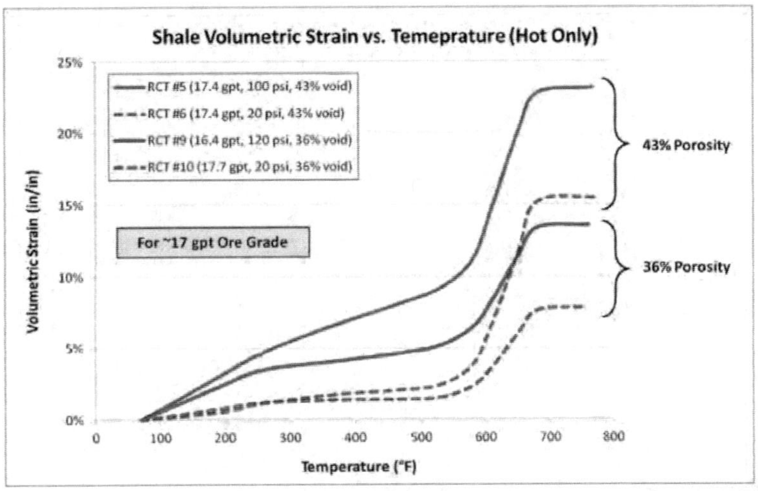

REDLEAF RESOURCES, INC.

Figure 11

Volumetric Strain vs. Void Fraction for 9 gpt Shale Ore

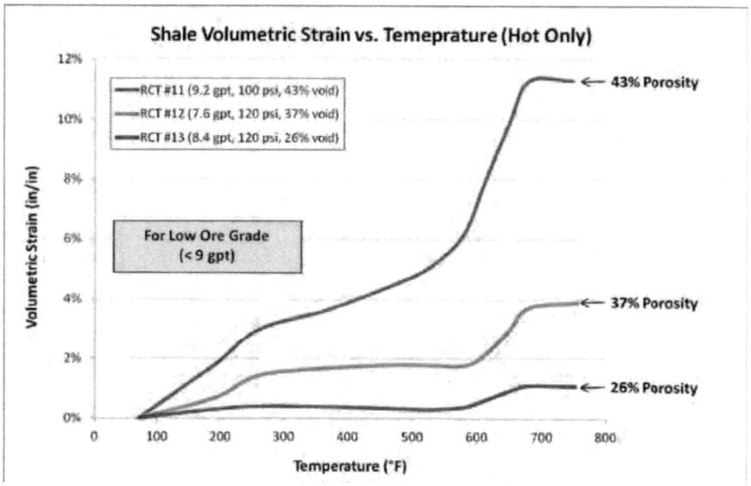

REDLEAF RESOURCES, INC.

Figure 12

Volumetric Strain vs. Ore (Oil Shale) Grade (Richness)

Rubble compressive strength (on retorting) and resulting settling are also a function of shale ore richness

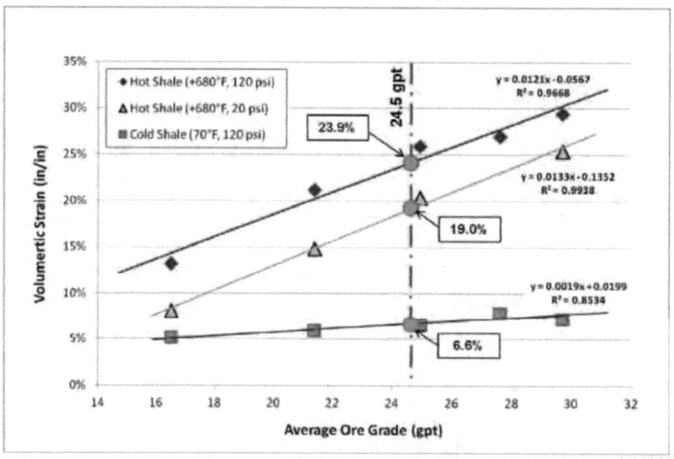

REDLEAF RESOURCES, INC.

- Engineering to reduce void fraction can therefore also be targeted to offset increased settling of richer shale ore, to allow richer shale ore to be processed in a given capsule geometry
- In addition, void fraction and a particle size distribution are extremely important in determining heating fluid (recycle gas) transport and heat transfer in the capsule (see also the presentation by Plikas et al this conference)
- Void fraction is therefore one of the more important parameters to be engineered for capsule structural design, maximum oil production and thermal efficiency from a given capsule size or surface footprint

REDLEAF RESOURCES, INC.

Some or all of the information contained in the presentation represents views, estimates and assumptions of Red Leaf Resources, Inc. and JWBA, Inc. (together, the "Companies"), any of which may change without notice. Neither the Companies nor any of their respective shareholders, affiliates, partners, members, directors, officers, management, employees or representatives makes any representation or warranty, express or implied, as to the accuracy or completeness of any of the information or any other information (whether communicated in written or oral form) transmitted or made available to you

REDLEAF RESOURCES, INC.

UMATAC Industrial Processes

Fushun ATP Project Update & Value Added Processing

Colorado School of Mines 35th Oil Shale Symposium (Salt Lake City)
October 5, 2015

UMATAC Industrial Processes
A company of ThyssenKrupp Industrial Solutions

ThyssenKrupp Industrial Solutions

ThyssenKrupp

Rev: 0

FMG Operation Update

UMATAC Industrial Processes
A company of ThyssenKrupp Industrial Solutions

Plant Operation Improvements:

→ 115 day run length achieved.
→ 80% overall plant availability (95% of design target).
→ 95% ATP Processor availability.

Challenges:

- **Equipment reliability.**
- **Bottlenecks.**
  → 87% of design feed rate.
- **Low ore grade from mine.**
  → Low grade = low oil production and low coke yield (primary fuel source).

Opportunities:

- **Used tire pyrolysis.**
  → Low CAPEX for tire shredding. High OPEX for tire collection.
  → No modifications to ATP.
  → Plant oil production +20%.

ATP Plant Main Flare

Oil Production

Page 2

FMG ATP Plant - 2014/15 Operations Update

UMATAC Industrial Processes
A company of ThyssenKrupp Industrial Solutions

Start Date	Run Length	Feed Rate	Notes
Jul. 2014	40 days	160-180 t/h	FMG sets goal of improving operability.
Oct. 2014	30 days	160-180 t/h	Operating error shuts down oil recovery system.
Dec. 2014	70 days	160-180 t/h	Train derailment disrupts shale supply, ATP burner modified to burn fuel gas.
Feb. 2015	50 days	120-160 t/h	Baghouse bag blinding, plant operates through minus 30°C weather.
Apr. 2015	**115 days**	**140-180 t/h**	**Equipment reliability challenges (i.e. 115 days of operation plus ~12 days of short unplanned outages).**
Sep. 2015	**> 100 days**	**180-230 t/h**	**Partial install of high capacity baghouse bags, new flue gas boiler.**

FMG ATP Plant Control Room

Page 3

FMG ATP Plant - Availability

UMATAC Industrial Processes
A company of ThyssenKrupp Industrial Solutions

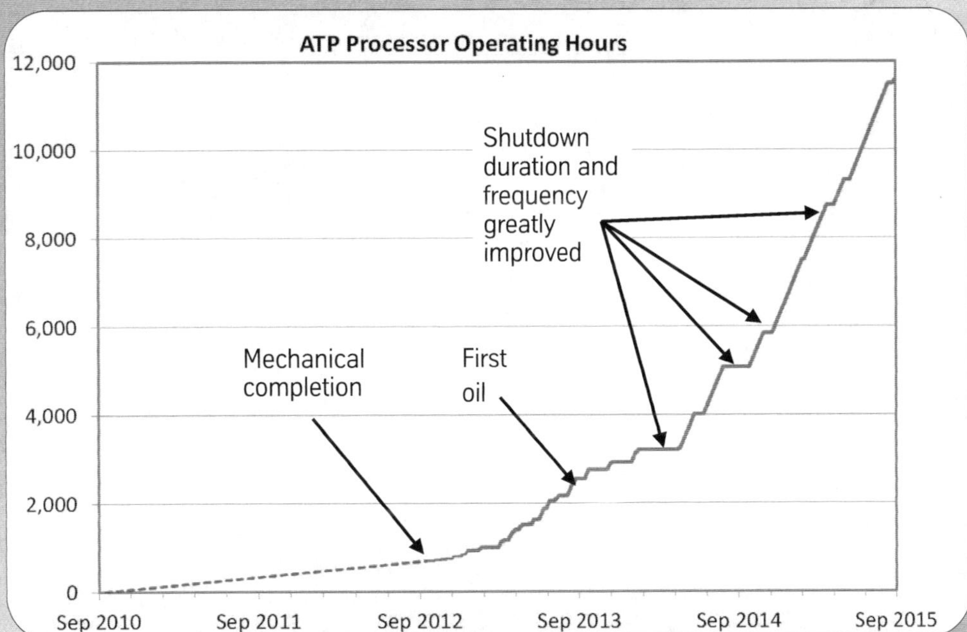

Shutdown target: 10 days, once every 3 months – achieved.

Page 4

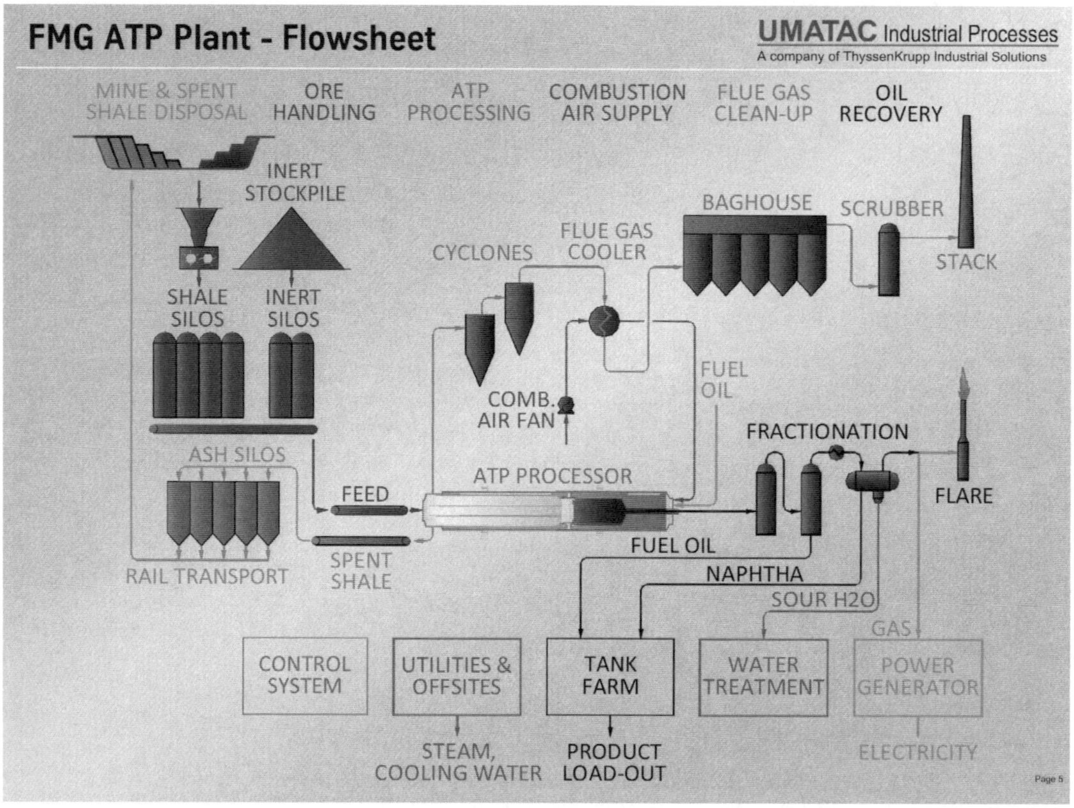
FMG ATP Plant - Flowsheet
UMATAC Industrial Processes
A company of ThyssenKrupp Industrial Solutions
MINE & SPENT SHALE DISPOSAL
ORE HANDLING
ATP PROCESSING
COMBUSTION AIR SUPPLY
FLUE GAS CLEAN-UP
OIL RECOVERY
INERT STOCKPILE
BAGHOUSE
SCRUBBER
FLUE GAS COOLER
CYCLONES
STACK
SHALE SILOS
INERT SILOS
FUEL OIL
COMB. AIR FAN
FRACTIONATION
ASH SILOS
ATP PROCESSOR
FEED
FLARE
FUEL OIL
SPENT SHALE
RAIL TRANSPORT
NAPHTHA
SOUR H2O
GAS
CONTROL SYSTEM
UTILITIES & OFFSITES
TANK FARM
WATER TREATMENT
POWER GENERATOR
STEAM, COOLING WATER
PRODUCT LOAD-OUT
ELECTRICITY
Page 5

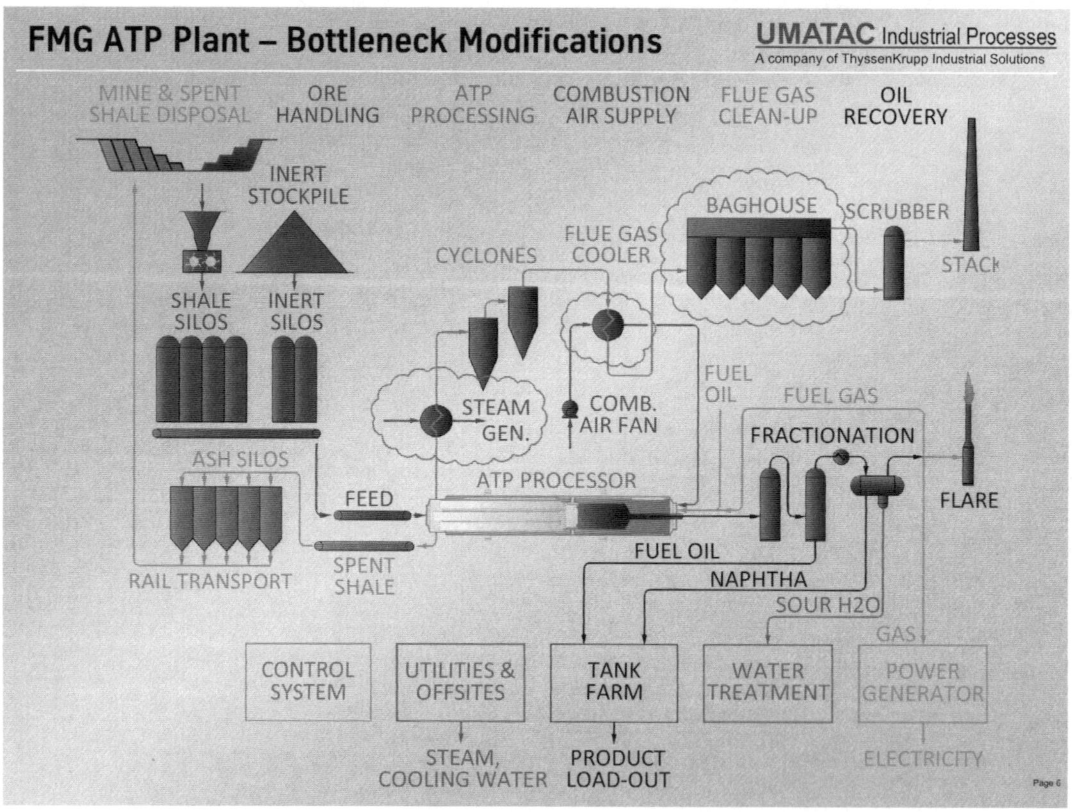
FMG ATP Plant – Bottleneck Modifications
UMATAC Industrial Processes
A company of ThyssenKrupp Industrial Solutions
MINE & SPENT SHALE DISPOSAL
ORE HANDLING
ATP PROCESSING
COMBUSTION AIR SUPPLY
FLUE GAS CLEAN-UP
OIL RECOVERY
INERT STOCKPILE
BAGHOUSE
SCRUBBER
FLUE GAS COOLER
CYCLONES
STACK
SHALE SILOS
INERT SILOS
FUEL OIL
FUEL GAS
STEAM GEN.
COMB. AIR FAN
FRACTIONATION
ASH SILOS
ATP PROCESSOR
FEED
FLARE
FUEL OIL
SPENT SHALE
RAIL TRANSPORT
NAPHTHA
SOUR H2O
GAS
CONTROL SYSTEM
UTILITIES & OFFSITES
TANK FARM
WATER TREATMENT
POWER GENERATOR
STEAM, COOLING WATER
PRODUCT LOAD-OUT
ELECTRICITY
Page 6

Flue Gas Steam Boiler & Baghouse Bags

UMATAC Industrial Processes
A company of ThyssenKrupp Industrial Solutions

Flue Gas Cooler / Air Preheater

- Fouling issues from 2014 solved, but solution reduced cooler capacity.

Flue Gas Steam Boiler

- New steam boiler installed – increases flue gas system capacity and produces utility steam.

Result

- 1 MPa steam produced and ATP feed rate increased to >200 t/h.
- High capacity baghouse bags to be installed. ATP rate to 100% of design.

Page 7

Unplanned Downtime Comparison

UMATAC Industrial Processes
A company of ThyssenKrupp Industrial Solutions

Stuart Demonstration (2004)

→Ash system.
→Duct fouling.
→Corrosion (2004).
→ATP support systems (e.g. lube oil cleanliness).

FMG ATP Plant (2014/15)

Avoidable downtime:

→Operating error.
→Sample valve failure.

Equipment performance:

→Cooler fouling (solved).
→Baghouse capacity.

Equipment quality:

→Sample valve, flue gas fines screw, MCC air conditioning, electrical issues.

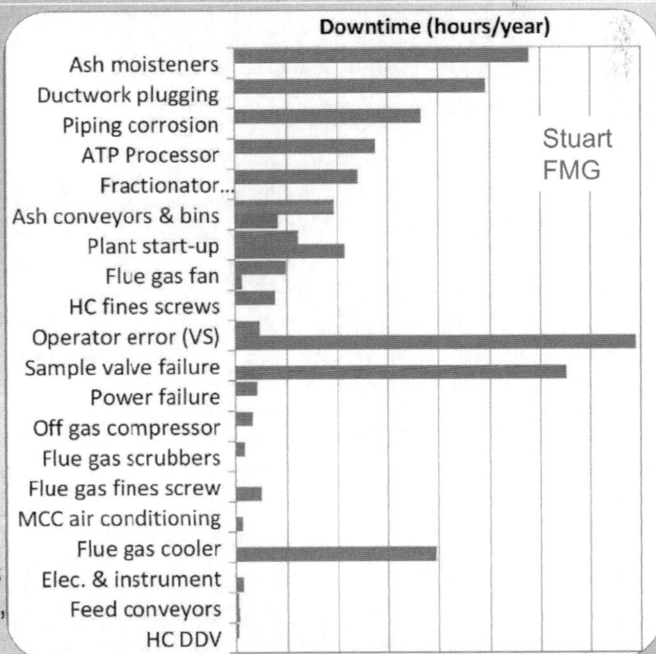

Learnings from Demonstration Plant Implemented; New Challenges at FMG.

Page 8

Total Downtime Comparison

UMATAC Industrial Processes
A company of ThyssenKrupp Industrial Solutions

Stuart Demonstration (2004)

- Lengthy planned shutdowns.
- Organizational constraints.
- ATP retort rear support routine maintenance usually the critical path.

FMG ATP Plant (2014/15)

- Turnaround duration greatly improved (target 10 days).
  - → FMG focus on turnaround performance!
  - → ATP retort rear support eliminated.
  - → Plant approaching design availability (85%).

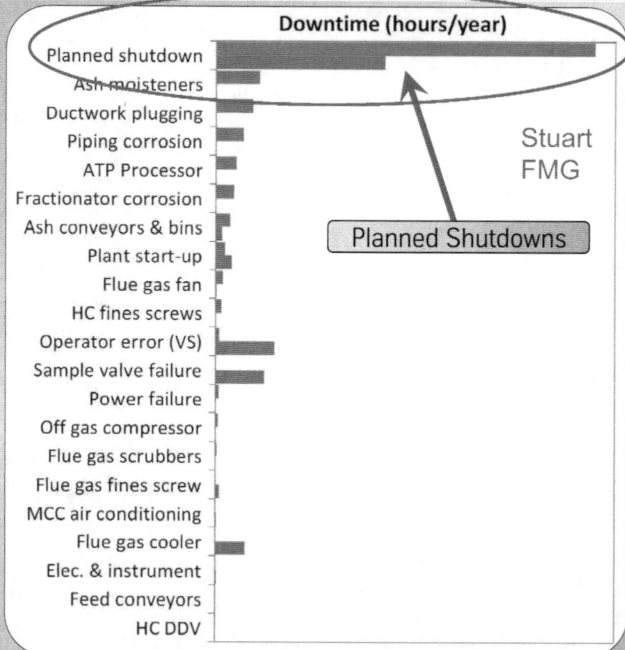

Turnaround Time Significantly Reduced with Good Engineering and Planning

Page 9

Value Added Processing

UMATAC Industrial Processes
A company of ThyssenKrupp Industrial Solutions

West Pit Mine is Closed.

- Pilot testing and design were done on West Pit ore: **7% oil yield, 5% free water.**
- Excellent particulate physical strength.

East Pit Opened – ore grade during early years of mine operation will be low.

- 2015 operations used East Pit ore. **4-7% oil, 5-8% water = low oil production.**
- Lower particulate physical strength, resulting in higher fines generation.

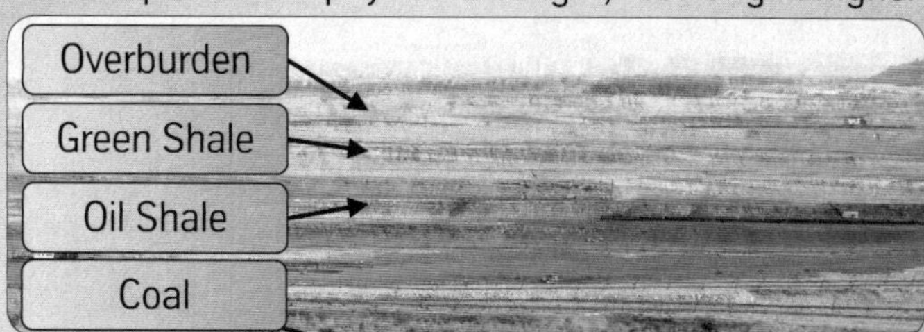

Page 10

Value Added Processing

UMATAC Industrial Processes
A company of ThyssenKrupp Industrial Solutions

Oil Shale Processing:

- Maximize revenue & profit:
  - → Operate at highest availability.
  - → Minimize waste disposal cost & OPEX.
  - → Overall yield vs. most valuable yield. Maximum yield ≠ maximum revenue.

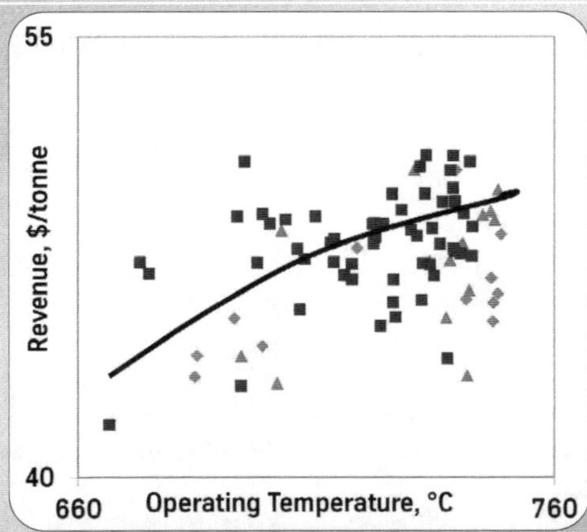

Other Opportunities:

- Large scale pyrolysis system, oil recovery system, infrastructure in place.
- Other readily available hydrocarbon bearing materials?
  - → Charge to process material (e.g. oilfield wastes): $/tonne feed processed.
  - → High oil yield feedstocks (e.g. rubber, plastic): $/bbl oil produced.
- Expend minimal capital, don't impact existing plant operability.

Used Rubber Tire

Page 11

Example – Used Tire Pyrolysis

UMATAC Industrial Processes
A company of ThyssenKrupp Industrial Solutions

Addition to Existing Oil Shale Pyrolysis Plant:

- Low grade East Pit ore:
  - → 2000 bbl/d from 2500 bbl/d plant.
- **Fushun oil shale ore yields 50 to 80 L/t**
- **Used rubber tire yields 575 L/t**
- Zillions of used tires in Shenyang/Fushun region.
- Process 120 t/d of tires = 3.7 million tires/year.

Tire Shredding

Revenue: $8 to 24 MM/a

- Environmental fee $4/tire = $15.0 million/a
- 20 t/d of steel @ $250/t = $ 1.5 million/a
- 375 bbl/d oil, @ $35/bbl = $ 4.0 million/a
- 7 MWt gas → 2 MWe = $ 1.8 million/a
- 11 MWt coke → 3 MWe = $ 2.2 million/a (displaces current fuel gas use).

Cost to collect tires is major expense. "Tire Tax" is incentive for recyclers in some jurisdictions.

Tire derived oil improves quality of Fushun raw shale oil product.

Page 12

Example – Used Tire Pyrolysis

UMATAC Industrial Processes
A company of ThyssenKrupp Industrial Solutions

Addition to Existing Oil Shale Pyrolysis Plant:

CAPEX: $5-10 MM

- Stockpiles, shredding, conveying systems.
- No additional CAPEX for pyrolysis system.

OPEX: $5-10 MM/a

- Tire collection: $6-10 MM/a (location specific)
- Shredding system: $2 MM/a

Revenue: $8-24 MM/a (location specific)

Environmental benefit: Small, polluting "Illegal tire oil extraction" pyrolysis plants displaced.

Tire Shred Feed Hopper

Standalone Tire Pyrolysis Plant:

CAPEX: $30 to 80 MM

- Shredding, pyrolysis, coke processing plants; product shipping, utilities, offices, maintenance shops, fire protection...)

OPEX: $12-30 MM/a

Revenue: $13- 29 MM/a (carbon +$9, gas -$4 MM/a)

Standalone facility = $30 to 80 MM CAPEX. Shale plant add-on < $10 MM

Page 13

Summary

UMATAC Industrial Processes
A company of ThyssenKrupp Industrial Solutions

Availability:

- ATP plant availability >80% (target 85%).
- Run length >100 days (target 90 days).
- Maintenance turnaround ~10 days.

Throughput:

- Plant has reliably achieved > 90% of plant design oil shale feed rate.
- ATP Processor is working well & handling ore much different than design.

Opportunities:

- Price of oil & low grade oil shale are economic challenges.
- Further improving plant operation.
- Co-processing alternate feedstocks to increase revenue.

TKIS and FMG Operating Team

Page 14

Questions?

UMATAC Industrial Processes
A company of ThyssenKrupp Industrial Solutions

谢谢 Thank You شكرا

ThyssenKrupp

Page 15

Contact Information

UMATAC Industrial Processes
A company of ThyssenKrupp Industrial Solutions

UMATAC Industrial Processes Inc.
Suite #110, 6835 Railway Street S.E.
Calgary, Alberta, Canada
T2H 2V6

Telephone: +1-403-910-1000
Facsimile: +1-403-910-1040

email: umatac@thyssenkrupp.com
web: www.umatac.ca

Page 16

Optimizing Reactor Parameters in an Ex-Situ Oil Shale Reactor

Anand Alembath, Hassan Golpour, Joseph D. Smith, Ph.D.

Energy Research & Development Center
Missouri University of Science and Technology
Rolla, MO

Contact Information:
Dr. Joseph D. Smith
Email: smithjose@mst.edu
Office: (573) 341-4294

Outline

- Current work describes Aspen process simulation of an ex-situ oil shale process using vertical moving-bed reactor
- Process Model was developed in which drying zone, combustion zone and reactor zone are simulated separately
- Identified key process parameters for the reactor
- Optimized production of syncrude from oil shale which enhanced process environmental and economic sustainability

Oil Shale Process Model

- Process model has four sections:
  1. Drying zone
  2. Reaction zone
     - Pyrolysis Reaction
     - Mineral Reactions
  3. Combustion zone
  4. Oil Recovery and Separation

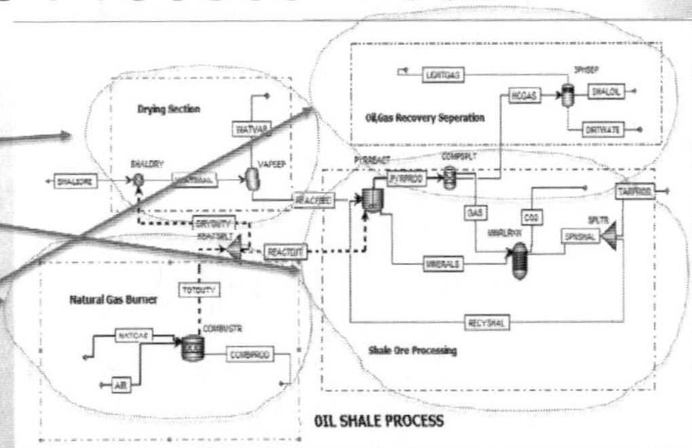

- Pyrolysis reaction is modeled as a CSTR which predicts extent of conversions as a function of temperature and volume
- Mineral reactions are thermodynamically modeled as Gibbs reactor
- Combustion zone provides duty required for reactor (Natural Gas Burner is heat source)

Case 1: Optimizing Temperature

- Operating Conditions
  - Fixed Volume, Fixed Feed Rate, Variable Bed Temperature
  - Pyrolytic reactor Volume = 0.05 m^3
  - Oil Shale Feed Rate = 26 TPD
- Objective: find optimum bed temperature to maximize crude production
- Optimum point = maximum kerogen yield, minimum CO_2 production
- Heat duty required for reactors directly related to natural gas consumption
- Sensitivity analysis performed using Aspen to record following responses to varying bed temperature:
  - Shale oil production rate
  - Light gas production rate
  - Natural gas consumption rate
  - Carbon dioxide production from both reactors

MISSOURI S&T

Case 1 Results

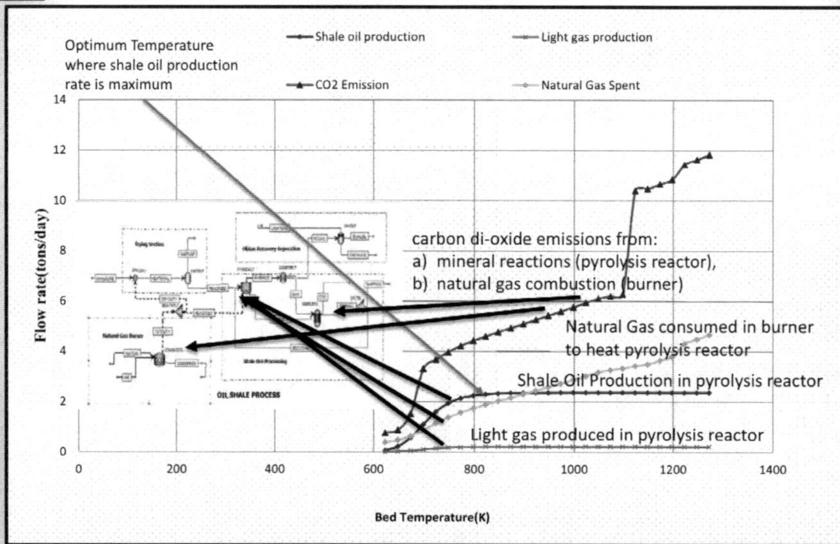

MISSOURI S&T

Carbon Dioxide Emission Analysis

- Two critical points at 700K and 1100K.
- 700 K is where the dolomite decomposition takes place.
- The dolomite decomposes to calcite which further decomposes to CaO and CO_2 at 1098K.

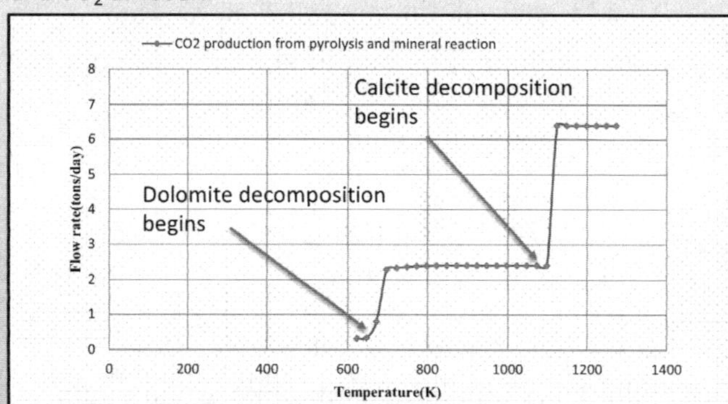

Case 2: Optimizing Flow Rate

- Operating Conditions:
  - Fixed Volume, Limited Heat Duty, Fixed Temperature, Variable Feed Rate
  - Volume of reactor: 0.05 m^3
  - Reactor temperature: 873 K
  - Limited reactor heat duty: 1.46×10^6 BTU/hr
- Optimum temperature used found in Case 1: 873K
- As feed rate increases:
  - residence time goes down
  - bed temperature remains constant as natural gas flow increases
  - natural gas flow limited so conversion decreases as temperature decreases
- Based on physical limitation of gas burners, heat duty limit assumed to be 1.46×10^6 BTU/hr

Missouri University of Science and Technology 7

Case 2 Results

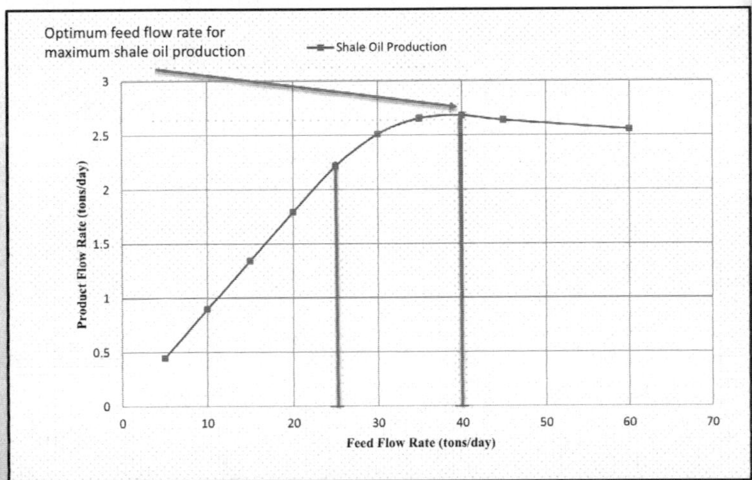

- Above 26 TPD (flow rate considered in first case), shale oil production increases with increased flow to reactor
- Above 40 TPD, shale oil production declines with increasing flow to reactor (insufficient heat duty available)

Missouri University of Science and Technology 8

Case 3: Optimizing Combination of Reactor Temperature and Shale Feed Rate

- Operating Conditions:
  - Fixed Volume, Variable Temperature, Variable Feed Rate, Limited Heat Duty
  - Reactor volume: 0.05 m^3
  - Limited reactor heat duty: 1.46E6 BTU/hr
- For specific reactor temperature, increasing feed rate to find optimum combination of temperature and feed rate to maximize maximum shale oil production

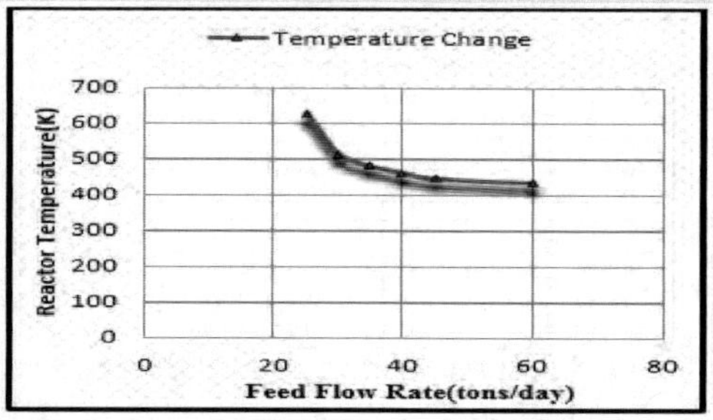

Reactor Temperature Change with Feed Flow Rate at Fixed Heat Duty

MISSOURI UNIVERSITY OF SCIENCE AND TECHNOLOGY 9

Optimum Temperature to maximize shale oil production rate with minimum CO2 production = 500°C

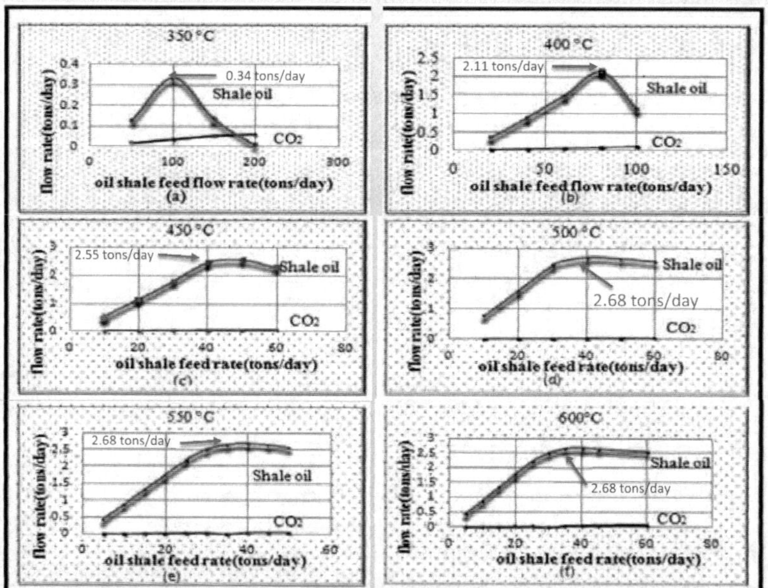

MISSOURI UNIVERSITY OF SCIENCE AND TECHNOLOGY 10

MISSOURI S&T

Case 4: Reactor Volume

- Operating Conditions:
  - Fixed Temperature, Fixed Feed Rate, Limited Heat Duty
- Reactor volume cannot be changed or optimized once reactor is built
- Shale oil yield increases as volume increases till maximum heat duty limit is reached

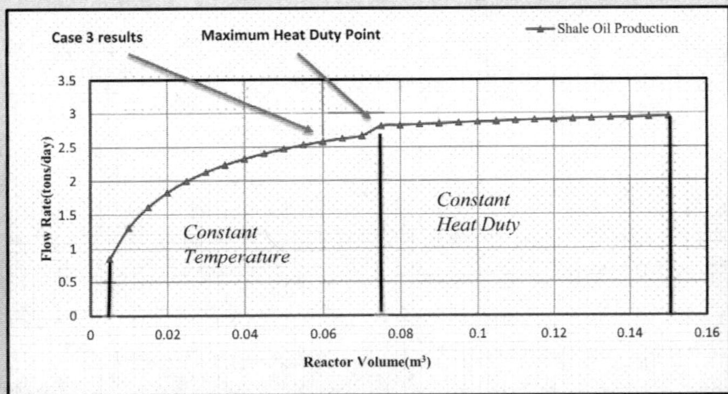

MISSOURI UNIVERSITY OF SCIENCE AND TECHNOLOGY 11

MISSOURI S&T

Conclusions and Recommendations

- Reactor operating parameters have been studied by:
  - Developing aspen model to simulate surface shale oil production process
  - Conducting sensitivity analysis composed of several cases using model
- Case 1 concluded that pyrolysis reaction enhanced with increasing reactor temperature until:
  - No further increase in shale oil production
  - Reactor temperature for increased carbon dioxide generation due to mineral decomposition
- Different cases concluded that during process scale up, bed temperature, flow rate and reactor volume had significant effect on maximizing shale oil yield
- Model allows investigation of various conditions difficult to test in lab (impossible in plant) to identify optimal system operating temperature and flow rate
- Recommend using Aspen model to conduct similar analysis for process design to establish system operating conditions

MISSOURI UNIVERSITY OF SCIENCE AND TECHNOLOGY 12

MISSOURI S&T

References

- Kleinberg, R. L. (2006). Oil Shale
- Diaz, J. C., & Braun, R. L. (1984). Process simulation model for a staged, fluidizedbed oil shale retort using lift pipe combustor.
- Sheritt, R., Jia, J., Meilani , P., & Schmidt, J. (2009). Advances in steady-state process modeling of oil shale retorting. 29th Oil Shale Symposium.
- Singleton, M. F., Koskinas, G. J., Burnham, A. K., & Raley, J. H. (1986). Assay Products from green river oil shale. Lawrence Livermore National Laboratory.
- Smoot, & Douglas. (2012). Demonstration of advanced technology for surface processing of oil shale.
- Hendrickson, T. A. (1974). Oil shale processing methods. 7th Oil Shale Symposium.
- Aspen Plus. (2011). Model for Oil Shale Retorting Process.
- Speight, J. G. (2012). Shale Oil Production Processes

MISSOURI UNIVERSITY OF SCIENCE AND TECHNOLOGY 13

MISSOURI S&T

- Questions?

Energy Research & Development Center
Missouri University of Science and Technology
Rolla, MO

Contact Information:
Email: smithjose@mst.edu
Office: (573) 341-4294

MISSOURI UNIVERSITY OF SCIENCE AND TECHNOLOGY 14

OCCURRENCE OF RARE EARTHS IN GREEN RIVER OIL SHALE

Alan K. Burnham
Jeremy Boak

PRESENTED TO: 35th Oil Shale Symp.
October 5-6, 2015

AMSO
AMERICAN SHALE OIL, LLC

Is Green River oil shale a potential source of Rare Earths?

AMSO

2

- Rare earth elements (REE) are important for many industrial uses
  - Magnets, batteries, lamps, phosphors, lasers, catalysts, medical imaging, high-index glasses, ceramics, etc.
- China now produces most of the REE
- Some have proposed Green River oil shale as a source of REE
- AMSO obtained analyses of many element through the Parachute Creek and Garden Gulch members for a core on its RD&D lease
- Some samples from other locations were also analyzed

Bottom Line: REE concentrations are near or below average crustal abundance

Background from USGS reports

3

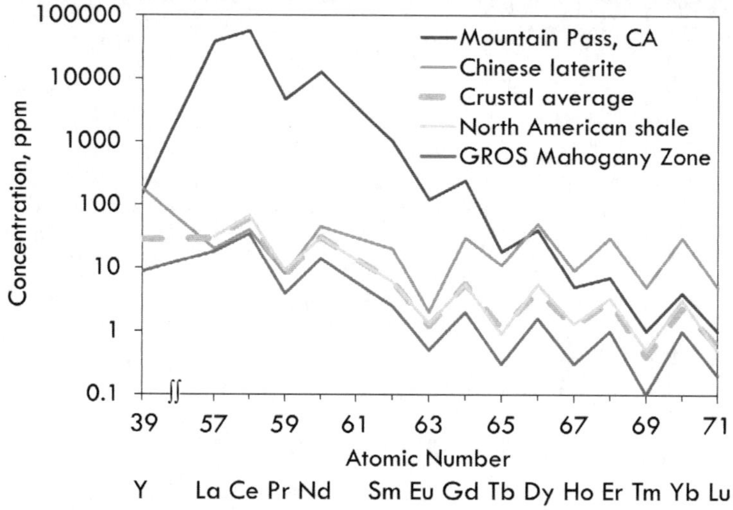

Long et al., USGS S.I. Rpt. 2010-5220; Birdwell, USGS O.F. Rpt. 2012-1016

The Garden Gulch Member is richer in REE

4

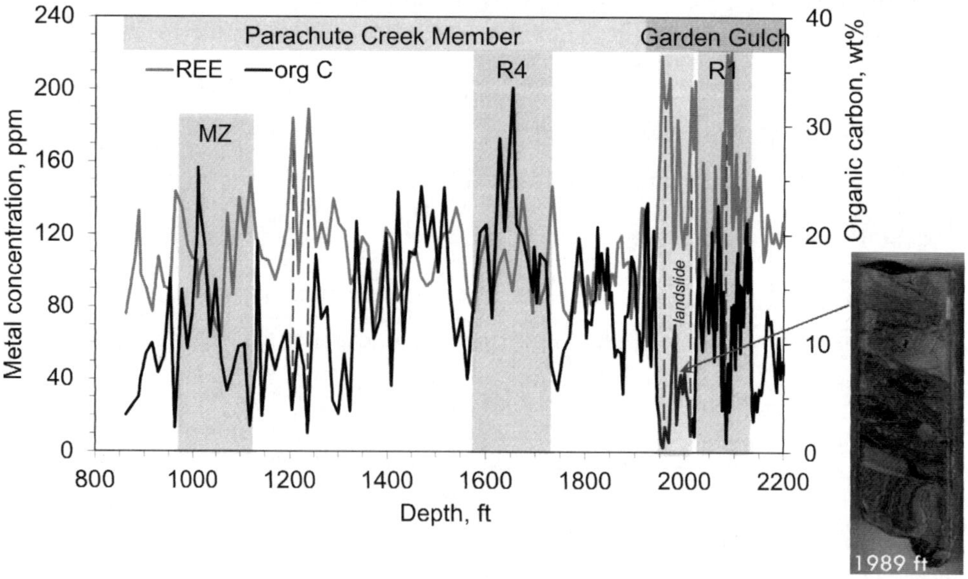

Rich REE interval near top of the L-1 interval

5

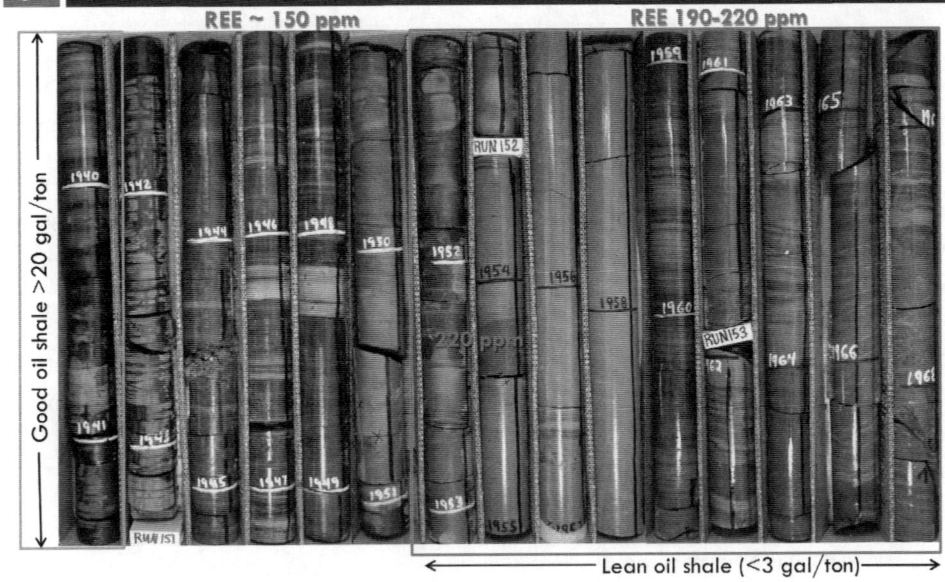

Rich REE – lean kerogen interval within R-1

6

Good oil shale (avg. 35 gal/ton)

Lean oil shale (avg. <10 gal/ton)

REE correlate strongly with each other and to varying degrees with Th, Al, Nb, Fe, and Si

7

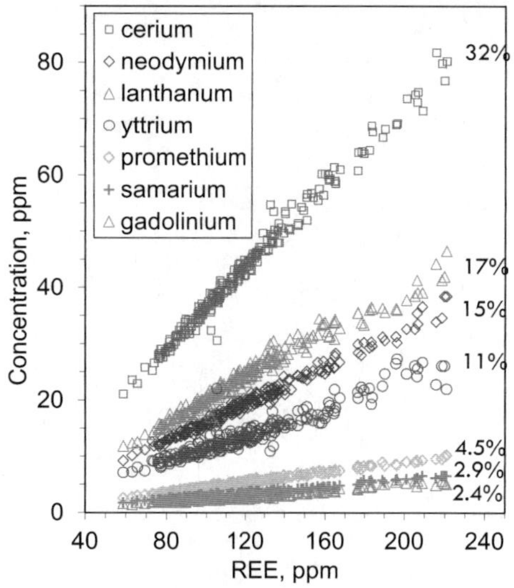

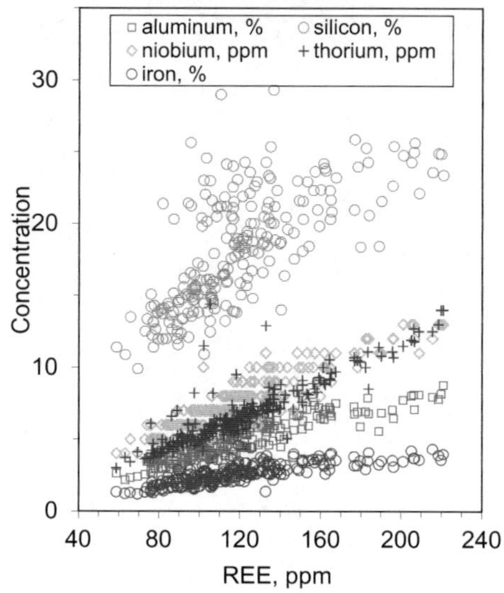

Distribution and maximum values are at crustal average!

REE concentrations are diluted by carbonates and kerogen

8

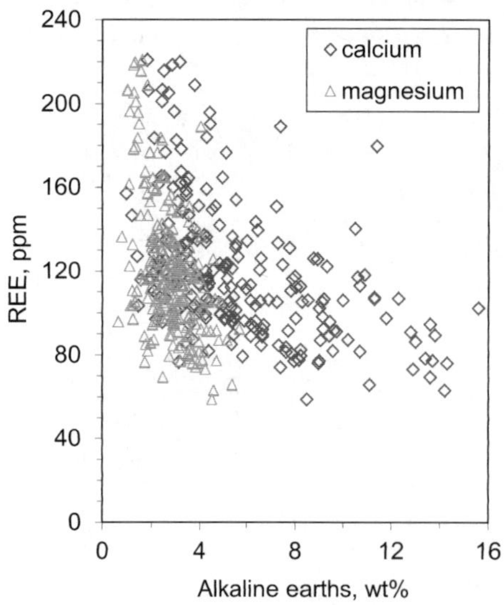

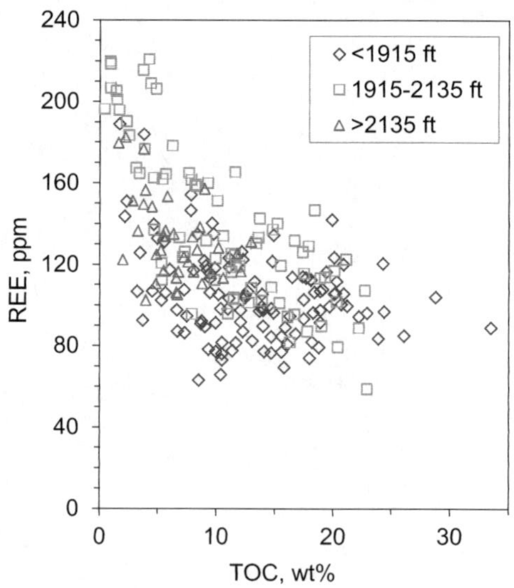

REE, Al_2O_3 and Zr correlate roughly with clay content in the Garden Gulch member

9

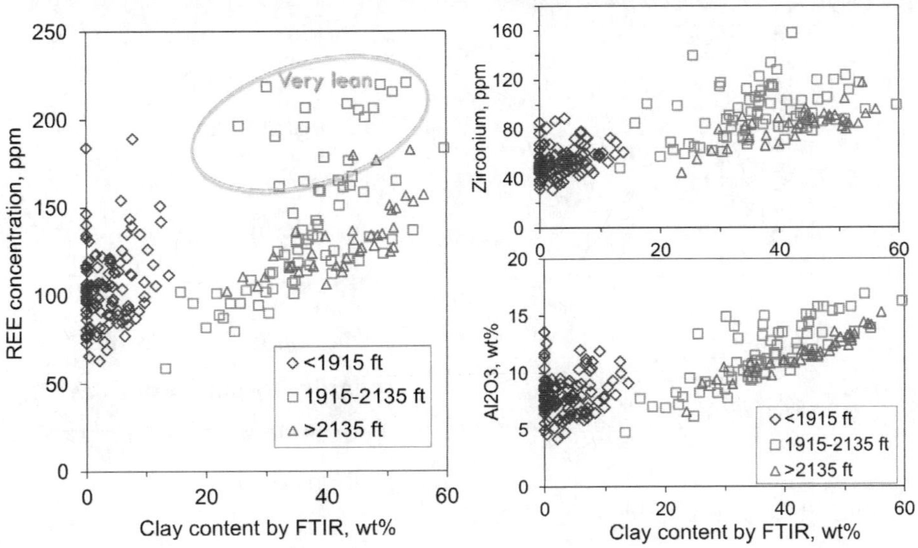

REE are elevated in the landslide and R1 intervals but U, Bi, and Pb only in the R1

10

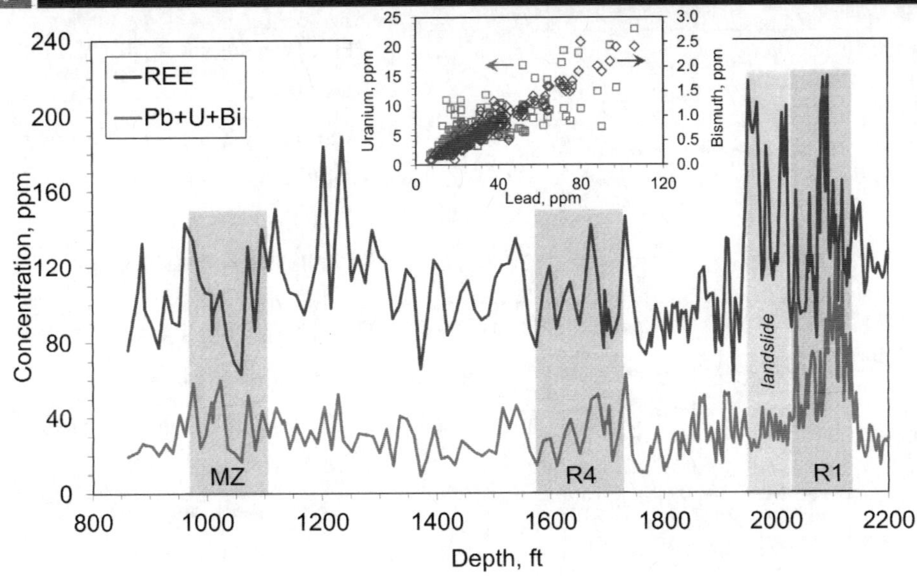

Similar results are obtained from the western portion of the Piceance Basin

11

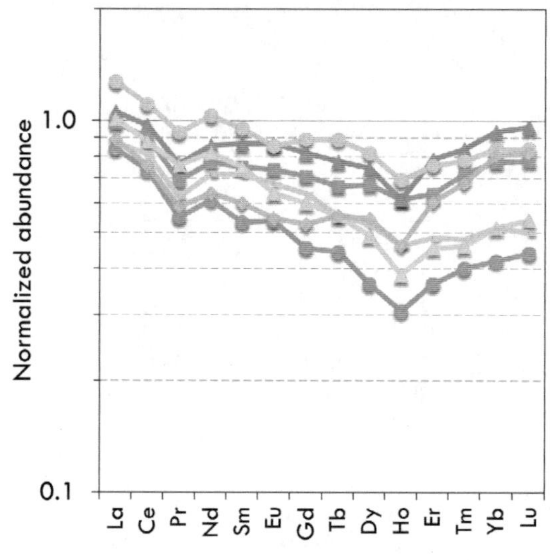

- Various samples from the lower Green River Form. (Douglas Creek)
- Normalized to North American shale composite
- No enrichment!
- Dilution by carbonates and kerogen means lower concentrations than crustal abundance

Samples richest in phosphate are not particularly enriched, either

12

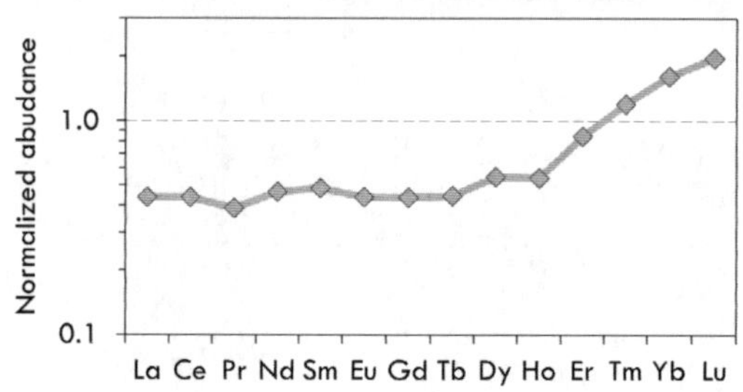

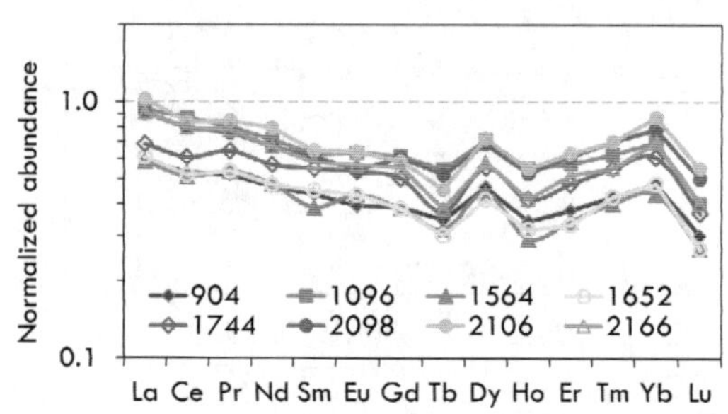

- REE often substitute for Ca in apatite
- Mahogany Zone sample near Cathedral Bluffs
  - 3.4% P_2O_5
  - Slightly enriched in the heaviest REE
  - 23% CaO explains depletion of light REE
- Eight samples from the AMSO core had 1-2% P_2O_5
  - None were enriched in REE

Green River oil shale is not a viable source for Rare Earth Elements

13

- The maximum concentrations of REE in the Green River Formation are approximately crustal abundance
  - The distribution among the elements reflects very large scale averaging through sediment recycling
- The highest REE concentrations are associated with very lean strata
- REE concentrations correlate strongly with other elements associated with clay minerals
- REE concentrations are diluted by carbonates and kerogen
- REE concentrations do not correlate with U, Pb and Bi concentrations, which correlate with each other

USGS
science for a changing world

Comparison of Mineral Occurrences in the Piceance and Uinta Basins of the Eocene Green River Formation

Justin E. Birdwell, Ronald C. Johnson, Tracey J. Mercier, and Michael E. Brownfield
U.S. Geological Survey, Denver, CO

U.S. Department of the Interior
U.S. Geological Survey

35th Oil Shale Symposium, October 5-6, 2015
Salt Lake City, Utah

Purpose

- Recover and assess data from large-scale XRD mineralogy studies by the U.S. Geological Survey and former U.S. Bureau of Mines of Uinta and Piceance Basin oil shale cores
- Improve our understanding of the geographic distribution of minerals in the principal basins of the Green River Formation
- Compare mineral distributions from the different basins and use variations in mineralogy as a proxy to examine differences in lake chemistry and how it changed through time

USGS

2

Previous studies

- Smith, Trudell and Robb, 1972, U.S. Bureau of Mines Report of Investigations 7693, Mineral occurrences with peak height relative abundance for ~1,100 samples from the WOSCO EX-1 Core.
- Desborough, 1976, U.S. Geological Survey Open-File Report 76-381, Mineralogy the upper Parachute Creek Member in the eastern Uinta Basin, Utah, 388 samples between the Curly Bed and Wavy Bed Tuffs (3 cores, includes Mahogany zone).
- Cole and Picard, 1978, Geological Society of America Bulletin, v. 89, p. 1441-1454, Nearshore and offshore comparison of Parachute Creek Member mineralogy in Piceance and eastern Uinta Basins, 22 Uinta samples (WOSCO core).

USGS

3

Overview of Lake Uinta

Highlighting differences in Uinta and Piceance Basins

USGS

4

Two large, long-lived saline lakes were present in the Rocky Mountain region in Eocene time, **Lake Gosiute** in the Greater Green River Basin and **Lake Uinta** in the Uinta and Piceance Basins.

Lake Uinta was a meromictic (density stratified) lake, preceded by a mostly freshwater lake that may have occasionally been brackish.

Major mineralogical transitions occurred over the history of lake and are well described for the Piceance Basin.

USGS

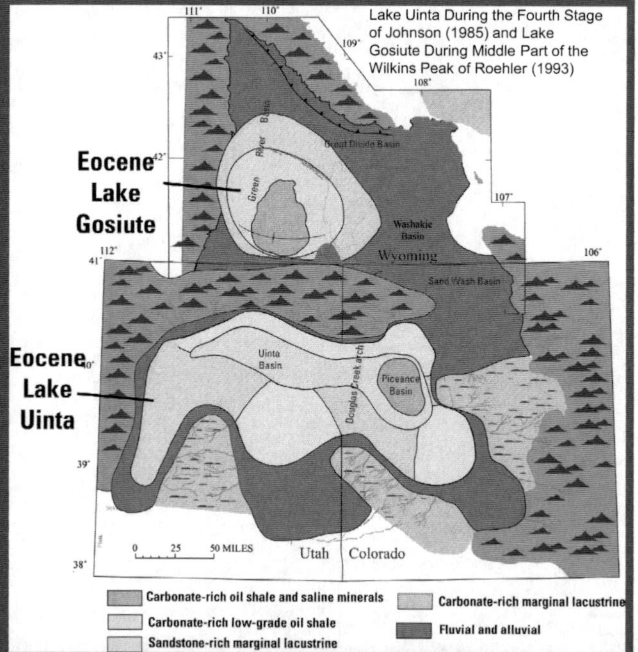

5

Fresh-water lakes were present in the Uinta and Piceance Basins prior to the Long Point Bed.

This was followed by deposition of the clay-rich (illite) Garden Gulch Member.

The R-2 zone marks the transition to the more dolomite- and feldspar-rich oil shale of the Parachute Creek Member, which includes a substantial saline interval in Piceance Basin and the Mahogany zone.

USGS

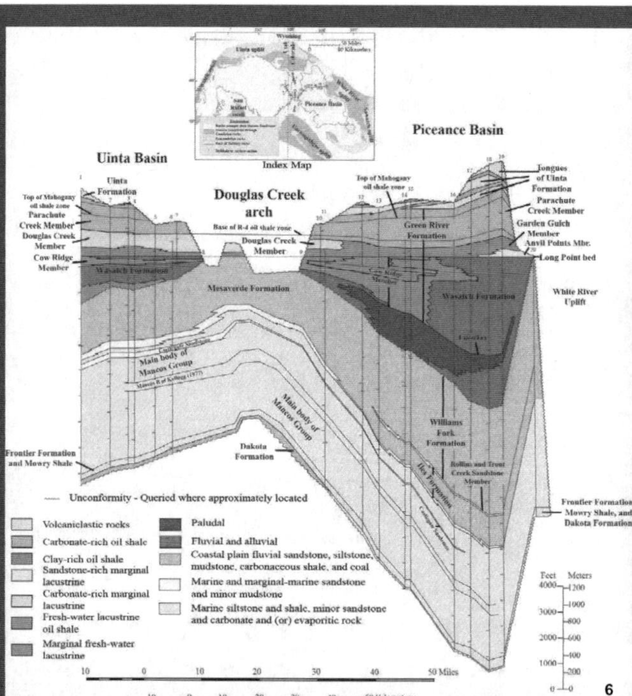

6

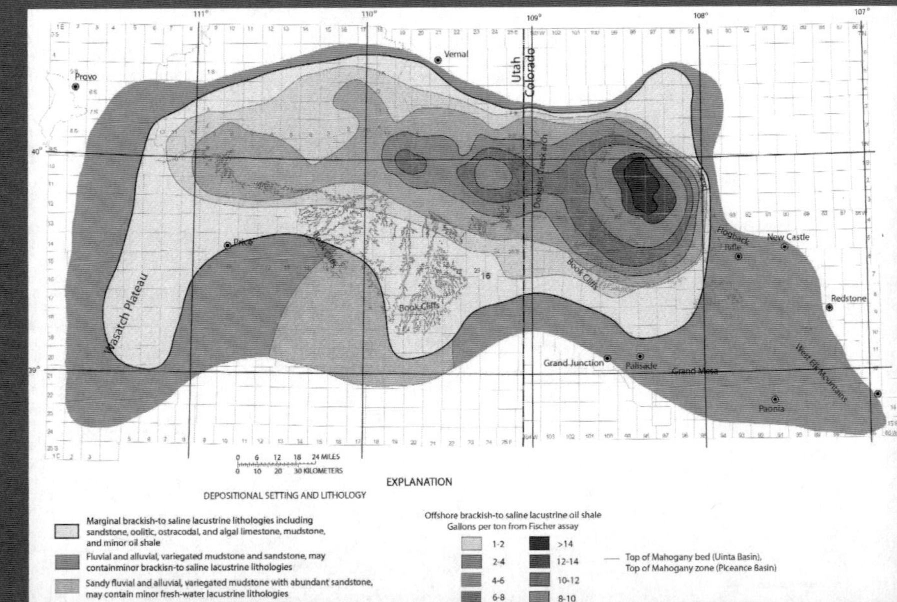

USGS Extent of Lake Uinta during R-0 deposition

7

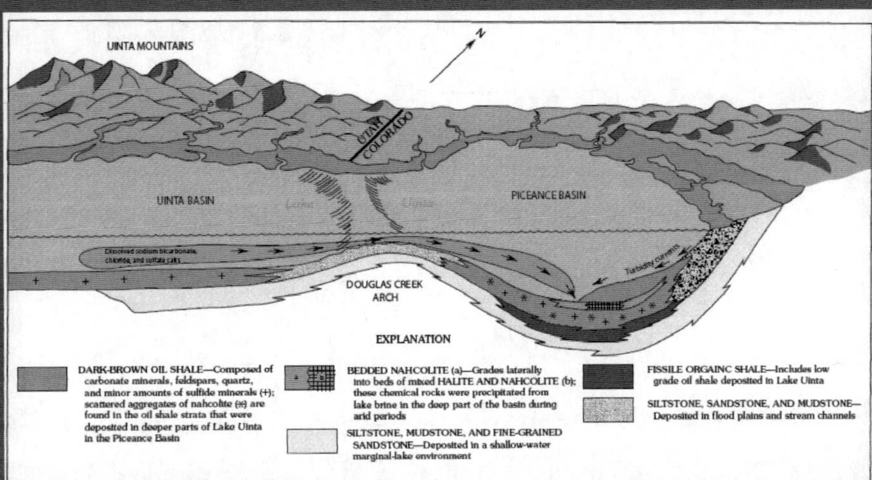

Saline minerals accumulated primarily in Piceance Basin in the lower Parachute Creek Member. Dyni's model (1987, fig. 12, above) showing net flow of water out of the Uinta Basin and into the Piceance facilitating nahcolite and halite deposition.

USGS

8

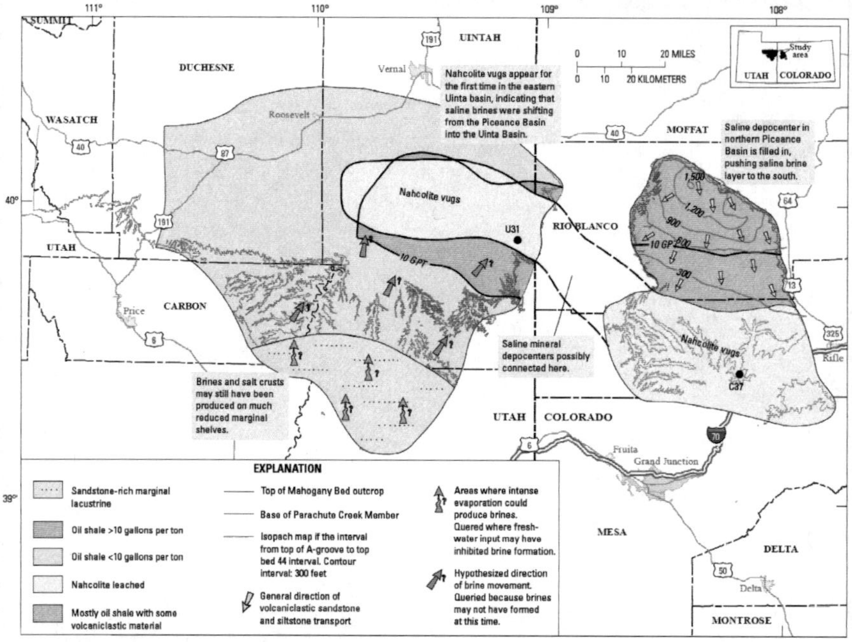

Figure 35. Paleogeographic map of the Uinta and Piceance Basins at the end of the A-groove to bed 44 interval showing (1) where nahcolite vugs occur in both basins, (2) where oil yields are greater than and less than 10 gallons per ton (GPT) (modified from Johnson and others, 2010 a, b), and (3) areas of marginal lacustrine deposition (modified from Johnson, 1985).

9

Methodology

- A database of X-ray diffraction (XRD) mineralogy results for approx. 7,500 samples from 39 coreholes was compiled from U.S. Bureau of Mines and USGS data.
- Sampling frequency varied from well to well: ~60% of samples represent < 5 ft. of core, 20% represent ~10 ft.
- Only the presence or absence of minerals is indicated. Detection limits are unknown but are likely to be ~5% or lower for most minerals.

USGS

10

Database details

UINTA BASIN

- Total samples: **1,159** from 14 cores
- 22 minerals identified; 14 occur in > 5% of samples

PICEANCE BASIN

- Total samples: **6,394** from 25 cores
- 36 minerals identified; 13 occur in > 5% of samples

USGS

11

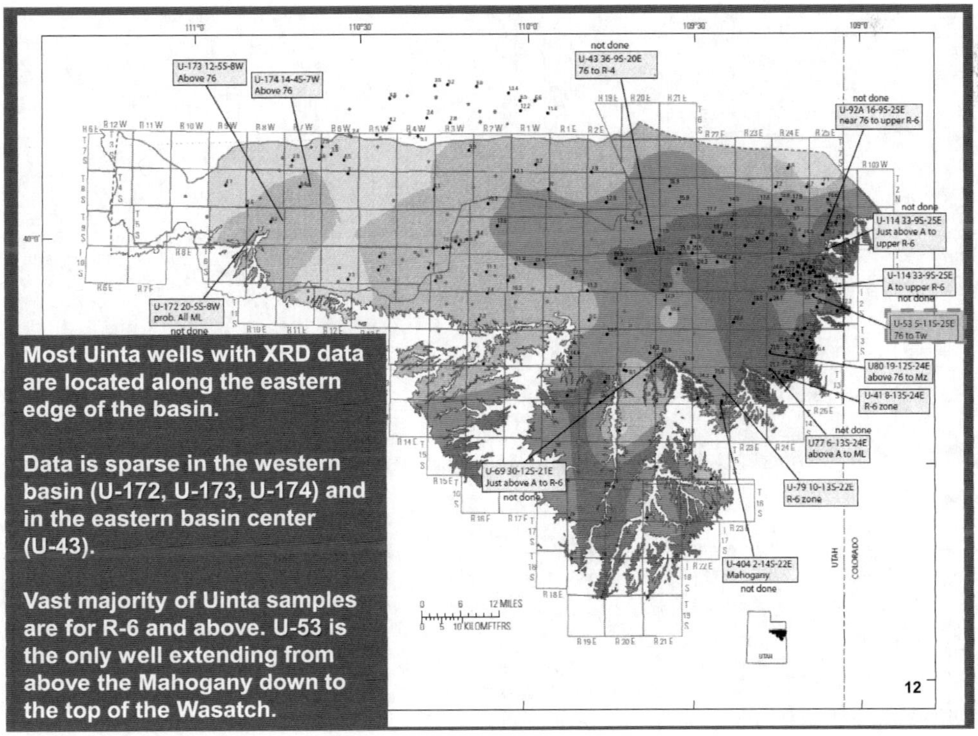

Most Uinta wells with XRD data are located along the eastern edge of the basin.

Data is sparse in the western basin (U-172, U-173, U-174) and in the eastern basin center (U-43).

Vast majority of Uinta samples are for R-6 and above. U-53 is the only well extending from above the Mahogany down to the top of the Wasatch.

12

Well locations

Most representative wells containing nearly all intervals between Cow Ridge to above A-groove: C-34, C-151, C-158, C-195, C-200, C-214, C-230, C-329

Other wells mostly represent Parachute Creek (R-2 through Mahogany)

C-162 is the only core focused on lower Parachute Creek and Garden Gulch (R-0 through R-4)

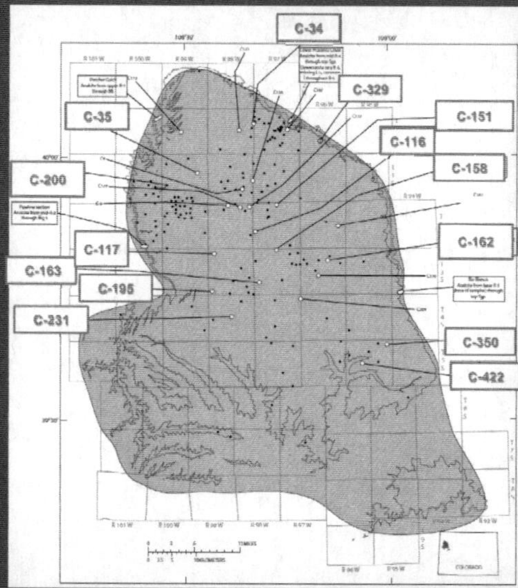

USGS

13

Differences in Organic Richness & Oil Properties

Based on Fischer assay results

USGS

14

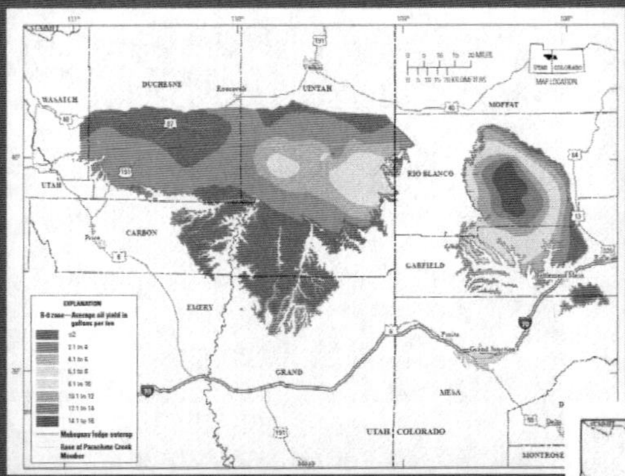

Average (acre basis) Fischer assay oil yield (gallons per ton) for the R-0 zone (Garden Gulch Member).

Organic shale concentrated in the central and central-eastern edge of the Uinta (6 to 10 GPT) and in the center (slightly north-western) part of the Piceance Basin (~15 GPT).

Distribution of average Fischer assay oil specific gravity for the R-0 zone.

Oil density is nearly uniform (~0.90) in the Piceance Basin with a few exceptions, but increases substantially from north to south in the Uinta (**thermal maturity effect**).

USGS

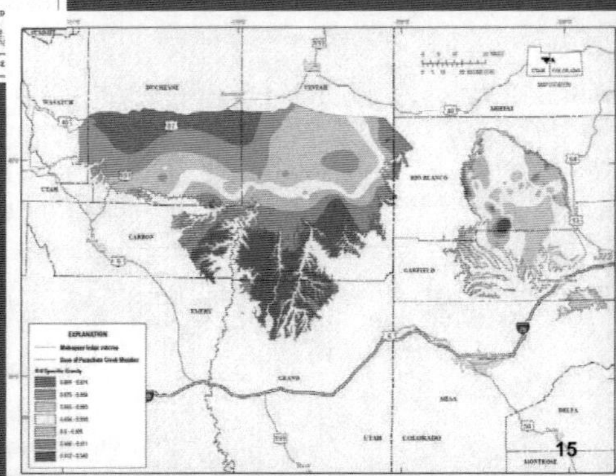

15

Deeply buried, thermally mature R-0 oil shale along the northern edge of the Uinta Basin.

Underlying freshwater lacustrine rocks are the major sources for oil fields (Altamont-Bluebell) and unconventional oil plays (Uteland Butte), but depths for R-0 should be deep enough for the rocks to be significantly more mature than those in other parts of the basin.

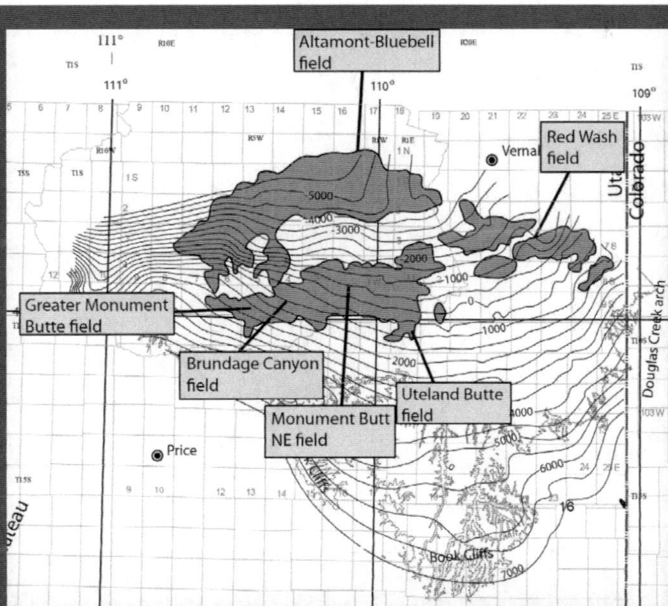

USGS

Structure contour horizon is base of Long Point Bed of the Green River Formation (Uteland Butte assessment, Johnson et al., 2015).

16

Average Fischer assay oil yield (gallons per ton) for the R-5 zone (Parachute Creek Member).

Although the distribution is more uniform, a similar organic richness pattern is observed in the Uinta (richer oil shale throughout on average, 5-10 GPT). The Piceance distribution is similar to R-0, but much richer (up to 40 GPT).

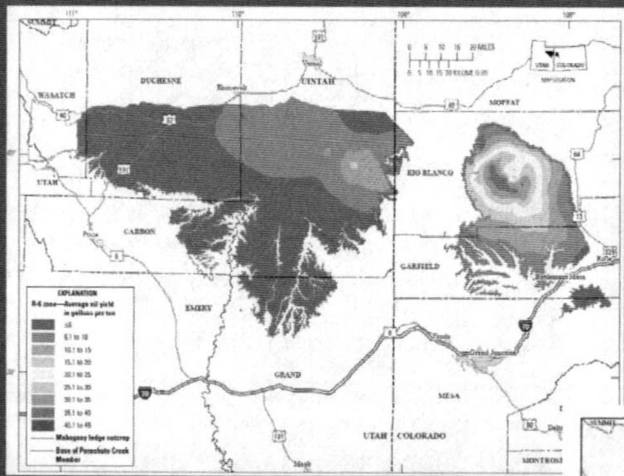

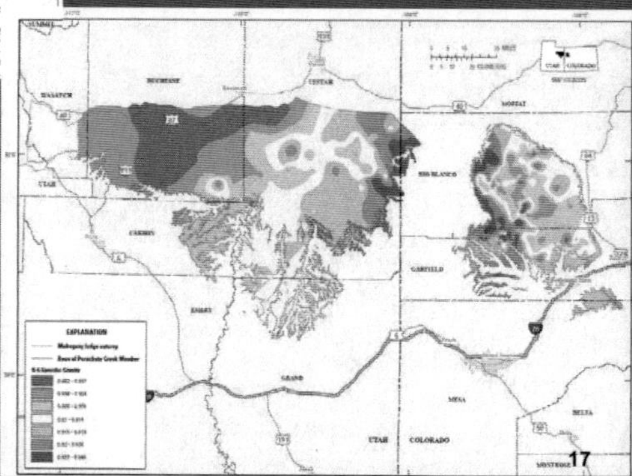

Distribution of average Fischer assay oil specific gravity for the R-5 zone.

Oil density is higher in the Uinta as you go from west to east and vice versa for Piceance (marginal/shallow deposition).

USGS

17

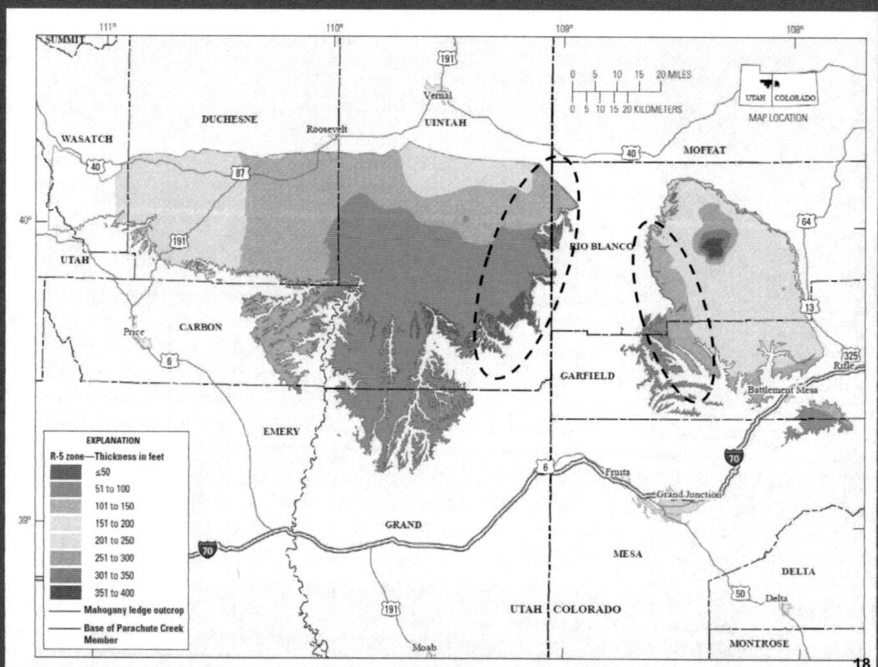

18

Average Fischer assay oil yield for the Mahogany zone.

Nearly uniform distribution of high grade (> 20 GPT) oil shale throughout the Piceance Basin. Rich oil shale (~20 GPT) concentrated along the eastern and southern edges of the Uinta.

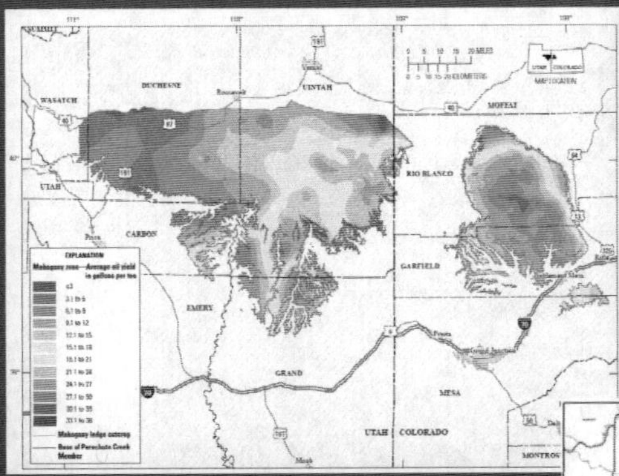

Distribution of average Fischer assay oil specific gravity for the Mahogany zone.

Uinta oils get heavier going from to east to west and south to north-east. Piceance Basin oil specific gravities vary significantly from the western edge to the basin center (preservation effect? Terrestrial inputs?).

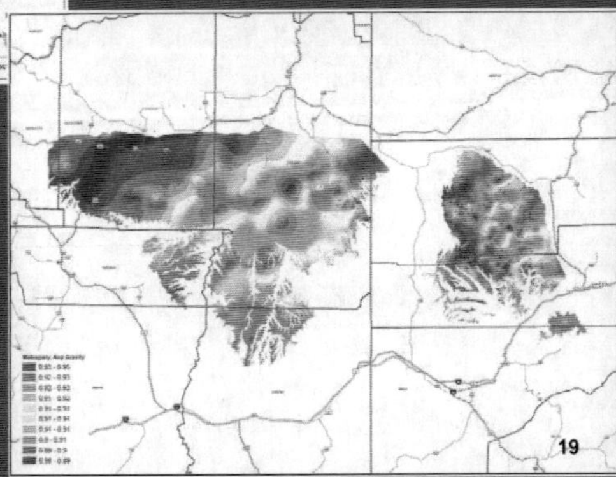

19

Mineral occurrence distributions by zone

For Uinta and Piceance Basins

USGS

20

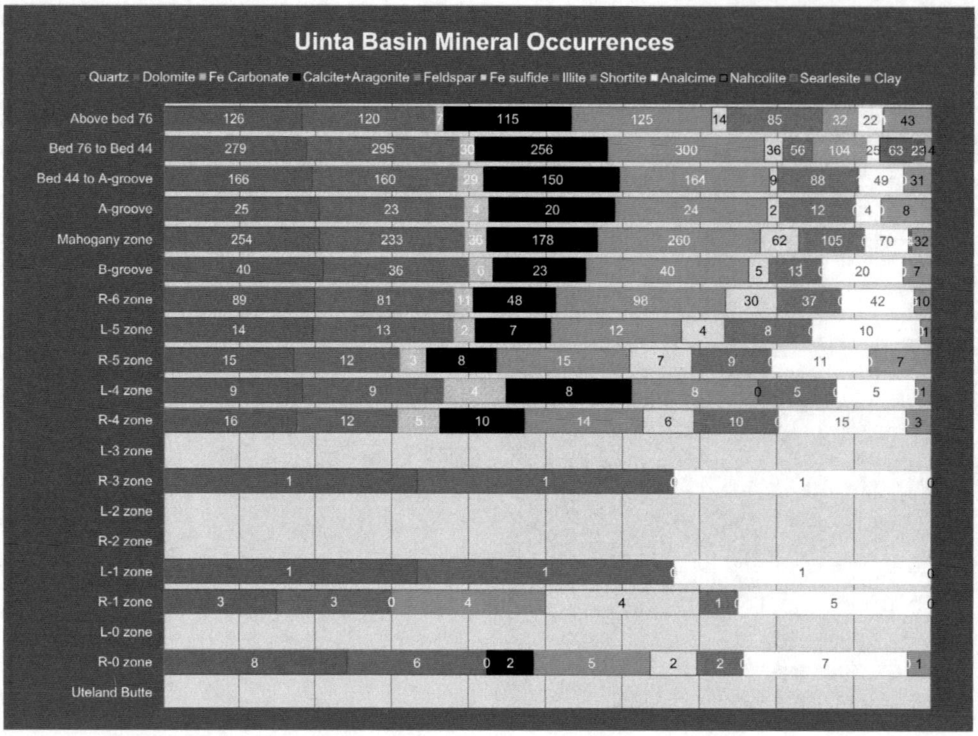
Uinta Basin Mineral Occurrences
Quartz Dolomite Fe Carbonate Calcite+Aragonite Feldspar Fe sulfide Illite Shortite Analcime Nahcolite Searlesite Clay
Above bed 76
Bed 76 to Bed 44
Bed 44 to A-groove
A-groove
Mahogany zone
B-groove
R-6 zone
L-5 zone
R-5 zone
L-4 zone
R-4 zone
L-3 zone
R-3 zone
L-2 zone
R-2 zone
L-1 zone
R-1 zone
L-0 zone
R-0 zone
Uteland Butte

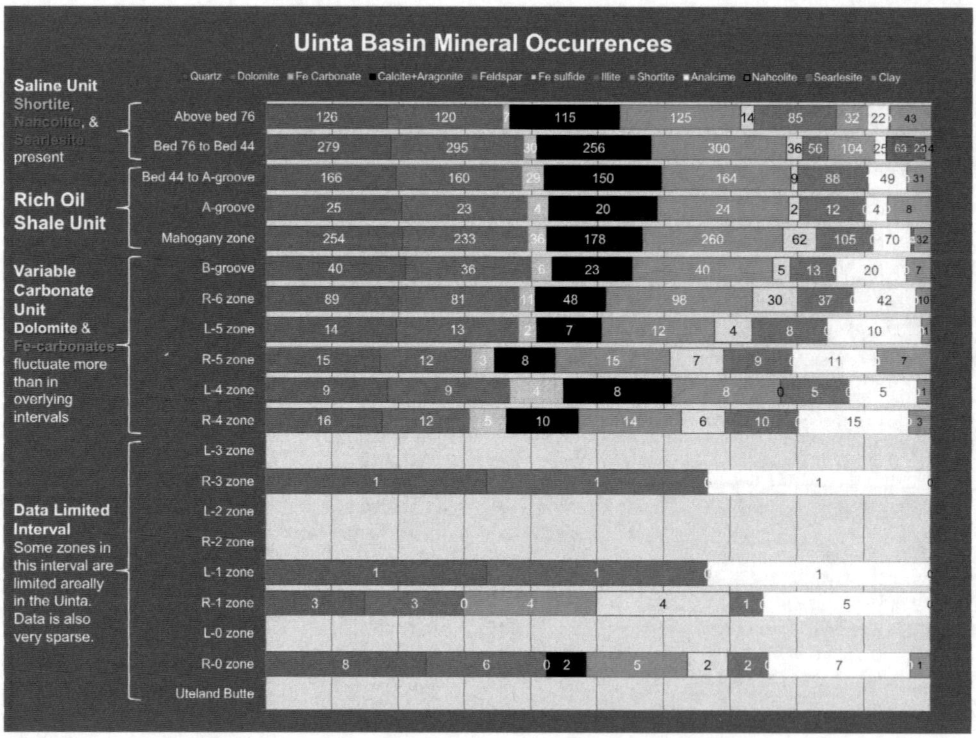
Uinta Basin Mineral Occurrences
Quartz Dolomite Fe Carbonate Calcite+Aragonite Feldspar Fe sulfide Illite Shortite Analcime Nahcolite Searlesite Clay
Saline Unit
Shortite, Nahcolite, & Searlesite present
Rich Oil Shale Unit
Variable Carbonate Unit
Dolomite & Fe-carbonates fluctuate more than in overlying intervals
Data Limited Interval
Some zones in this interval are limited areally in the Uinta. Data is also very sparse.
Above bed 76
Bed 76 to Bed 44
Bed 44 to A-groove
A-groove
Mahogany zone
B-groove
R-6 zone
L-5 zone
R-5 zone
L-4 zone
R-4 zone
L-3 zone
R-3 zone
L-2 zone
R-2 zone
L-1 zone
R-1 zone
L-0 zone
R-0 zone
Uteland Butte

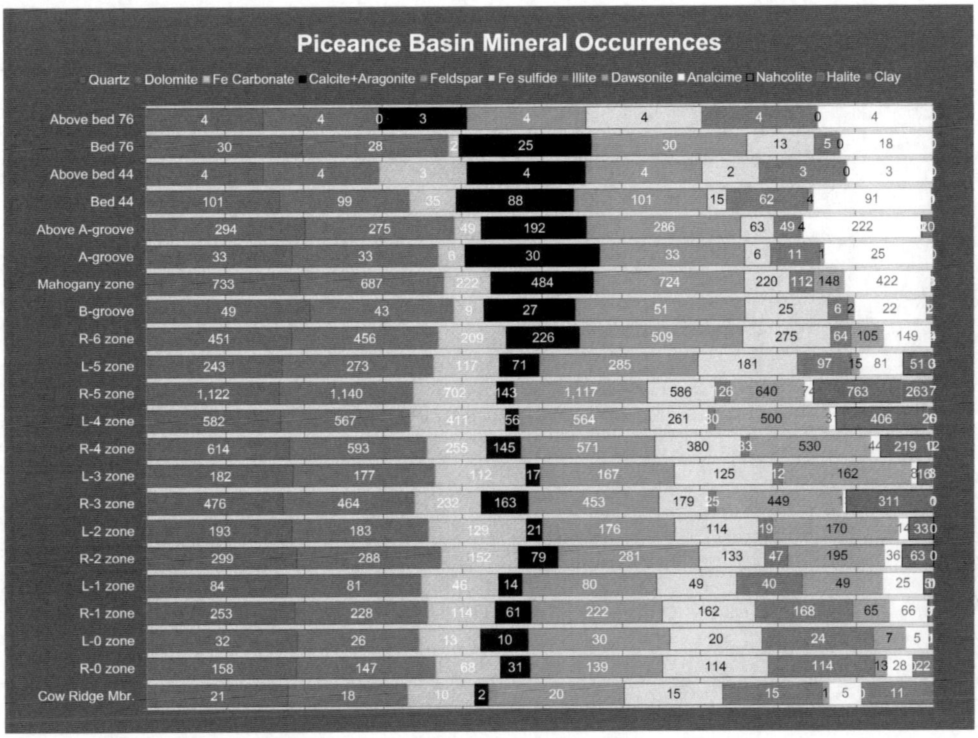
Piceance Basin Mineral Occurrences
Quartz Dolomite Fe Carbonate Calcite+Aragonite Feldspar Fe sulfide Illite Dawsonite Analcime Nahcolite Halite Clay
Above bed 76
Bed 76
Above bed 44
Bed 44
Above A-groove
A-groove
Mahogany zone
B-groove
R-6 zone
L-5 zone
R-5 zone
L-4 zone
R-4 zone
L-3 zone
R-3 zone
L-2 zone
R-2 zone
L-1 zone
R-1 zone
L-0 zone
R-0 zone
Cow Ridge Mbr.

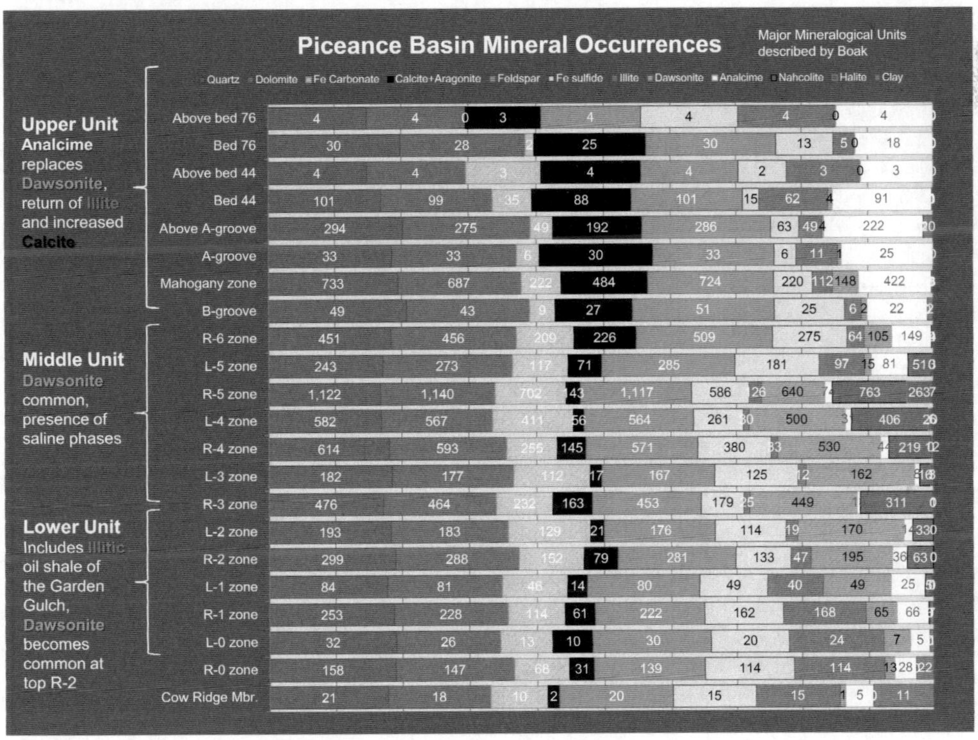
Piceance Basin Mineral Occurrences
Major Mineralogical Units described by Boak
Quartz Dolomite Fe Carbonate Calcite+Aragonite Feldspar Fe sulfide Illite Dawsonite Analcime Nahcolite Halite Clay
Upper Unit
Analcime replaces Dawsonite, return of Illite and increased Calcite
Middle Unit
Dawsonite common, presence of saline phases
Lower Unit
Includes Illitic oil shale of the Garden Gulch, Dawsonite becomes common at top R-2
Above bed 76
Bed 76
Above bed 44
Bed 44
Above A-groove
A-groove
Mahogany zone
B-groove
R-6 zone
L-5 zone
R-5 zone
L-4 zone
R-4 zone
L-3 zone
R-3 zone
L-2 zone
R-2 zone
L-1 zone
R-1 zone
L-0 zone
R-0 zone
Cow Ridge Mbr.

Indicator minerals by core and basin location

Sodium minerals in Uinta and Piceance Basins

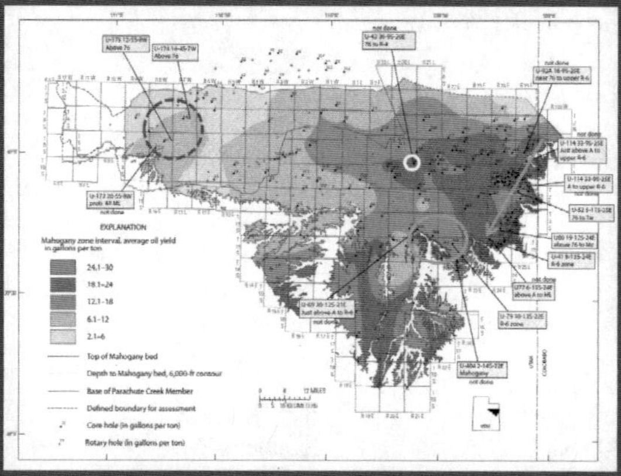

USGS

25

Eastern Edge – Uinta Basin (*proceeds from north to south*)

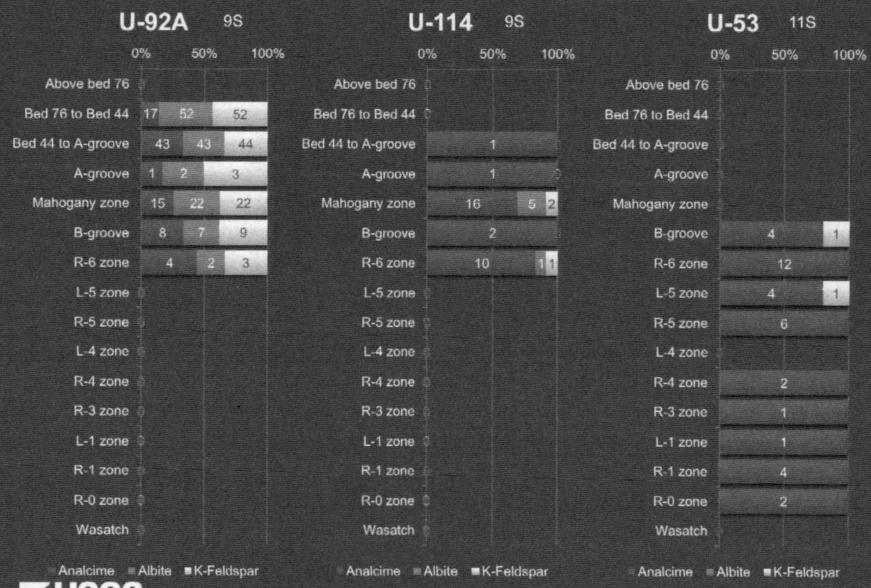

USGS

More offshore?

In U-53, Albite may be lumped into undifferentiated feldspar

26

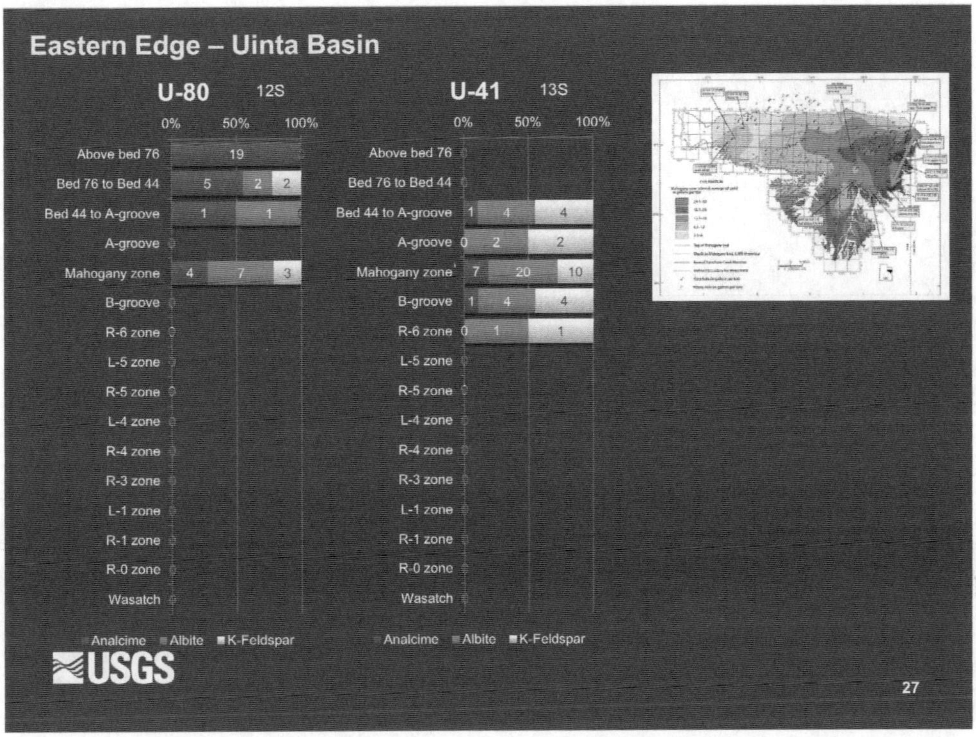
Eastern Edge – Uinta Basin
U-80 12S
U-41 13S
0% 50% 100%
Above bed 76
Bed 76 to Bed 44
Bed 44 to A-groove
A-groove
Mahogany zone
B-groove
R-6 zone
L-5 zone
R-5 zone
L-4 zone
R-4 zone
R-3 zone
L-1 zone
R-1 zone
R-0 zone
Wasatch
19
5 2 2
1 1
4 7 3
1 4 4
0 2 2
7 20 10
1 4 4
0 1 1
Analcime Albite K-Feldspar
USGS
27

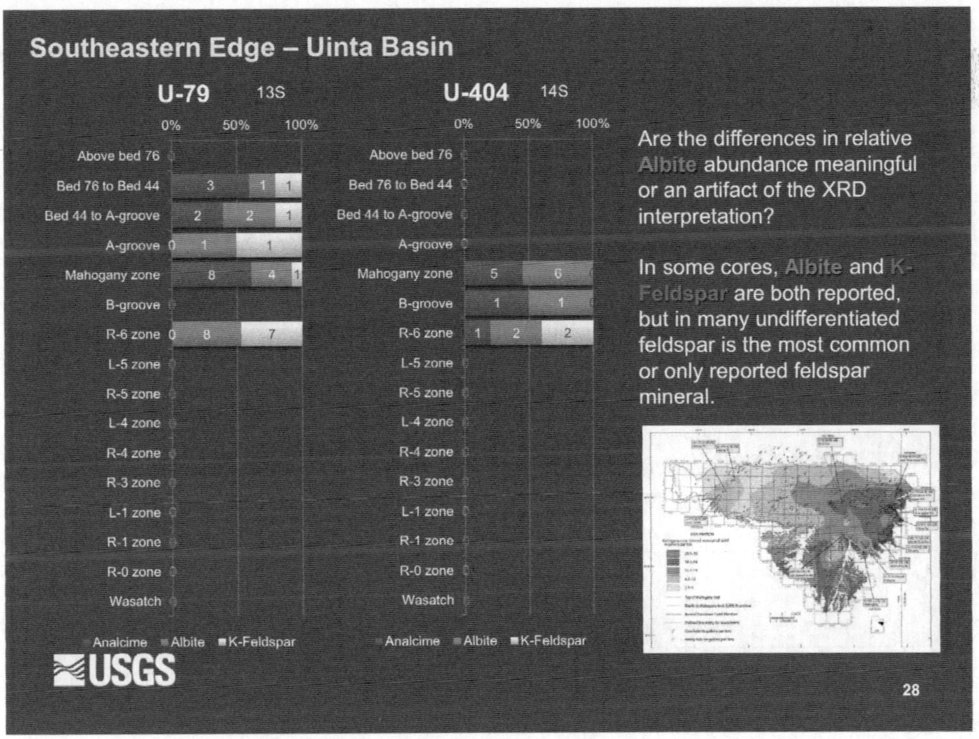
Southeastern Edge – Uinta Basin
U-79 13S
U-404 14S
0% 50% 100%
Above bed 76
Bed 76 to Bed 44
Bed 44 to A-groove
A-groove
Mahogany zone
B-groove
R-6 zone
L-5 zone
R-5 zone
L-4 zone
R-4 zone
R-3 zone
L-1 zone
R-1 zone
R-0 zone
Wasatch
3 1 1
2 2 1
0 1 1
8 4 1
0 8 7
5 6
1 1
1 2 2
Analcime Albite K-Feldspar
Are the differences in relative Albite abundance meaningful or an artifact of the XRD interpretation?
In some cores, Albite and K-Feldspar are both reported, but in many undifferentiated feldspar is the most common or only reported feldspar mineral.
USGS
28

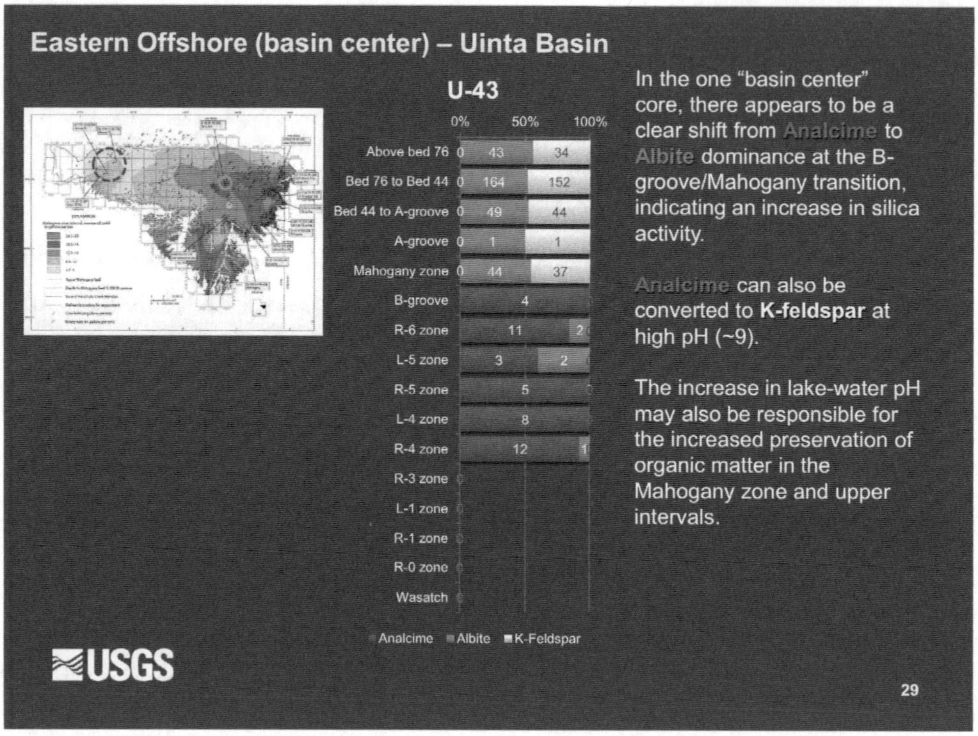
Eastern Offshore (basin center) – Uinta Basin
U-43
0%
50%
100%
Above bed 76
Bed 76 to Bed 44
Bed 44 to A-groove
A-groove
Mahogany zone
B-groove
R-6 zone
L-5 zone
R-5 zone
L-4 zone
R-4 zone
R-3 zone
L-1 zone
R-1 zone
R-0 zone
Wasatch
Analcime
Albite
K-Feldspar
In the one "basin center" core, there appears to be a clear shift from Analcime to Albite dominance at the B-groove/Mahogany transition, indicating an increase in silica activity.
Analcime can also be converted to K-feldspar at high pH (~9).
The increase in lake-water pH may also be responsible for the increased preservation of organic matter in the Mahogany zone and upper intervals.
USGS
29

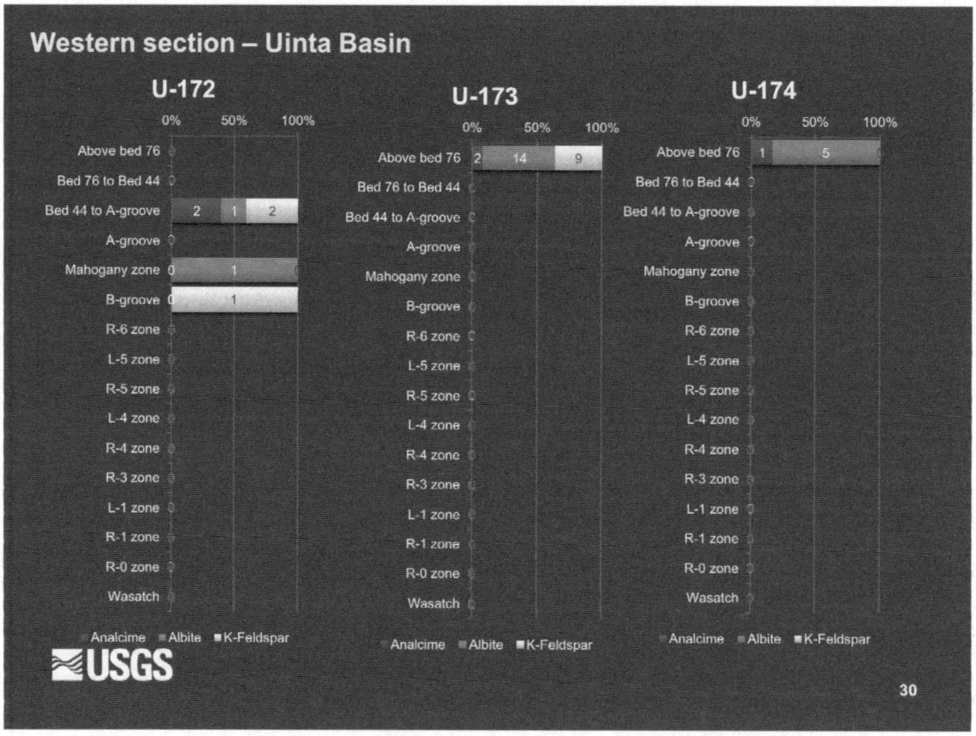
Western section – Uinta Basin
U-172
U-173
U-174
0%
50%
100%
Above bed 76
Bed 76 to Bed 44
Bed 44 to A-groove
A-groove
Mahogany zone
B-groove
R-6 zone
L-5 zone
R-5 zone
L-4 zone
R-4 zone
R-3 zone
L-1 zone
R-1 zone
R-0 zone
Wasatch
Analcime
Albite
K-Feldspar
USGS
30

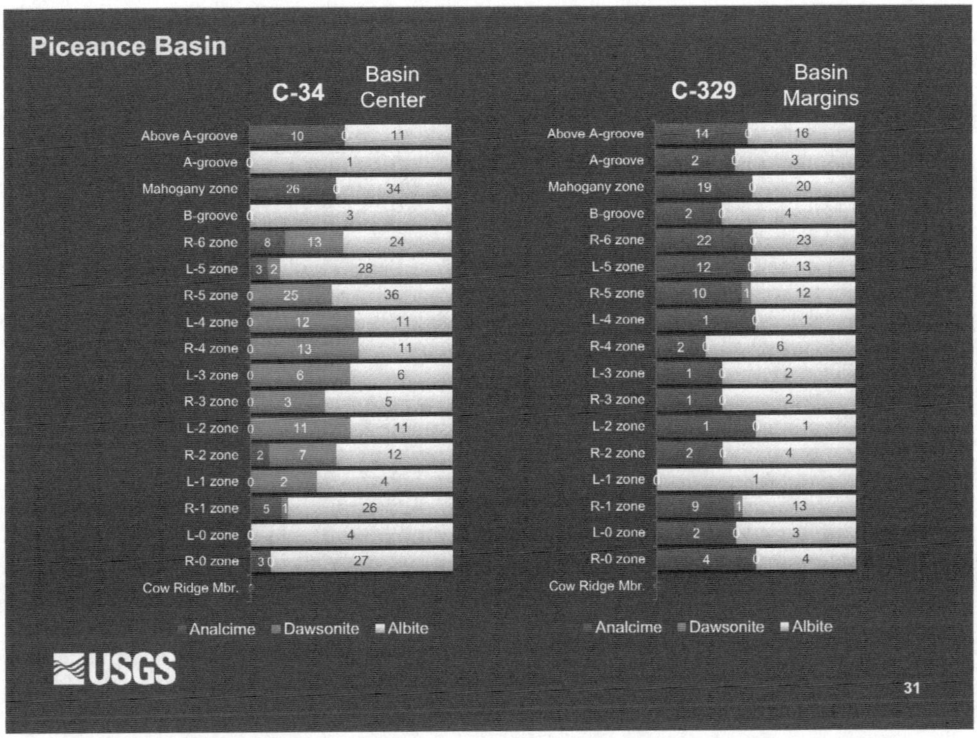

Summary

- Organic matter and mineral distributions appear to be linked – both relate to lake history and chemistry, but this connection is currently clearer in the Piceance Basin
- Distinctive mineralogical units are clearly delineated in Piceance Basin; aside from the saline interval at the top of the Uinta Basin, clear mineralogical units cannot be identified based on the available data
- The chemistry in the Uinta Basin portion of Lake Uinta was distinctly different and less OM preserving and chemically aggressive than in the Piceance.

USGS

32

Questions?

USGS

STEMS
Stanford Total Enhanced Modeling of Source rock

Multiscale Characterization of Geologic Properties of Green River Oil Shale

OIL SHALE SYMPOSIUM, OCT 2015

Y. Mehmani[1], A.K. Burnham[1], M.D. Vanden[2] Berg, H. Tchelepi[1]

[1]Stanford University [2]Utah Geologic Survey

TOTAL
COMMITTED TO BETTER ENERGY

Stanford | SCHOOL OF EARTH, ENERGY & ENVIRONMENTAL SCIENCES | Energy Resources Engineering

Photo: US Bureau of Mines (reprinted in Russell, 104)

STEMS
Stanford Total Enhanced Modeling of Source rock

OUTLINE

- Motivation
- Setting and Hypothesis
- Optical method for mapping kerogen content
- Field validation
- Ongoing lab validation
- Conclusions
- Future work

2

STEMS
Stanford Total Enhanced Modeling of Source rock

Motivation

3

SEM Backscattered Electron Image

Physical properties are:

- Strong functions of grade
- Non-linearly dependent
- Micro-structure dependent

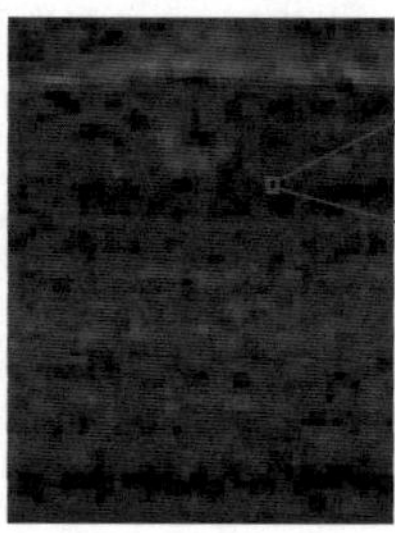

- Binary mix approx.
- Mixing ratio known

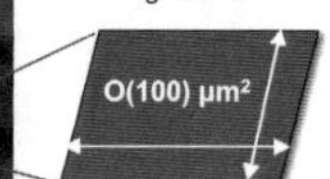

Want:
- Thermal Conductivity
- Porosity/Permeability
- Young's Modulus
- Poisson's Ratio

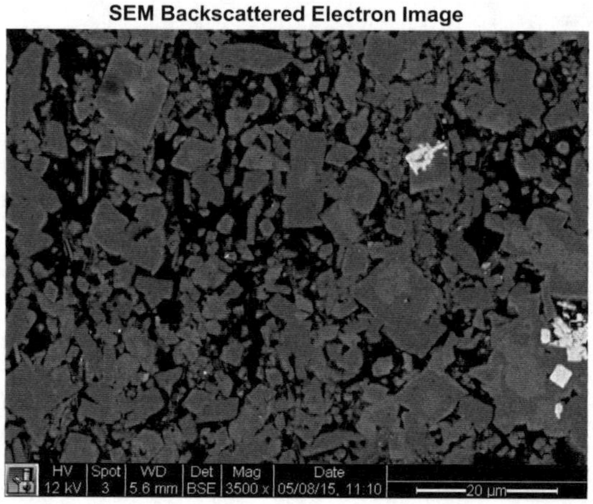

4

STEMS
Stanford Total Enhanced Modeling of Source rock

Setting and Hypothesis

5

Geology of First Application:

SHALE OIL RESOURCES	
Zone	(10^9 bbls)
R-8	(ND)
Mahogany	172.94
R-6	159.09
L-5	52.42
R-5	178.72
L-4	60.85
R-4	107.78
L-3	18.72
R-3	58.38
L-2	20.08
R-2	53.07
L-1	10.70
R-1	115.35
Total	1008.10

17% (6% global)

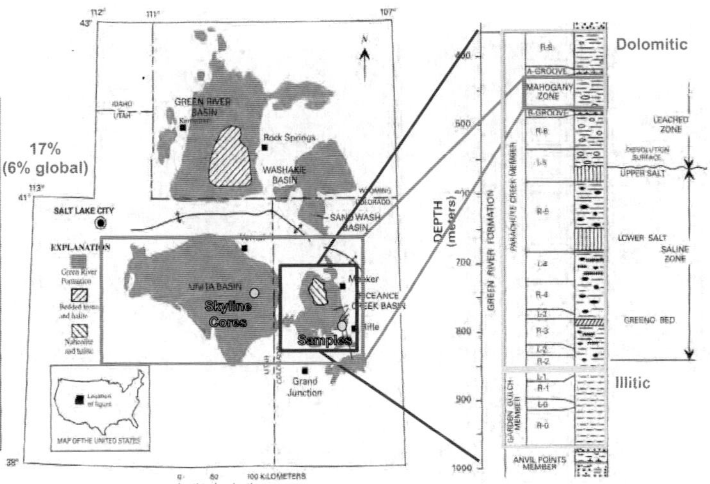

- *Graphics after Dyni (2006)*
- *Table from Pitman et al (1989)*

6

STEMS
Stanford Total Enhanced Modeling of Source rock

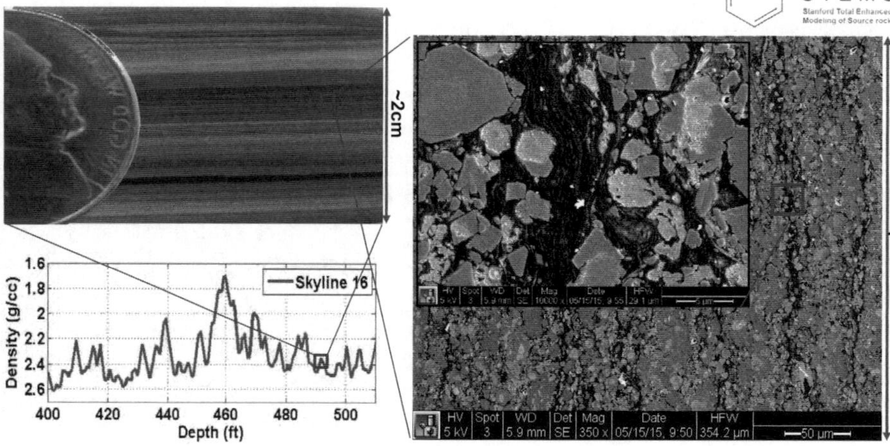

7

STEMS
Stanford Total Enhanced Modeling of Source rock

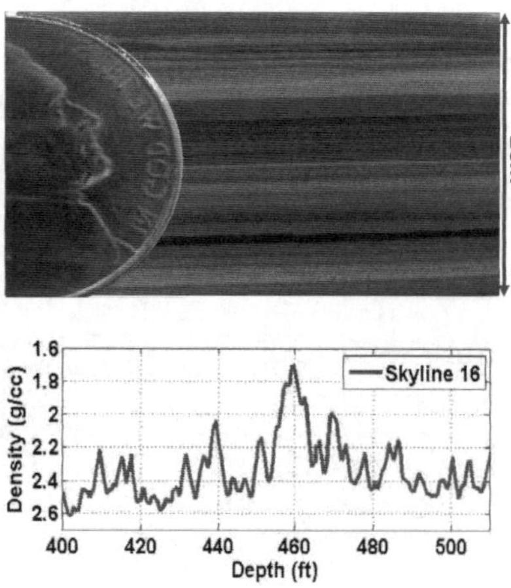

Observations:

- Rock banded at all scales
- Bands have different color intensities

Hypothesis:

Optical images contain <u>quantifiable</u> information about spatial distribution of kerogen content

Implication:

Physical properties (thermal/mechanical/etc.) can be derived from camera images

8

STEMS
Stanford Total Enhanced Modeling of Source rock

Optical Method for Mapping Kerogen Content

9

STEMS
Stanford Total Enhanced Modeling of Source rock

Optical Images:

Color mapping
White balance
Gamma correction
Compression

GSI:
Red
Green
Blue

Kubelka-Munk Theory of Reflectance:

$$F(R_\infty) = \frac{(1-R_\infty)^2}{2R_\infty} = \frac{\sum v_i K_i}{\sum v_i S_i}$$

Kubelka-Munk (1931)
Duncan (1940,1949)

$$v_{kr} = \gamma \frac{F(R_\infty)}{F(R_\infty)+\beta} \propto FA_{gal/ton} \qquad R_\infty = \frac{GSI}{GSIW}$$

$$R_\infty = \frac{I}{I_0}$$

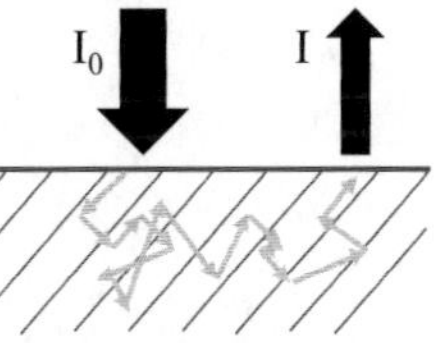

10

Red, Green, or Blue?

Agilent Cary 6000i UV-VIS-NIR

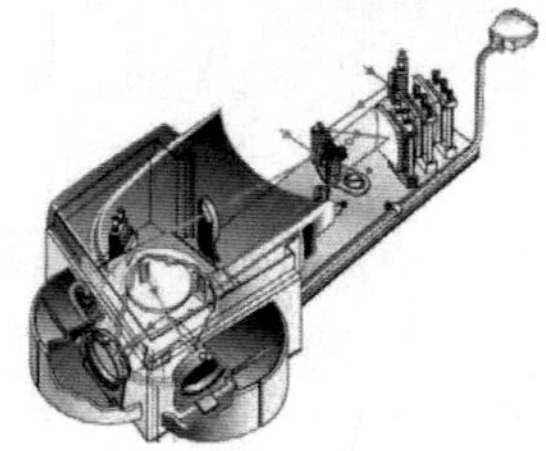

Integrating Sphere

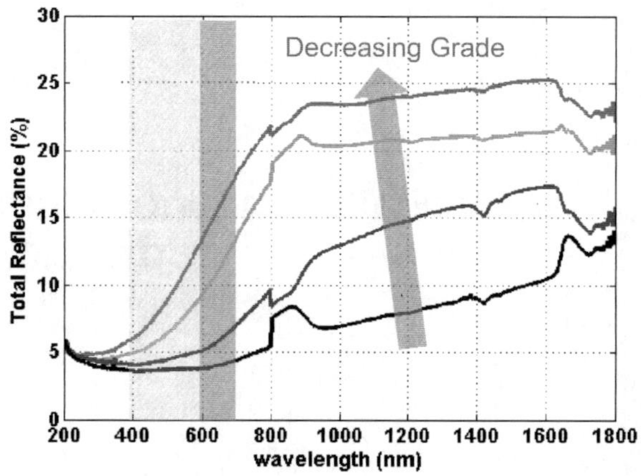

Red channel preferred in visual range

11

STEMS
Stanford Total Enhanced Modeling of Source rock

Field Validation

12

Dataset:

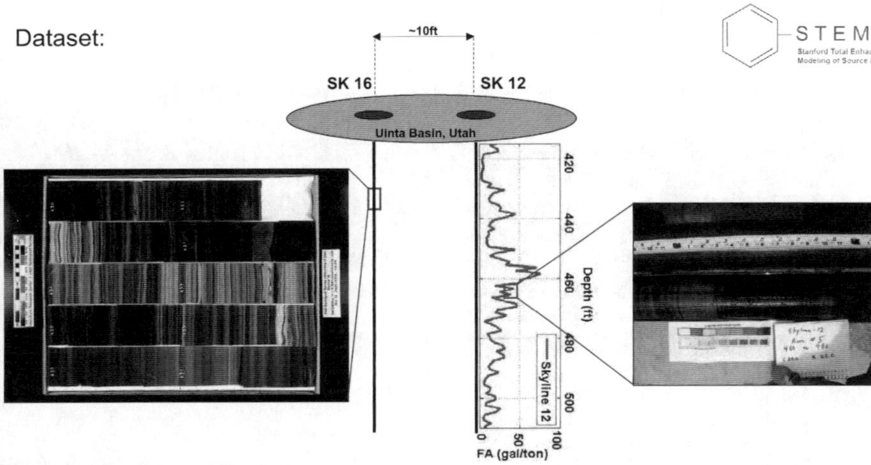

* Data courtesy of Michael Vanden Berg, Utah Geologic Survey, and Lauren Birgenheier, University of Utah

13

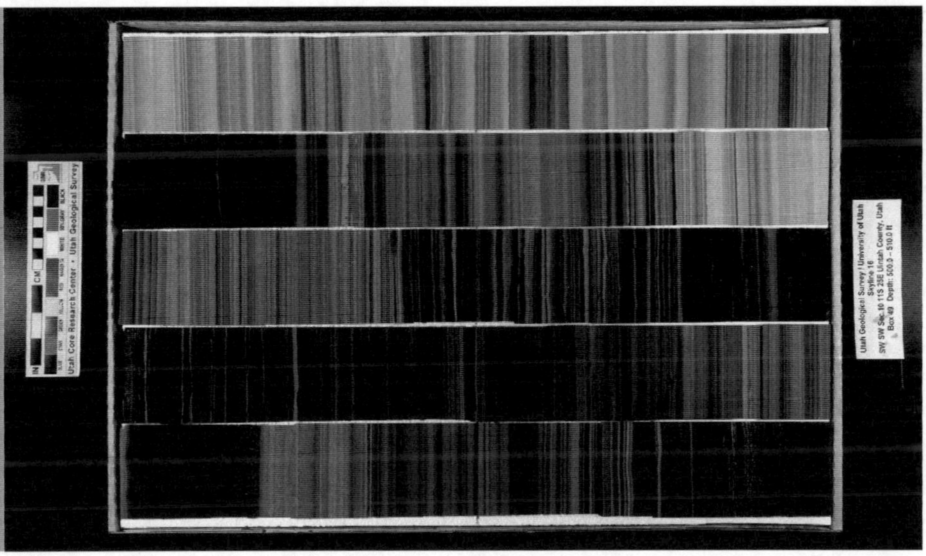

Optical Log:

STEMS
Stanford Total Enhanced Modeling of Source rock

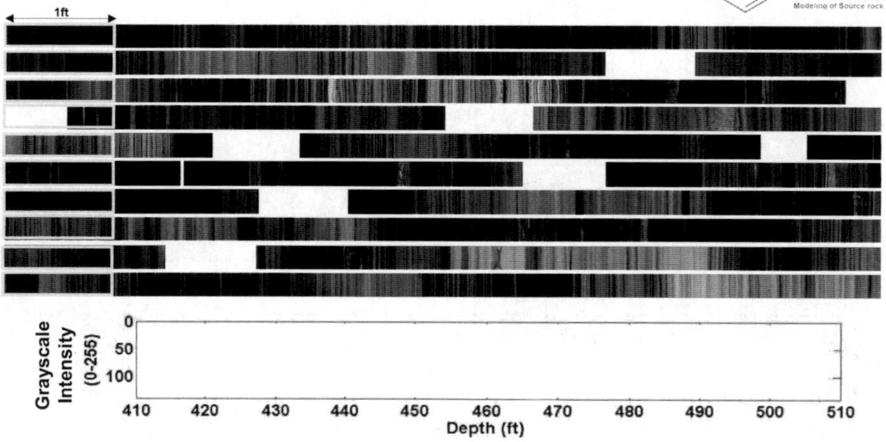

15

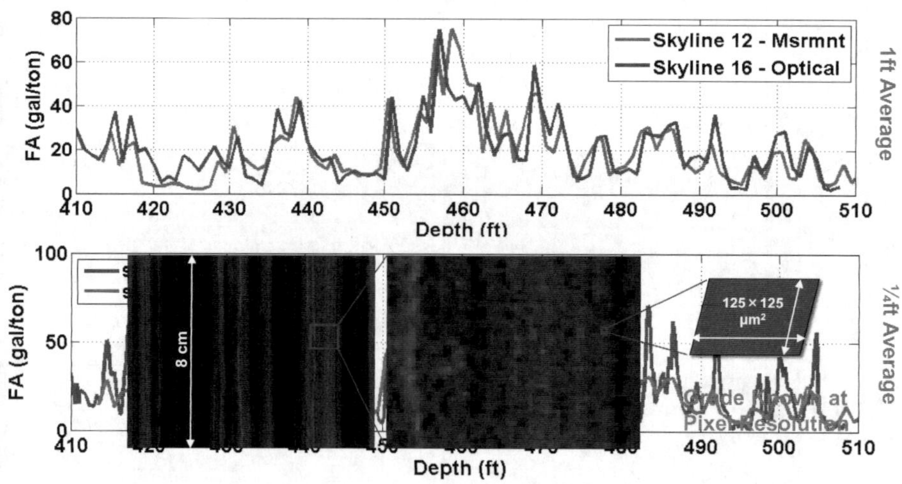

16

STEMS
Stanford Total Enhanced Modeling of Source rock

Ongoing Lab Validation

17

STEMS
Stanford Total Enhanced Modeling of Source rock

Anvil Points Mines, CO

18

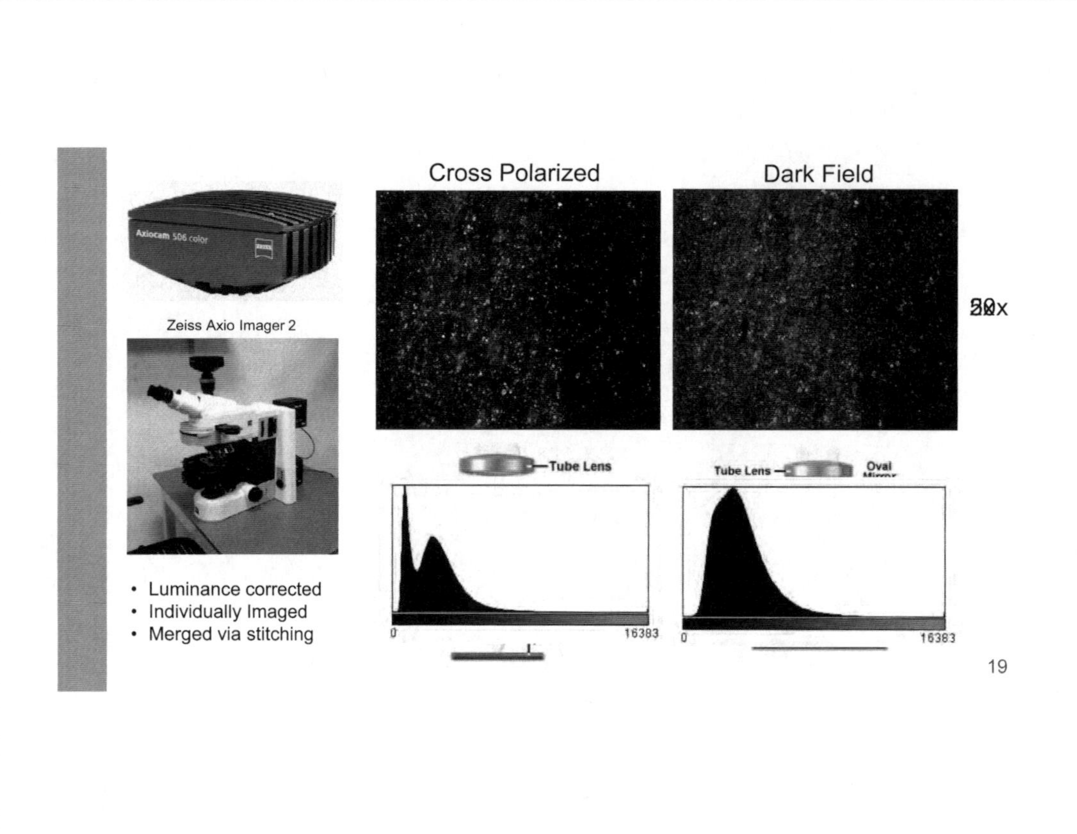
Cross Polarized
Dark Field
Zeiss Axio Imager 2
Tube Lens
Tube Lens
Oval
0
16383
0
16383
• Luminance corrected
• Individually Imaged
• Merged via stitching
19

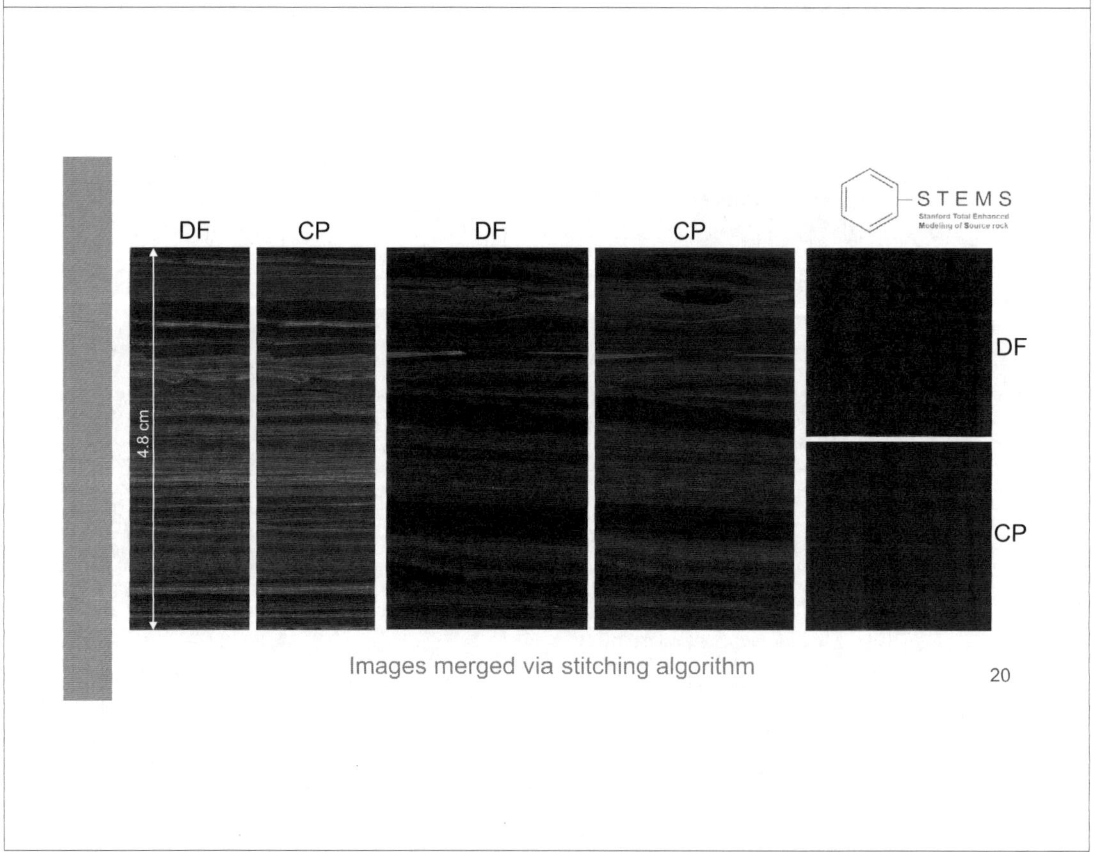
STEMS
DF
CP
DF
CP
DF
CP
4.8 cm
Images merged via stitching algorithm
20

Minimal cut loss
(8-25mg)

21

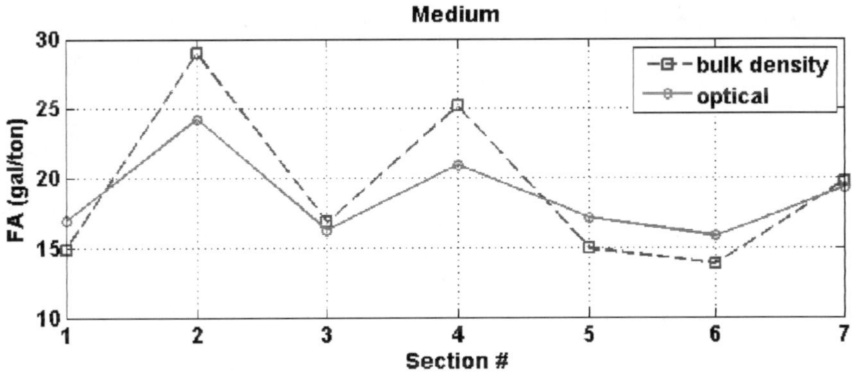

Accurate TOC measurements underway

22

CONCLUSIONS

STEMS
Stanford Total Enhanced Modeling of Source rock

- Spatial distribution of kerogen seems quantifiable from common photographs
- Best results expected from red channel of cross-polarized or dark-field images
- Images calibrated to low resolution measurements (~2ft) could provide high resolution (~150μm) information needed for bridging multiple spatial scales

23

FUTURE WORK

STEMS
Stanford Total Enhanced Modeling of Source rock

- Scale identification and geostatistical characterization of variability
- Derivation and upscaling of thermo-mechanical properties for field simulations
- Pore-scale reconstruction and modeling of fundamental processes involved in pyrolysis

24

STEMS

Stanford Total Enhanced
Modeling of Source rock

26

Using ^{13}C NMR Chemical Structure Data to Model Oil Shale Pyrolysis for Green River and Estonian Oil Shale

Thomas H. Fletcher
Chemical Engineering Dept., Brigham Young University

Ronald J. Pugmire
Chemistry and Chemical Engineering Depts., University of Utah

35th Oil Shale Symposium
October 5-6, 2015

BYU | CHEMICAL ENGINEERING
IRA A. FULTON COLLEGE

INSTITUTE FOR
Clean and Secure Energy
THE UNIVERSITY OF UTAH

Topics

1. Implications of advanced solid-state ^{13}C NMR data for Green River oil shale
2. Advance solid-state ^{13}C NMR data for kukersite (Estonian oil shale)
3. Modeling oil shale pyrolysis based on chemical structure

2

2D Kerogen Model

Siskin's 1995 2D model for a type I kerogen from a Green River oil shale

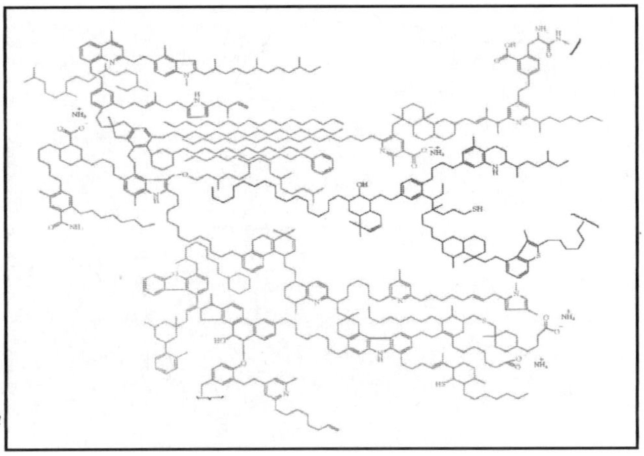

Siskin et al, in *Composition, Geochemistry and Conversion of Oil Shales (NATO ASI Series C* volume 455*)* **1995**

NMR Studies of Green River Kerogen

- **Goal: Characterize the solid hydrocarbons (kerogen) without reaction**
- **Analytical Technique: Solid-State C-13 NMR**
- **14 Structural parameters and 8 lattice parameters were developed for coal**
  - **Applied to oil shale kerogen and other hydrocarbons**

FUNCTIONAL GROUP DIVISION

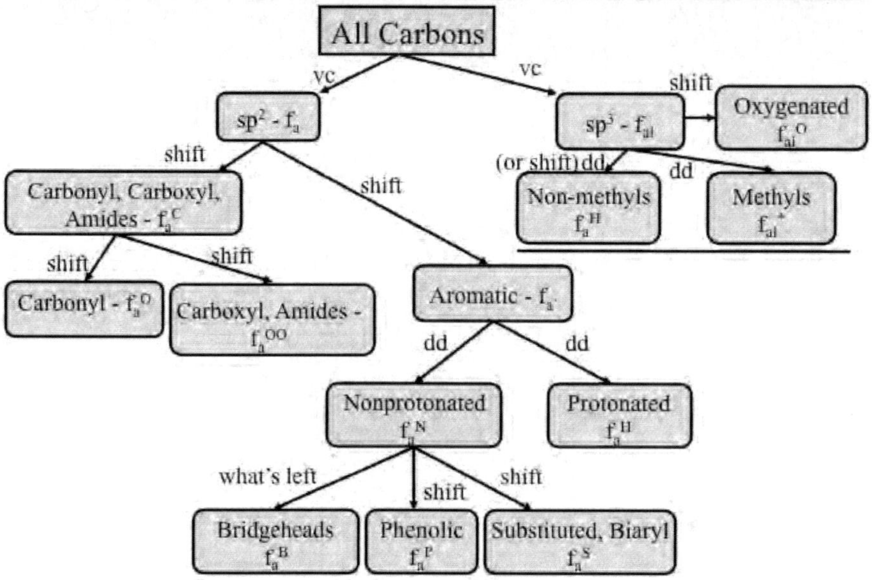

Empirical Relationship to get # of C/Cluster

Fraction of Bridgehead Carbons

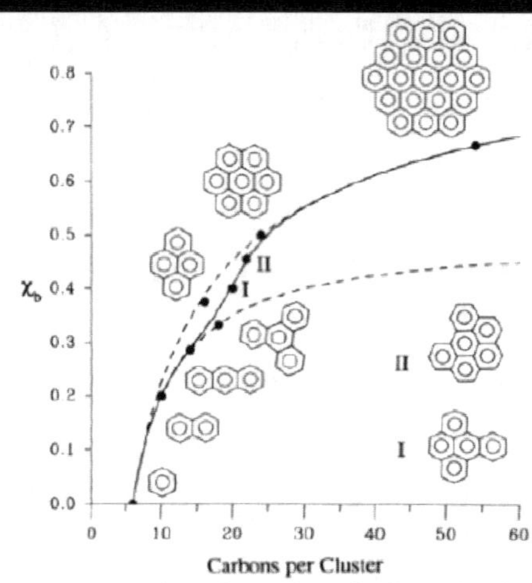

Figure 6. Plot of the mole fraction of bridgehead carbons, χ_b, vs C where C is the number of carbon atoms per aromatic cluster. The solid curve is for the combined model, the upper dashed curve is for the circular catenation model, and the lower dashed curve is for primary catenation model.

From Solum et al., Energy & Fuels (1989)

Representative Hydrocarbon Molecule

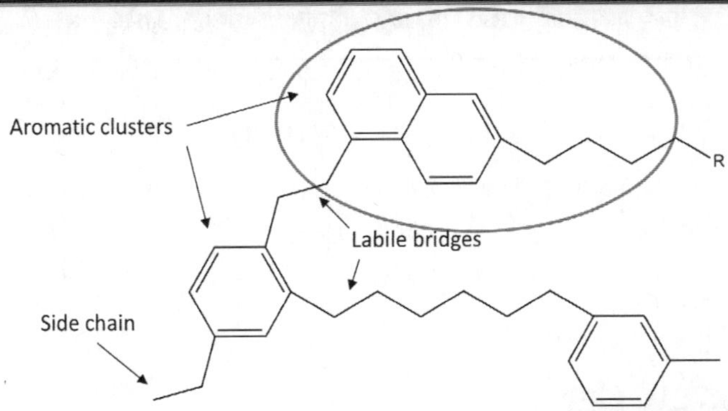

- Aromatic clusters
- Bridges
- Side Chains

- Bridges break during heating
- Lattice statistics tell fraction of detached clusters
- Vapor pressure determines if detached cluster will evaporate

Solid State NMR on Demineralized Green River Kerogens

(GR1,2,3)

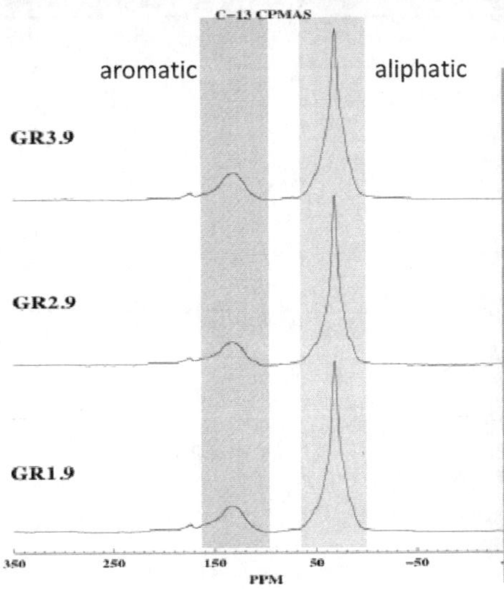

- Aromaticity ($f_{a'}$) = 0.193
- Average Aromatic Cluster Size (C) = 10.3 ~ (naphthalene)
- Average Aliphatic Chain Length = 11.5
- This means bridge length = 23
- Fraction of aromatics with attachments = 0.5
- Average cluster molecular weight = 832 dalton
  - This includes aromatic cluster plus side chains and half the bridges
- Average attachment molecular weight = 138 dalton

Structure Parameters of Char

Structural Parameter	**GR1.9**	**300°C**	**375°C**	**410°C**	**445°C**	**495°C**
aromatic carbon, $f_a = f_{a'} + f_a^C$	0.24	0.23	0.28	0.28	0.35	0.73
carbonyl, f_a^C	0.04	0.04	0.03	0.01	0.03	0.04
aldehydes and ketones, f_a^O	0.01	0.02	0.015	0.005	0.01	0.01
acids and esters, f_a^{OO}	0.03	0.02	0.015	0.005	0.02	0.03
aromatic carbon, carbonyl subtracted, f_a'	0.2	0.19	0.25	0.27	0.32	0.69
protonated aromatic carbon, f_a^H	0.06	0.05	0.08	0.09	0.12	0.11
nonprotonated aromatic carbon, $f_a^N = f_a^P + f_a^S + f_a^B$	0.14	0.14	0.17	0.18	0.2	0.58
aromatic carbon with oxygen attachment (phenolic), f_a^P	0.03	0.03	0.04	0.03	0.04	0.08
aromatic carbon with alkyl attachment, f_a^S	0.07	0.06	0.08	0.07	0.1	0.19
aromatic bridgehead and inner carbon f_a^B	0.04	0.05	0.05	0.08	0.06	0.31
aliphatic carbon, f_{al}	0.76	0.77	0.72	0.72	0.65	0.27
aliphatic CH and CH_2, f_{al}^H	0.65	0.66	0.63	0.61	0.54	0.19
aliphatic CH_3 and nonprotonated carbon, f_{al}^*	0.11	0.11	0.09	0.11	0.11	0.08
aliphatic with oxygen attachment, f_{al}^O	0	0.01	0.01	0.01	0.01	0.01
aromatic bridgehead carbons, χ_b	0.2	0.263	0.2	0.296	0.188	0.449

9

Lattice Parameters of Char

Lattice Parameter	**GR1.9**	**300°C**	**375°C**	**410°C**	**445°C**	**495°C**
average number of carbons per cluster, *C*	10	12.6	10	14.4	9.6	21.9
total attachments per cluster, $\sigma + 1$	5	6	4.8	5.3	4.2	8.4
fraction of intact bridges per cluster, p_0	-0.1	-0.22	0.25	-0.01	0.21	0.7
bridges and loops per cluster, *B.C.*	--	--	1.2	--	0.9	5.9
side chains per cluster, *S.C.*	--	--	3.6	--	3.3	2.5
molecular weight per cluster, MW_{cl}	776	1010	591	797	442	448
molecular weight per side chain, m_δ	131	143	98	117	77	22
Ratio f_{al}/f_a^S	10.86	12.83	9.00	10.29	6.50	1.42
Mass Release (wt% of parent kerogen)	0.0%	7.1%	7.1%	17.2%	41.0%	77.6%

10

Comparison of NMR Data with Siskin Model

- Counted # of bridges and attachments per cluster in Siskin model
- Averaged value of σ+1 was 3.3 in the Siskin model
- ^{13}C NMR characterization showed values of σ+1 ranging from 4.5 to 5.9
- Bridges and loops per cluster was 2.0 for Siskin model, but NMR technique could not get this number due to challenges with branched aliphatics
- **Lesson: Advanced NMR characterization has new information that must be incorporated into models of chemical structure!**

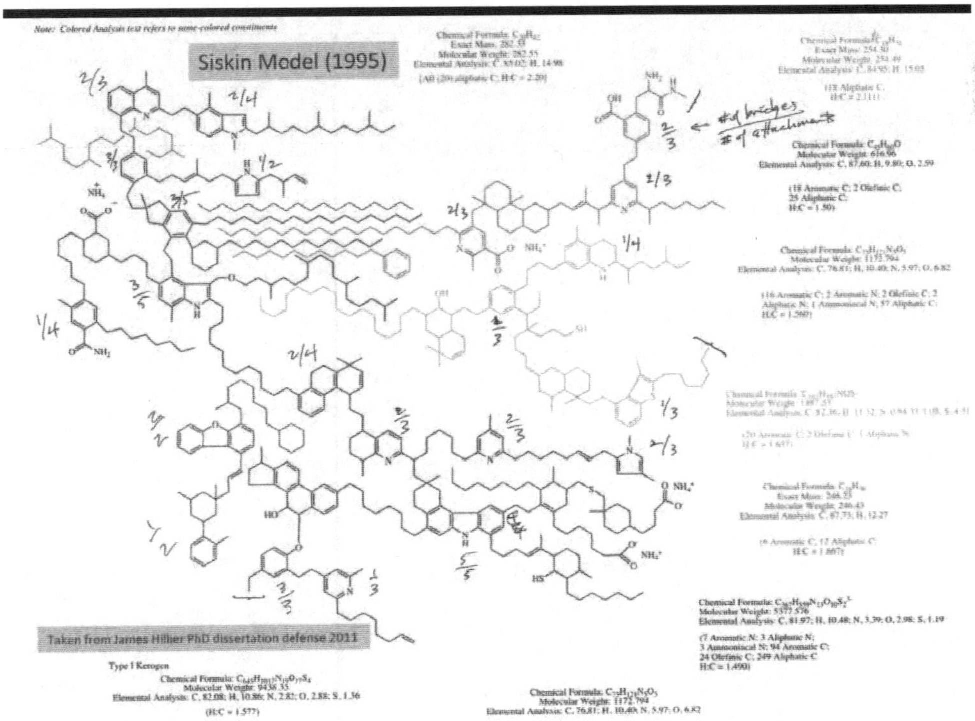

CPD model
(Chemical Percolation Devolatilization)

- Originally developed for coal pyrolysis
- Based on chemical structure of coal
  - Lattice of aromatic clusters connected by aliphatic bridges
  - Thermal decomposition of aliphatic bridges chops up the lattice
  - Percolation statistics relate fraction of broken bridges to the amount of separated fragments
  - Vapor pressure used to determine if fragments evaporate
  - Large fragments crosslink back into the char

Vapor-Liquid Equilibrium and Crosslinking

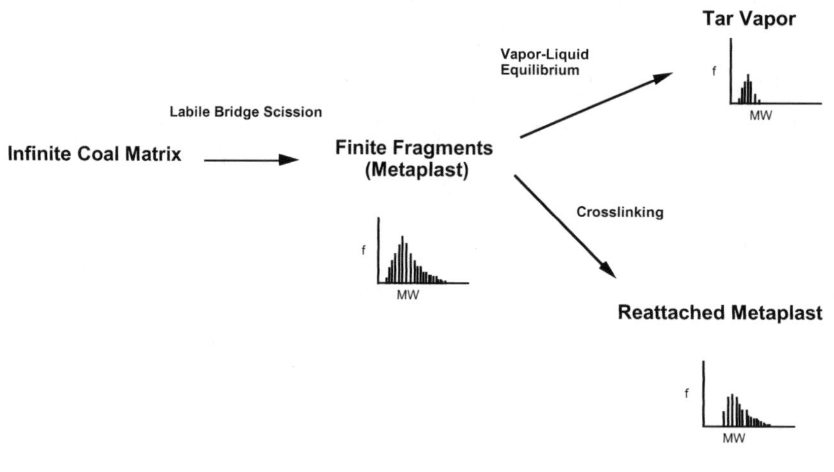

Input Parameters Required by the CPD Model
(values given for GR2.9 Kerogen)

Measured by NMR

- Number of attachments per cluster ($\sigma+1$) (i.e., coordination number)
  ------- **4.5**
- Fraction of attachments that are bridges (p_0) (bridges/bridges+side chains)
  ------ **0.5** (not measured for kerogen)
- Molecular weight per aromatic cluster (M_{cl})
  ------ **775**
- Molecular weight per side chain (M_δ)
  ------ **148**
- Fraction of bridges that are initially stable (c_0)
  ------ **0.0**

Other Parameters
(not usually adjusted)

- Rate coefficients
  - One set for coals, one set for biomass, one set for kerogen
  - Set based on extensive comparisons with data
  - Uses sequential (not parallel) distributed activation energy
  - A_b, E_b, σ_b, A_g, E_g, σ_g, A_{cr}, E_{cr}, ρ (ratio of 2 A's)
- Vapor pressure coefficients
  - Assumed to be MW-dependent

Changed Bridge Scission Mechanism

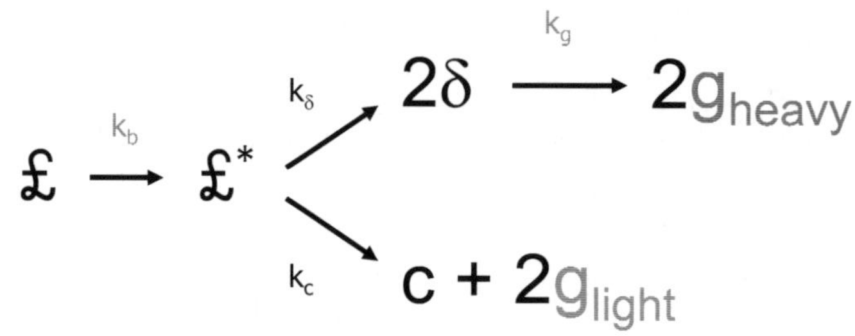

Best-fit CPD Model

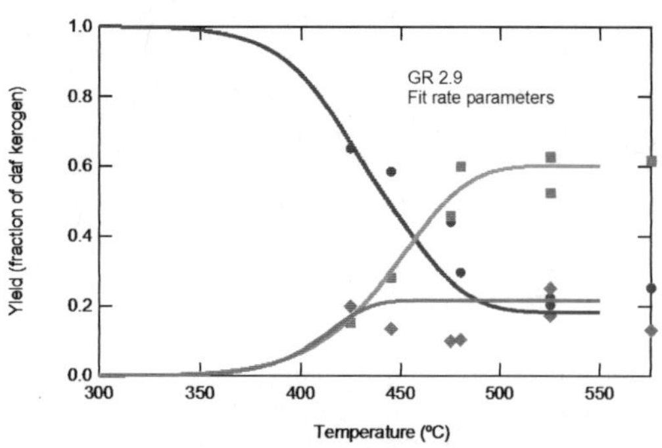

- Very good agreement for GR1.9 and GR3.9
- Side chain release (E_g) lower than bridge breaking (E_b)

- Curve-fit values of
  - Bridge-breaking rate
  - Side chain release rate
  - Ratio of cleavage to crosslinking
  - These are bridge variables, not mass variables
- Assumed $MW_{light\ gas}$ = 20 amu
- $MW_{heavy\ gas}$ estimated from
  - $MW_{light\ gas}$,
  - $MW_{side\ chain}$, and
  - 80/20 ratio of heavy gas to light gas

	GR 1.9	GR 2.9	GR 3.9
ab	9.8E+12	9.8E+12	1.8E+15
eb	47500	47500	51500
ebsig	1	1	1
ag	1.58E+10	2.58E+10	1.58E+14
eg	41800	41800	54800
egsig	600	600	600
ac=rho	0.9	0.9	1.8

Possible Extension to Other Oil Shales

- Measure chemical structure characteristics
  - Example: Kukersite
- Use chemical structure-based pyrolysis model
- Are the bridge-breaking kinetics similar???

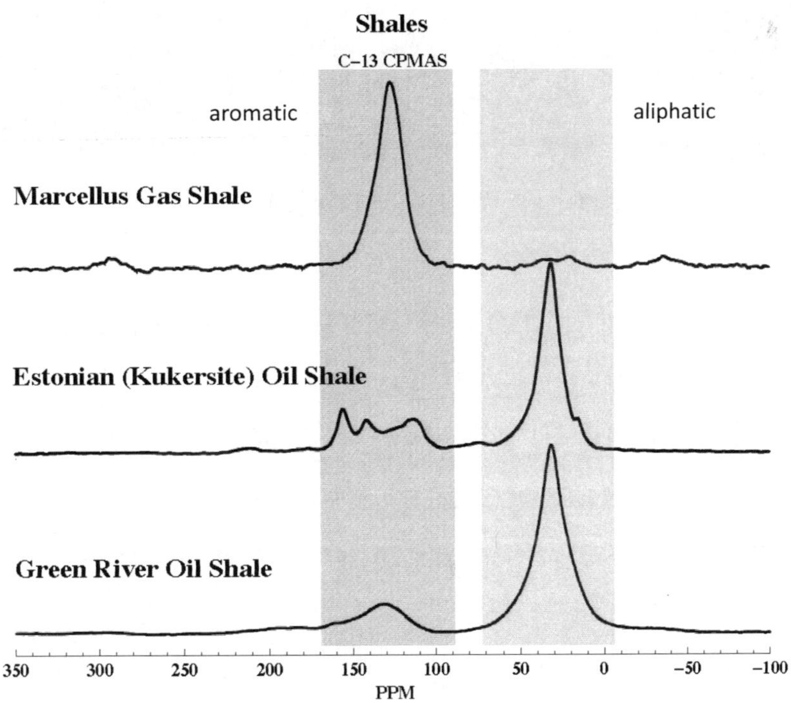

Kukersite Data

Structural Parameters														
Compound	f_a	f_a^C	f_a^O	f_a^{OO}	$f_{a'}$	f_a^H	f_a^N	f_a^P	f_a^S	f_a^B	f_{al}	f_{al}^H	f_{al}^*	f_{al}^O
Kukersite	0.38	0.04	0.03	0.01	0.34	0.10	0.24	0.08	0.08	0.08	0.62	0.51	0.11	0.07

Lattice Parameters								
Compound	χ_b	C	$\sigma+1$	P_0	$B.L.$	$S.C.$	$M.W.$	M_δ
Kukersite	0.235	11.3	5.3	0.31	1.6	3.7	--	--

	Kukersite	GR1.9	GR2.9	GR3.9
C	78.0	77.4	77.5	76.2
H	9.8	9.8	10	9.5
N	0.4	2.8	2.6	2.5
O	10.2	8.2	8	8.1
S	1.6	2	2	3.7
fa'	0.34	0.2	0.18	0.2
C_cl	11.3	10	9	12
Sig+1	5.3	5	4.5	5.9
MW_cl	511.1	775.2	774.2	944.9
M_del	70.8	131.0	148.0	135.7

Kukersite vs Green River Oil Shale

- Higher aromaticity in kukersite
  - 34% vs 18-24%
- Side chain length in kukersite is only about half that of Green River oil shale
- Attachments per cluster are similar (~5.3)
- Aromatic carbons per cluster are similar (~11)
- Molecular weight per cluster varies due to side chain length

22

Use GROS Rate Coefficients for Kukersite in CPD Model

10 K/min

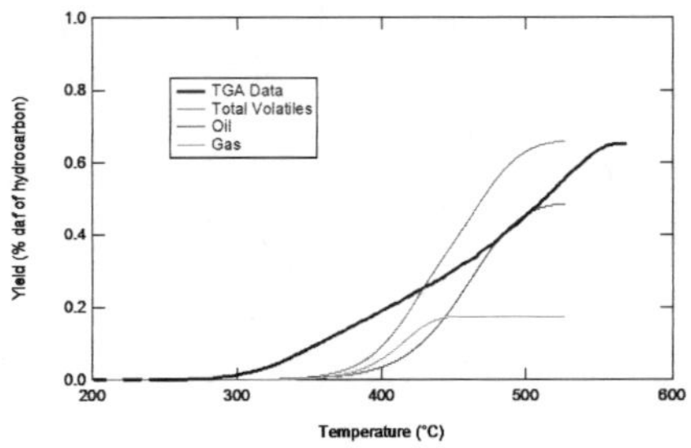

- Yields normalized to same value for comparison
- GROS pyrolysis occurs in a narrower T range than Kukersite pyrolysis
- Rates of oil and gas evolution unknown for Kukersite
- Kinetic rates may need to be related to chain length

TGA analyses originating from Kukersite from the Republic of Estonia

23

Conclusions

1. Advanced solid-state ^{13}C NMR data for Green River oil shale show that kerogen is more connected than indicated by Siskin model
   - More refinement now possible based on advanced analysis
2. Kukersite (Estonian oil shale) significantly differs from GROS
   - Higher aromaticity (34% vs ~22%)
   - Shorter side chain lengths (~70 vs 140)
3. Kukersite pyrolysis not solely influenced by chemical structure
   - Kinetics of bridge breaking seem different

24

Thank you

25

26

CPD model

(Chemical Percolation Devolatilization)

- Originally developed for coal pyrolysis
- Based on chemical structure of coal
  - Lattice of aromatic clusters connected by aliphatic bridges
  - Thermal decomposition of aliphatic bridges chops up the lattice
  - Percolation statistics relate fraction of broken bridges to the amount of separated fragments
  - Vapor pressure used to determine if fragments evaporate
  - Large fragments crosslink back into the char

9-Step Demineralization Process

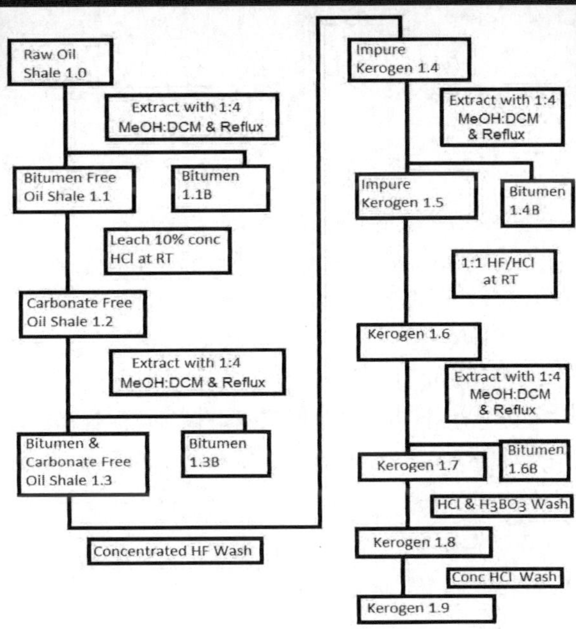

After Vandegrift, Winans, Scott, Horowitz, Fuel, 1980

Representative Hydrocarbon Molecule

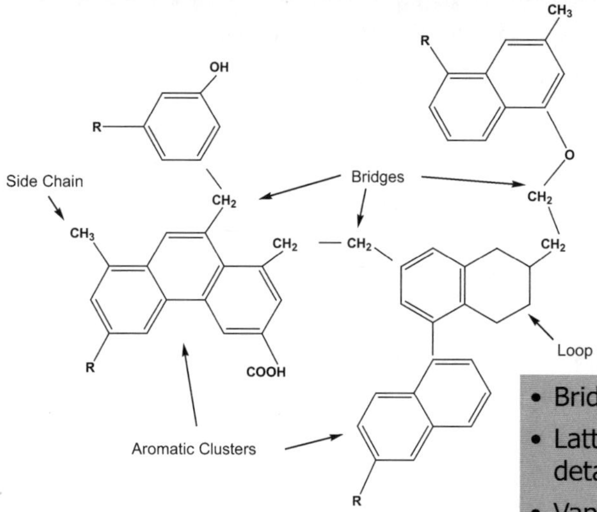

- Aromatic clusters
- Bridges
- Side Chains

- Bridges break during heating
- Lattice statistics tell fraction of detached clusters
- Vapor pressure determines if detached cluster will evaporate

The Chemical Percolation Devolatilization (CPD) Model

Includes:

- NMR for coal structure
- Chemical mechanism for bridge scission
- Percolation lattice statistics
- Vapor pressure model
- Crosslinking

Predicts tar and light gas yields as a function of:

- Coal type
- Heating rate
- Temperature
- Pressure

Reaction Chemistry

- Aromatic clusters do not break apart at pyrolysis temperatures
  - Normal coal pyrolysis temperatures are ~900-1000 K at high heating rates
  - Aromatic compounds break apart at >2500 K
- Bridges between clusters break during pyrolysis
  - Not all bridges break (or else no char)
  - Crosslinking (new stable bridges formed)

Bridge Scission Mechanism

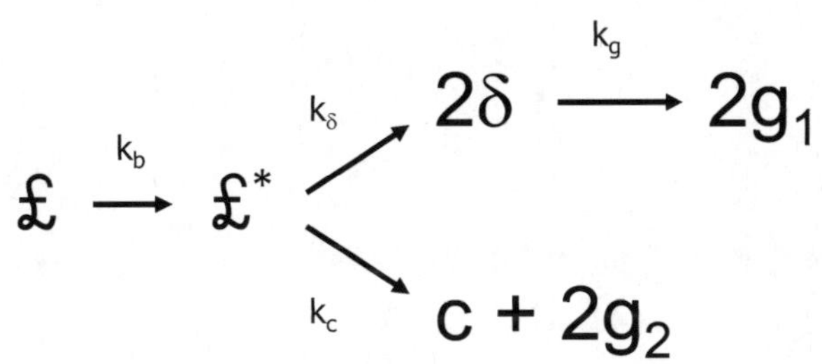

How Does Bridge-Breaking Relate to Mass Release?

⇨ Lattice structure (also called network)

Types of Lattices

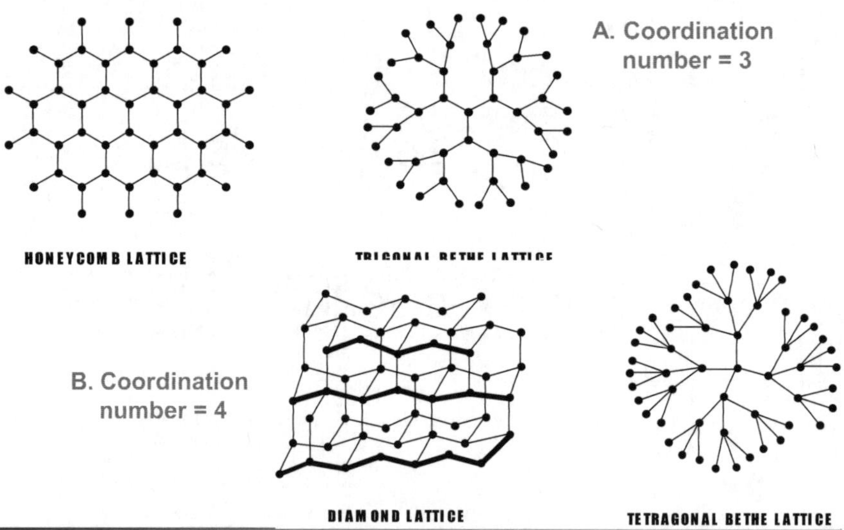

Relationship Between Broken Bridges and Finite Clusters

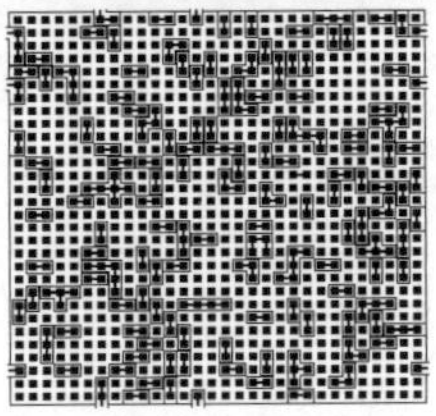

- 90% broken bridges
- 900 of 900 in fragments

a. p = 0.1

- 20% broken bridges
- 3 of 900 in fragments

b. p = 0.8

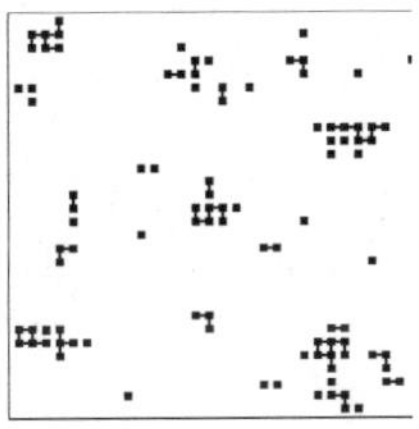

c. p = 0.55, finite fragments

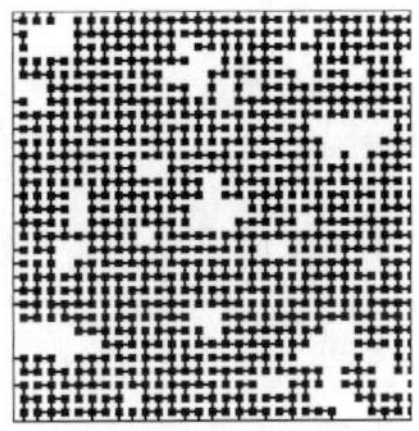

- 45% broken bridges
- 100 of 900 in fragments

d. p = 0.55, infinite lattice

Closed-Form Solution of Percolation Lattice Statistics

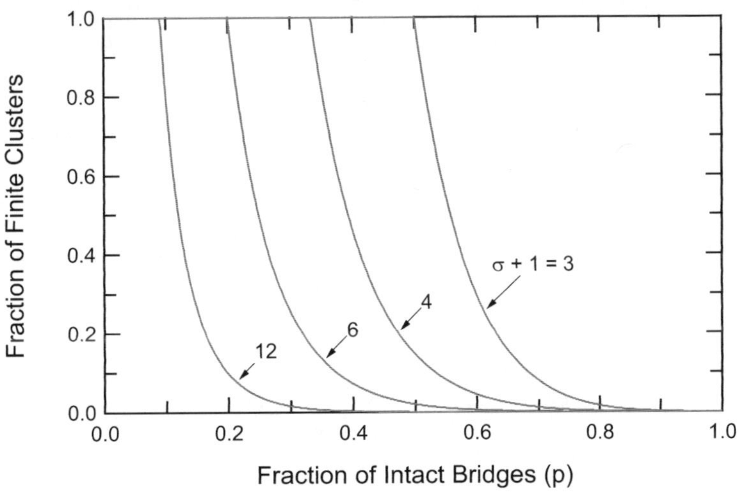

Vapor-Liquid Equilibrium and Crosslinking

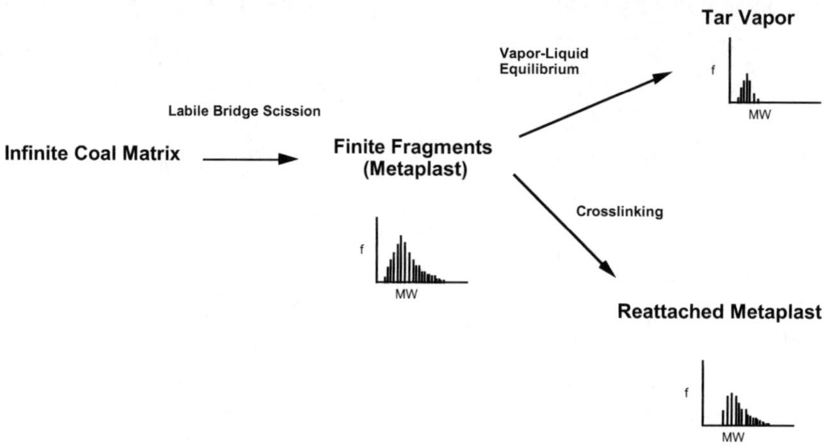

6.1 Irati Energy Development of the Oil Shale Resources Near Curitiba, Brazil

Karen Payrazyan

Irati Energy, Canada

When rubblizing oil shale to create a retort bed, the particle size distribution (PSD) will govern the void fraction (VF) in the bed. Void fraction is an important performance and economic variable, and process optimization requires that voids be controlled. Because oil shale particles fracture in shapes not readily described mathematically, an empirical approach has been taken to understand the relationship between PSD and VF. Particle shapes typically exhibit a long axis, a short axis and an intermediate dimension that define aspect ratios. Interestingly, for particles crushed only one or two times, the relationship of these axes dimensions are similar, irrespective of particle size. The consistency in aspect ratios allows for scaling of results from laboratory to field practice. Repeated crushing, however, changes these relationships and should be avoided. Particles of uniform size exhibit void fractions of 50% or greater. Lower void fractions can only be achieved through a proper recipe of a wider range of particle sizes. Particles varying in size by a factor of 100, mixed in ideal proportions yield void fractions as little as 20%. Results describing the PSD required to minimize void fraction have been developed and will be presented. The effect of void fraction on bed settling properties will also be illustrated.

JORDAN 2015 UPDATE

SHELL INTERNATIONAL EXPLORATION & PRODUCTION INC. (SIEP)

JORDAN OIL SHALE COMPANY (JOSCO)

Mariela Araujo
Jack Emmen
Tom Fowler
Thomas Meijssen

"The Treasury" – Petra, Jordan
UNESCO World Heritage Site

DEFINITIONS AND CAUTIONARY NOTE

The companies in which Royal Dutch Shell plc directly and indirectly owns investments are separate entities. In this presentation "Shell", "Shell group" and "Royal Dutch Shell" are sometimes used for convenience where references are made to Royal Dutch Shell plc and its subsidiaries in general. Likewise, the words "we", "us" and "our" are also used to refer to subsidiaries in general or to those who work for them. These expressions are also used where no useful purpose is served by identifying the particular company or companies. "Subsidiaries", "Shell subsidiaries" and "Shell companies" as used in this presentation refer to companies over which Royal Dutch Shell plc either directly or indirectly has control. Companies over which Shell has joint control are generally referred to "joint ventures" and companies over which Shell has significant influence but neither control nor joint control are referred to as "associates". In this presentation, joint ventures and associates may also be referred to as "equity-accounted investments". The term "Shell interest" is used for convenience to indicate the direct and/or indirect (for example, through our 23% shareholding in Woodside Petroleum Ltd.) ownership interest held by Shell in a venture, partnership or company, after exclusion of all third-party interest.

This presentation contains forward-looking statements concerning the financial condition, results of operations and businesses of Royal Dutch Shell. All statements other than statements of historical fact are, or may be deemed to be, forward-looking statements. Forward-looking statements are statements of future expectations that are based on management's current expectations and assumptions and involve known and unknown risks and uncertainties that could cause actual results, performance or events to differ materially from those expressed or implied in these statements. Forward-looking statements include, among other things, statements concerning the potential exposure of Royal Dutch Shell to market risks and statements expressing management's expectations, beliefs, estimates, forecasts, projections and assumptions. These forward-looking statements are identified by their use of terms and phrases such as "anticipate", "believe", "could", "estimate", "expect", "goals", "intend", "may", "objectives", "outlook", "plan", "probably", "project", "risks", "schedule", "seek", "should", "target", "will" and similar terms and phrases. There are a number of factors that could affect the future operations of Royal Dutch Shell and could cause those results to differ materially from those expressed in the forward-looking statements included in this presentation, including (without limitation): (a) price fluctuations in crude oil and natural gas; (b) changes in demand for Shell's products; (c) currency fluctuations; (d) drilling and production results; (e) reserves estimates; (f) loss of market share and industry competition; (g) environmental and physical risks; (h) risks associated with the identification of suitable potential acquisition properties and targets, and successful negotiation and completion of such transactions; (i) the risk of doing business in developing countries and countries subject to international sanctions; (j) legislative, fiscal and regulatory developments including regulatory measures addressing climate change; (k) economic and financial market conditions in various countries and regions; (l) political risks, including the risks of expropriation and renegotiation of the terms of contracts with governmental entities, delays or advancements in the approval of projects and delays in the reimbursement for shared costs; and (m) changes in trading conditions. All forward-looking statements contained in this presentation are expressly qualified in their entirety by the cautionary statements contained or referred to in this section. Readers should not place undue reliance on forward-looking statements. Additional risk factors that may affect future results are contained in Royal Dutch Shell's 20-F for the year ended December 31, 2014 (available at www.shell.com/investor and www.sec.gov). These risk factors also expressly qualify all forward looking statements contained in this presentation and should be considered by the reader. Each forward-looking statement speaks only as of the date of this presentation, October 5, 2015. Neither Royal Dutch Shell plc nor any of its subsidiaries undertake any obligation to publicly update or revise any forward-looking statement as a result of new information, future events or other information. In light of these risks, results could differ materially from those stated, implied or inferred from the forward-looking statements contained in this presentation.

We may have used certain terms, such as resources, in this presentation that United States Securities and Exchange Commission (SEC) strictly prohibits us from including in our filings with the SEC. U.S. Investors are urged to consider closely the disclosure in our Form 20-F, File No 1-32575, available on the SEC website www.sec.gov. You can also obtain these forms from the SEC by calling 1-800-SEC-0330.

PRESENTATION OUTLINE

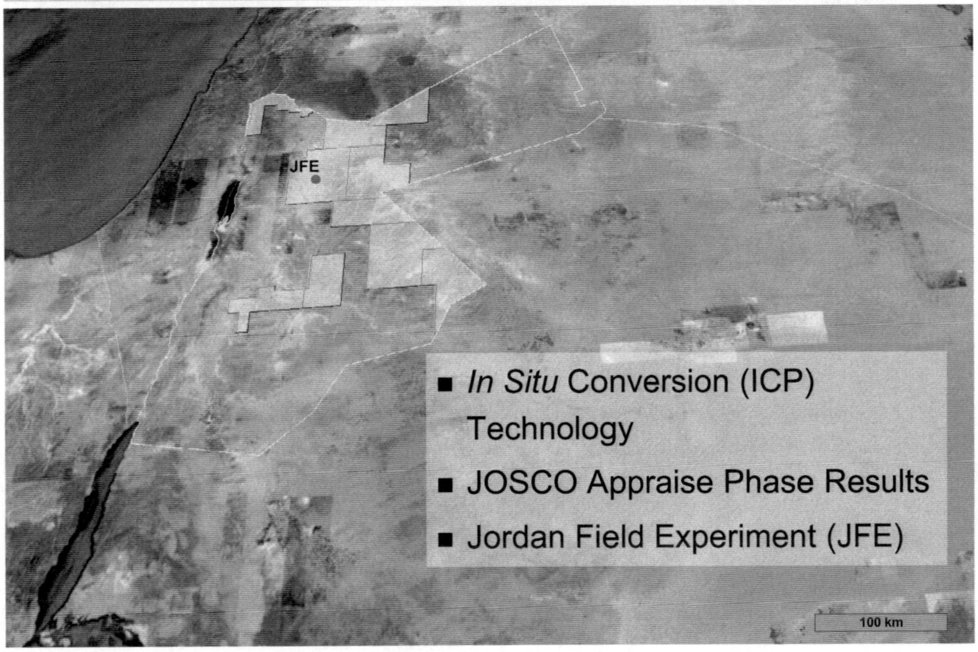

- *In Situ* Conversion (ICP) Technology
- JOSCO Appraise Phase Results
- Jordan Field Experiment (JFE)

IN-SITU CONVERSION PROCESS (ICP) TECHNOLOGY

- ICP enhances the natural maturation of kerogen by *in situ* heating resulting in:
  - Thermal cracking
  - *In situ* hydrogenation
- High Sweep (thermal conduction not impacted by heterogeneity)
- High Recovery (vapor-phase in reservoir)
- High Value Products (heavy ends left behind)

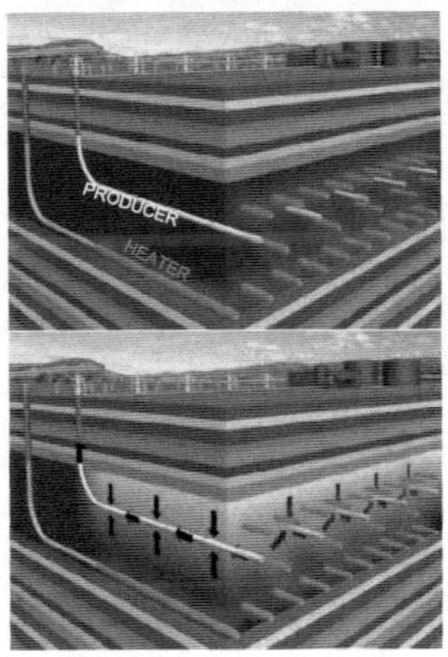

- *Heaters inserted into drilled holes*
- *Resource heated to ~330°C.*
- *Kerogen thermally cracked to lighter components*
- *Coke left behind*

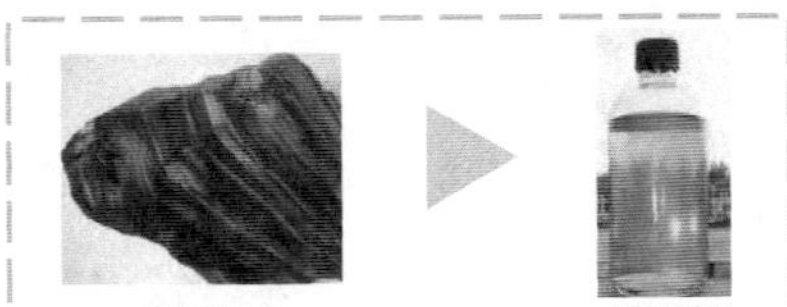

JOSCO – OIL SHALE CONCESSION AGREEMENT (OSCA)

- OSCA ratified by Jordan Parliament Aug 2009:
  - Initial acreage: 22,270 km²
  - Final acreage: 1,000 km²
- OSCA frames the path to a Commercial Project
- Assess Phase drilling began in 2010; 340 wells drilled
- Appraise Phase began May 2013; reduced holding to 15,000 km². Focus: Define final 1,000 km² and execute JFE
- Excellent Governmental support (Jordan imports 97% of its energy)

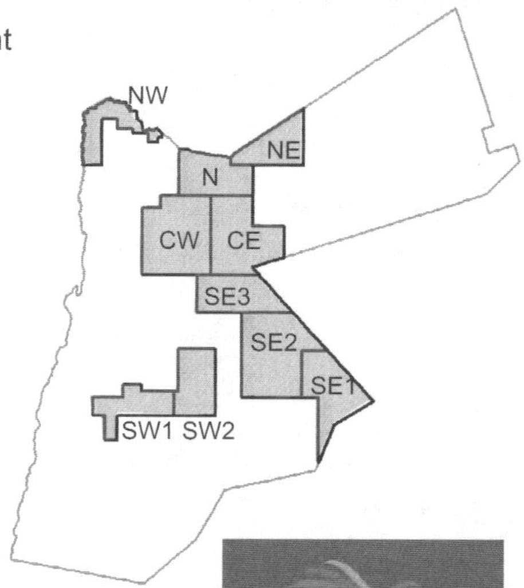

JOSCO
JORDAN OIL SHALE COMPANY

OSCA - PATH TO COMMERCIALITY

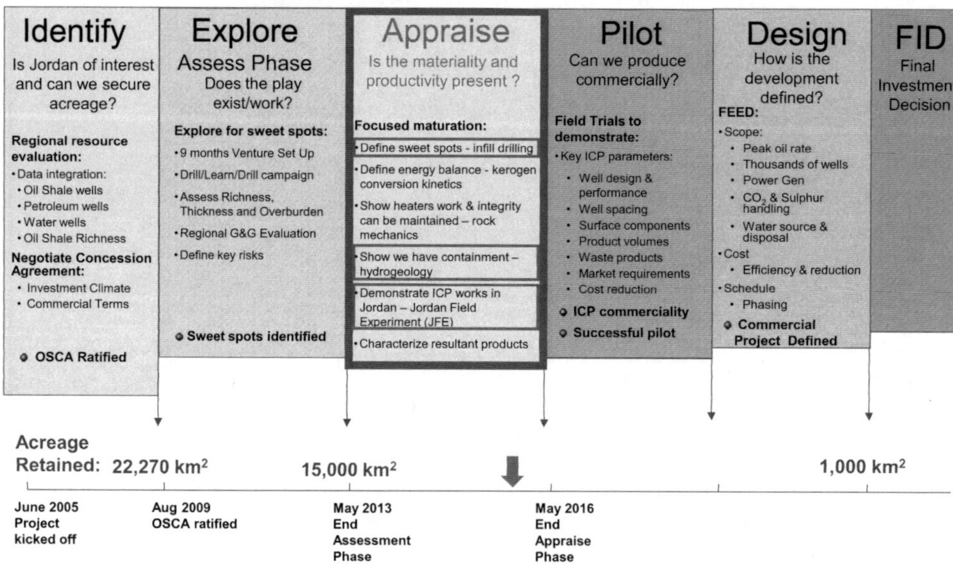

APPRAISE PHASE ACTIVITIES

- The Appraise Phase, begun in May 2013, is the back end of the 5 yr Assessment and Appraisal Effort.
- Drilled 340 wells to underpin:
  - Geology
  - Geophysics
  - Hydrology
  - Geomechanics
  - Reservoir Modeling

JOSCo Drilling Operations

JOSCo Drilling Rig Layout

ASSEMBLING AN INTEGRATED PICTURE

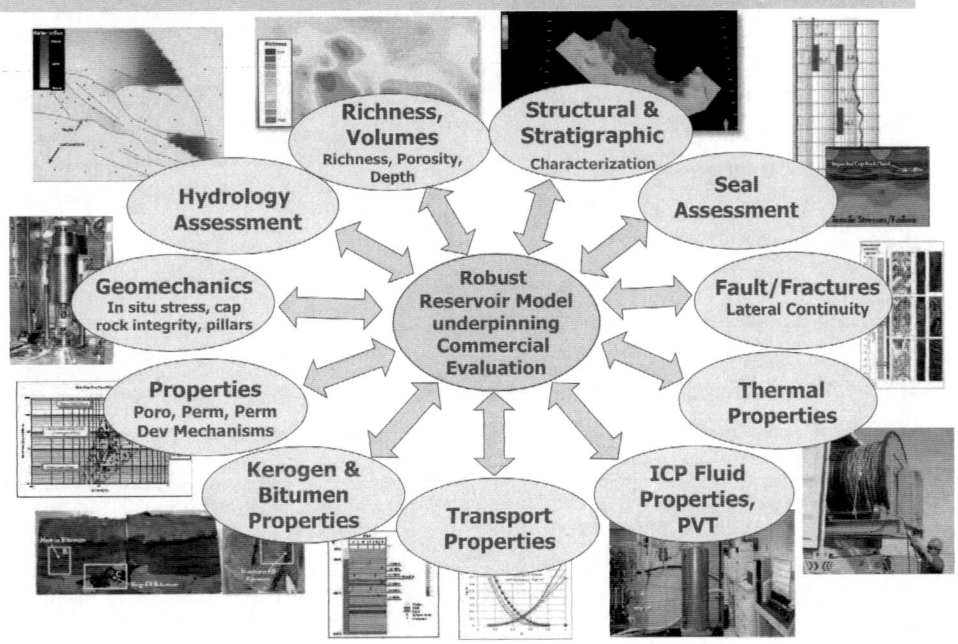

HYDROLOGY AND GEOMECHANICS

Hydrology Campaign

- Successful campaign, >85 wells & >1100 pressure transient tests; water quality samples for baseline in 42 wells.
- Large areas with acceptable hydrology confirmed. Good top and bottom seals.

Appraise Phase Hydrology Testing

Geomechanics

- Rock mechanics test program: measurements from multiple cores at ambient and high temperatures
- In situ stress and regional stress analysis
- Modeling to determine geomechanical pillars dimensions and buffer zone requirements

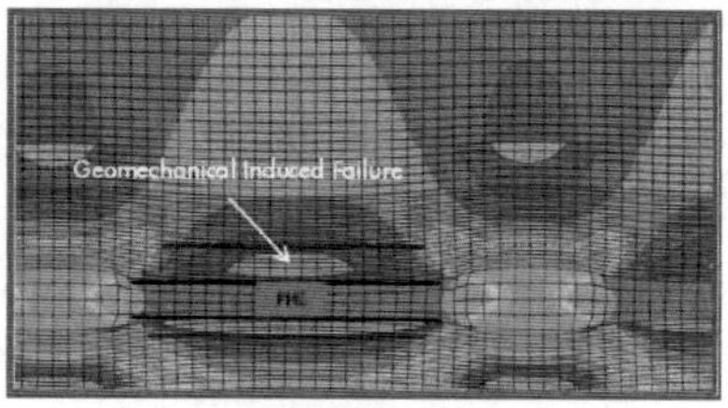

Appraise Phase Geomechanical Modeling

APPRAISE PHASE SUMMARY RESULTS

Appraise Phase subsurface work significantly progressed the JOSCO venture:

- Conducted extensive field and lab activities:
  - Drilled and logged 340 wells (150-1100 m deep)
  - Cut and analyzed 12 cores
  - Hydrotested more than 85 wells, >1100 individual tests
  - Performed Hot Wire & Stress Tests
  - Conducted >50 pyrolysis experiments to develop kinetic/PVT models
- Mapped large in-country resource.
- Identified three "sweet spot" areas using richness, thickness, porosity, containment potential and target depth as criteria.
- Developed Common Risk Segment (CRS) maps to quickly investigate / compare economic potential for ICP.
- Made good progress in the identification of the final 1,000 km^2 lease hold.

JORDAN FIELD EXPERIMENT (JFE)

- Overview
- Wells/Heaters
- Facilities
- Start-up

JFE – Drilling Operations

JFE – Pipe Rack to Flare KO and Flare Tower

JFE OVERVIEW

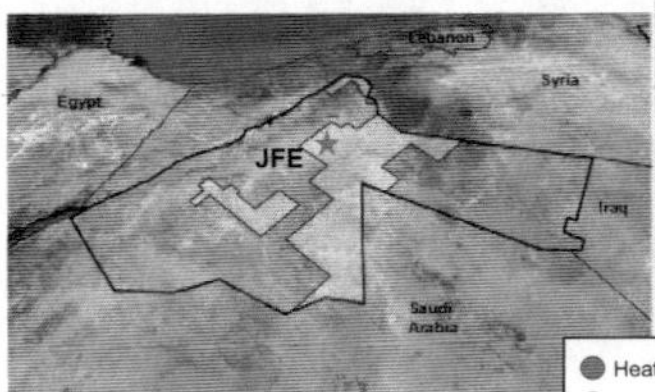

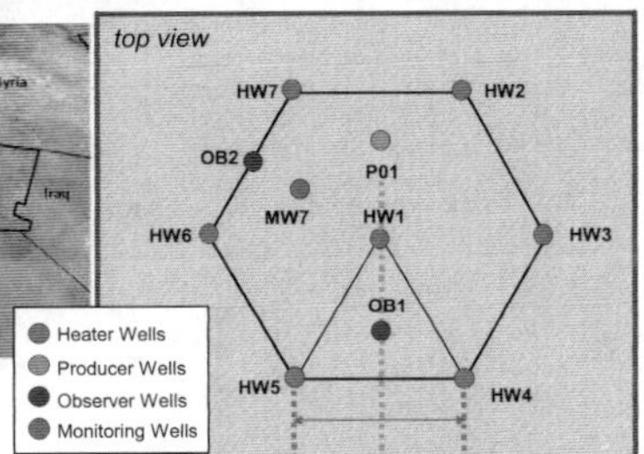

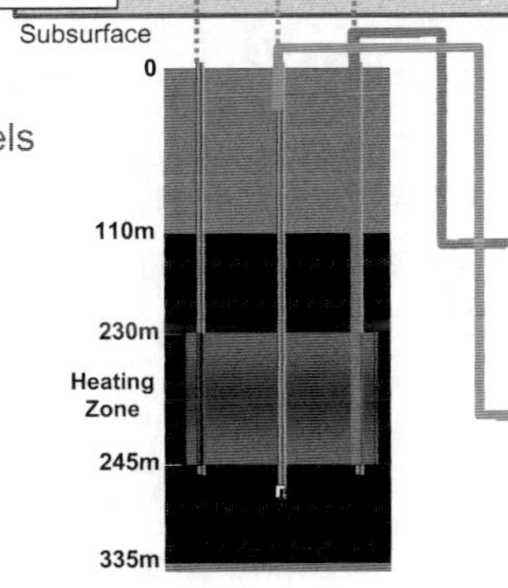

Objectives:
- Produce light, high value hydrocarbons using ICP
- Calibrate subsurface models
- Assess in-country capabilities

Dimensions:
- 7 heater wells (17 total wells)
- 15 m heated interval (230-245m depth)
- Final Temperature 350°C

Timeline:
- Permitting 2011–13
- Select Phase 2012
- Define Phase & FID 2013
- Skid Fab 2014–15
- Onsite Drilling and Constr. Apr 2014 – Sep 2015
- Heaters energized Sep 26 2015
- Heat until mid-2016

Surface Facilities:

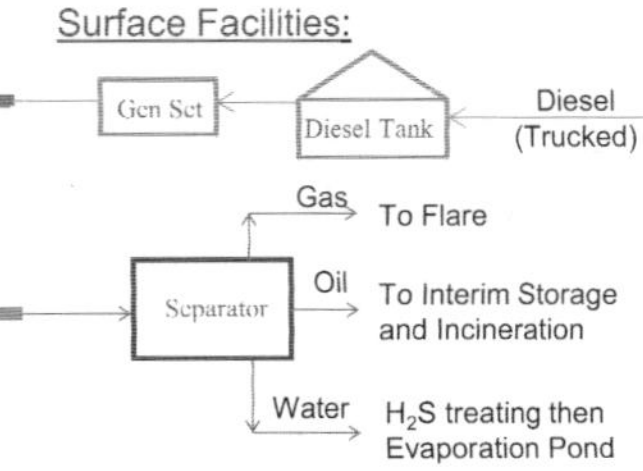

WELLS / HEATERS

- Drilling/Completion
  - 7 heaters, 17 total wells (directionally drilled / vertical).
  - Wells instrumented for pressure and temperature measurement.
  - Tight tolerance pushed the capability of the rig to its limit. All wells drilled within 2.0 ft radial tolerance.
  - Specialist skills and equipment imported from outside Jordan.
- Heaters
  - Three phase, Mineral Insulated (MI) electrical heaters
  - Vertical deployment, 15m heated interval for pilot application
  - Heater powered by diesel generators
  - Note: A full length, long life, high power commercial heater has been designed, built, and tested

JFE – Well Completions

JFE – Heater Deployment

FACILITIES

- Basic Oil, Gas and Water separation, sampling, and measurement. Focus on high reliability data collection
- Skid fabricated in USA, Europe, and Jordan. Engineering contractor: W. S. Nelson (Houston)
- Drilling conducted concurrently with Construction. SIMOPS effectively managed.
- Construction contractor: CCC- Jordan. Excellent safety record during construction (over 600 day “Goal Zero” attainment during JFE construction).
- Extensively instrumented. Real time data available in Amman and Houston with back-up for power outages.

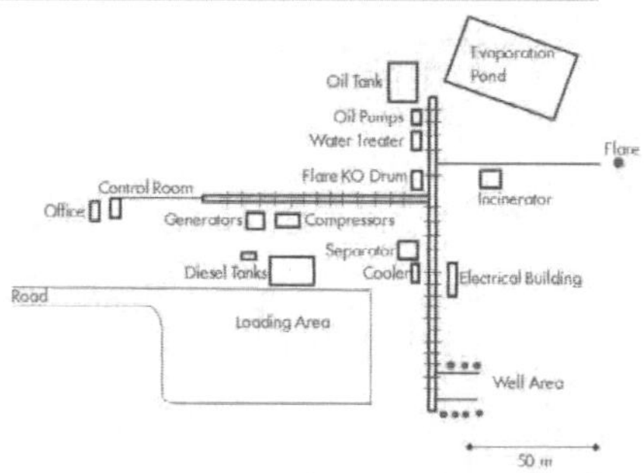

JFE – Overhead Layout

JFE – Separator Skid

START-UP – SEPTEMBER 26, 2015

- Heater low power "bump test" performed on Sep 18-19, 2015 to test the control sequence and instrumentation end elements.
- Heater start-up was achieved on Sep 26, 2015, ahead of P50 schedule established in 2014. Heaters were ramped up to 100% power during the first week of operation.
- Official opening by H.E. Dr. Ibrahim Saif, Minister of Energy and Mineral Resources of Jordan, Sep 29, 2015.
- The heating will be completed in mid-2016.

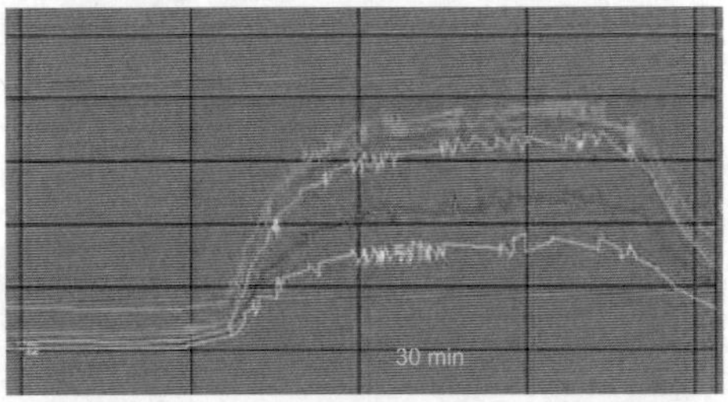

JFE – Temp readings from Heater "Bump Test"

Official opening of JFE – Sep 29,. 2015

ACKNOWLEDGEMENTS

- Co-authors: Mariela Araujo, Jack Emmen, Thomas Meijssen
- JOSCO Amman staff and Shell Houston staff
- Photos by Raj Bharambe

Jordan Field Experiment, September 2015

6.3 Jordan's oil shale development projects: Current status & way forward

Hazem Al-Ramini

Ministry of Energy & Mineral Resources, Jordan

At the present time, the government of Jordan has an agreement with one company for an in-situ oil shale project, three concessions with companies planning surface retorting of oil shale, and one agreement with a company for an oil shale-fired power plant. Economic feasibility studies conducted on different oil shale localities showed that the resource is economically viable for the proposed methods of utilization. The political unrest surrounding the region and the decline in oil prices have not impacted oil shale development projects yet in Jordan but slight delays are predicted pertaining to project financing and hence perhaps tarnishing the glamorous earlier image of early oil and/or electricity generation from oil shale. Several Memoranda of Understanding "MOU's" have been signed and Letters of Interest are placed to share with us the excitement to explore and exploit new oil shale localities of Jordan.

Forward Looking Statements

Some or all of the information contained in the presentation represents Red Leaf Resources, Inc.'s (the "Company") views, estimates and assumptions, any of which may change without notice. Neither the Company nor any of its shareholders, affiliates, partners, members, directors, officers, management, employees or representatives makes any representation or warranty, express or implied, as to the accuracy or completeness of any of the information or any other information (whether communicated in written or oral form) transmitted or made available to you. This presentation contains forward-looking information relating to, without limitation, the Company and its business, intentions, plans, expectations and anticipated financial performance or condition. Readers are cautioned not to place undue reliance on forward-looking information. The actual results, performance or achievements of the Company, the EcoShale® process and the joint venture with Total E&P USA Oil Shale, LLC, could differ materially from those expressed in or implied by forward-looking information contained in this presentation. This presentation is for information purposes only and does not constitute an offer to sell or a solicitation to buy securities.

REDLEAF RESOURCES, INC.

3

Red Leaf's Strategy

- Develop projects with partners in the Green River Formation of Utah, Colorado and Wyoming
  - 45,000 acres of leases on state lands in Utah
  - Two projects: Seep Ridge and Holliday Block
- License to other operators internationally
  - Currently have six license agreements
  - Licenses for projects in Jordan, Morocco, Australia and Utah
  - No country exclusive licenses other than Morocco
- Partner with other operators internationally
  - Likely to be on a non-operating, carried interest basis

REDLEAF RESOURCES, INC.

4

Red Leaf Licensees & Partners

- Total (France): 50/50 partner in Utah; owns a perpetual world-wide license
- Questerre (Canada): Owns an option for a license; has an MOU in Jordan
- Tomco (UK): Site specific license in Utah
- Whitehorn (Canada): Option for a non-site specific license and a license in Jordan
- TAQA (Abu Dhabi): Licenses in Morocco (exclusive) and Jordan
- OilCorp (Australia): Site specific license in Queensland, Australia

REDLEAF RESOURCES, INC.

5

Initial Steps for a License

1. Desktop review of potential resource (no field work)
   - How rich and deep is the resource? What is the stripping ratio?
   - What is the ownership of, and access to, the resource? Who has the surface rights?
2. Secure access to the resource
3. Plan and execute core drilling program
4. Chemical and physical characterization of the shale
   - Fischer Assay, chemical analysis, density analysis, ...*etc.*
5. 10-lb reactor test
   - Characterizes volume and quality of oil produced
6. Submit a pre-development plan

REDLEAF RESOURCES, INC.

6

License Requirements

To be issued a license, one must provide the following to Red Leaf:

- Site-specific resource identified and access secured
- Pre-development plan, including:
  - Permitting overview and timeline
  - Timeline to first production
  - Preliminary resource assessment
  - Financial model
  - Financing plan
  - Personnel plan
- A license will require the payment of a license fee plus a revenue royalty

REDLEAF RESOURCES, INC.

7

How Red Leaf can help

- Resource work:
  - Introductions to Geotech firms
  - Provide sample resource reports
  - Access to our Oil Shale Database (in development)
  - Consulting with our lab staff
- Shale analysis – see next slide
- Introductions to current licensees
- Introductions to engineering support companies
- "Licensee Path Forward"
  - Outlines steps a licensee may consider in moving a project forward
  - Provides rough estimates of time and cost to advance a project

Our objective is to support licensees in determining their own efficient path forward that minimizes initial capital investment and provides multiple decision points for moving forward or walking away from potential projects.

REDLEAF RESOURCES, INC. 8

Red Leaf Laboratory

Suite of Tests and Services (in-house and contracted):

- **Modified Fischer Assay - ore richness**
- Mining Intercept Optimization
- **Ore Properties**
  - **ore chemistry, minerology, organic carbon content, and density**
- **10-lb. Reactor Test – yield and product composition**
- 100-lb. Reactor Test – effects of time and temperature on yield and quality
- 500-lb. Reactor Test – produces samples for downstream testing
- Retort Compression Testing – tests mechanical properties of shale
- Oil Analysis
- Water Analysis
- Gas Analysis
- Spent Shale Testing – leach testing
- Containment Layer Testing - gas diffusion and permeability

*The tests in **bold** are critical steps for initial license application.

REDLEAF RESOURCES, INC. 9

Engineering Support
from Strategic Contractors

- Engineering Support
  - Conceptual
  - FEED
  - Feasibility
  - Final Engineering
- Instrumentation
- Operating Procedures
- Technology Updates

HATCH™

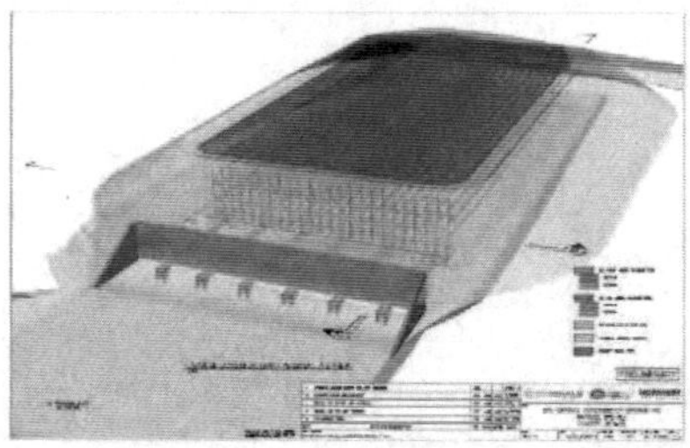

REDLEAF RESOURCES, INC.

10

REDLEAF
RESOURCES, INC.

CONTACTS

Alex Bocock
VP Business Development
abocock@redleafinc.com
801-878-8421

Red Leaf Resources
10808 South River Front
Parkway, Ste. 200
South Jordan, UT 84095
Telephone: 801-878-8100
Fax: 801-878-8101

Buried Heating Pipe Design for Ecoshale Process

H. GHORBANI (HATCH), M. AL-DOJAYLI (HATCH), M. MALEKI (HATCH), K. JOHNS (RED LEAF), J. PATTEN (RED LEAF)

Outline

- Introduction
- Capsule Piping System
  - Indirect vs. Direct Capsule Heating
- Capsule Piping Design
  - Cross Section Design
  - Flexibility Analysis
  - Component Design
- Pipe Embedment Requirements
- Pipe Fabrication & Installation Requirements
- Conclusions

Introduction

- EcoShale In-Capsule Technology is an ex-situ process to more environmentally safe production of liquid fuels from oil shale.
- The EcoShale process uses large gas-tight capsules heated to the pyrolysis temperature using a heating piping network

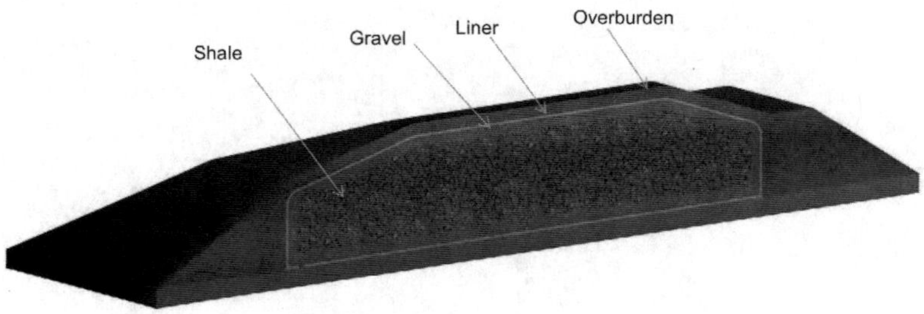

Capsule Piping System

Indirect Heating using Flue Gas

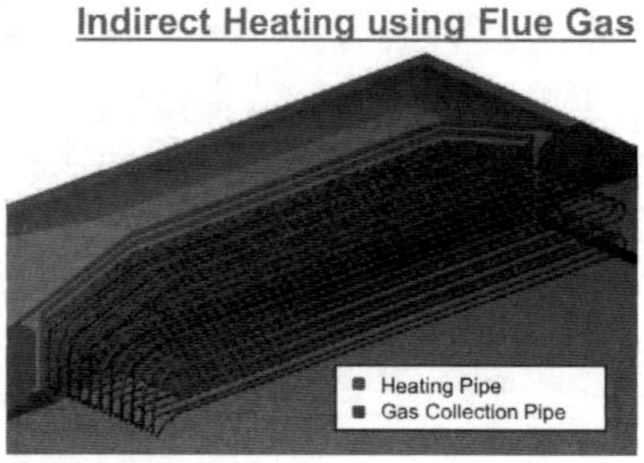

- **Heating Pipes:**
  - Multi-layer
  - gas tight
  - Steel corrugated pipes
- **Gas Collection Pipes:**
  - Multi-layer
  - Perforated
  - Smooth steel pipes

Direct Heating

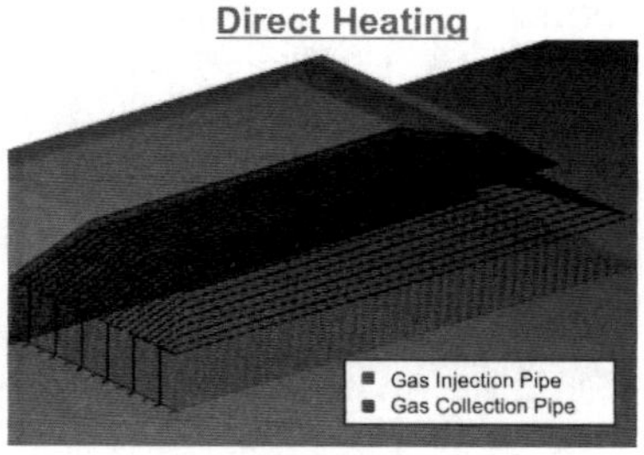

- **Gas Injection Pipes:**
  - Single layer
  - Perforated
  - Smooth steel pipes
- **Gas Collection Pipes:**
  - Single layer
  - Perforated
  - Smooth steel pipes

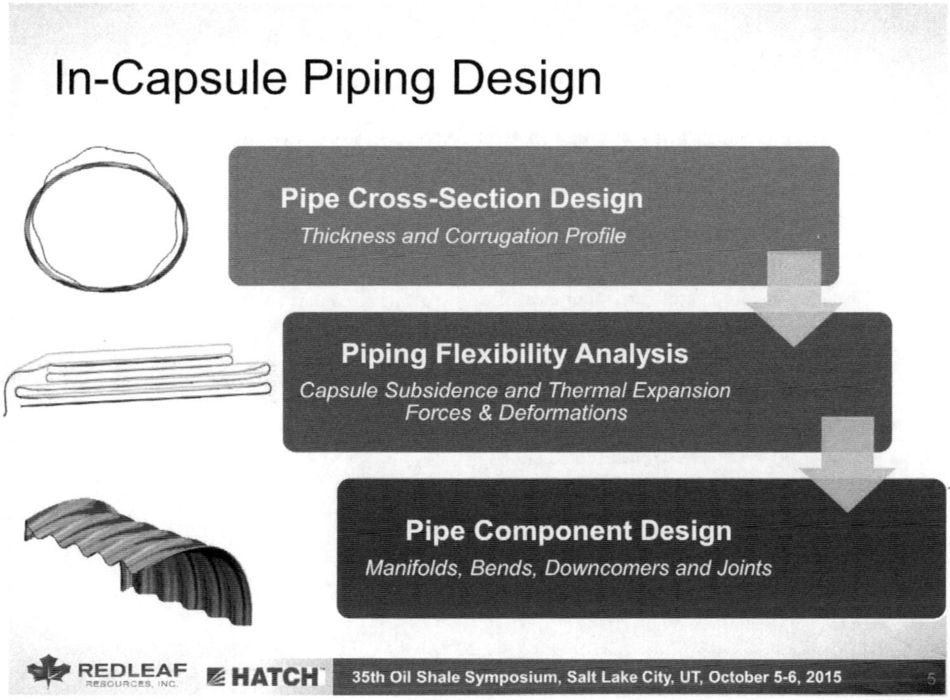
In-Capsule Piping Design
Pipe Cross-Section Design
Thickness and Corrugation Profile
Piping Flexibility Analysis
Capsule Subsidence and Thermal Expansion
Forces & Deformations
Pipe Component Design
Manifolds, Bends, Downcomers and Joints
REDLEAF RESOURCES, INC.
HATCH
35th Oil Shale Symposium, Salt Lake City, UT, October 5-6, 2015
5

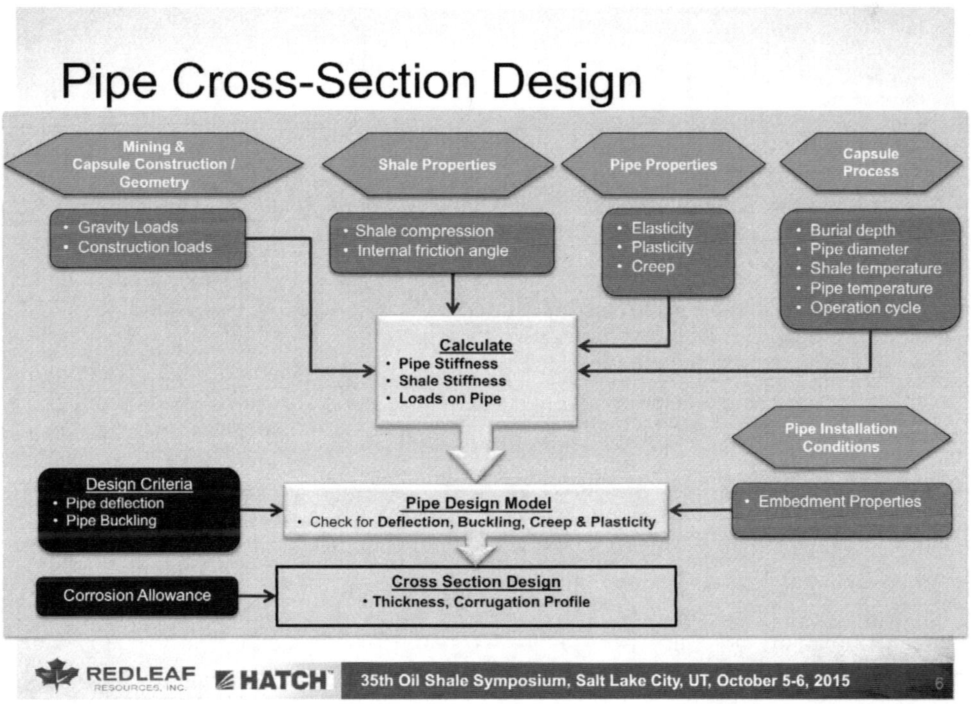
Pipe Cross-Section Design
Mining & Capsule Construction / Geometry
Shale Properties
Pipe Properties
Capsule Process
• Gravity Loads
• Construction loads
• Shale compression
• Internal friction angle
• Elasticity
• Plasticity
• Creep
• Burial depth
• Pipe diameter
• Shale temperature
• Pipe temperature
• Operation cycle
Calculate
• Pipe Stiffness
• Shale Stiffness
• Loads on Pipe
Pipe Installation Conditions
Design Criteria
• Pipe deflection
• Pipe Buckling
Pipe Design Model
• Check for Deflection, Buckling, Creep & Plasticity
• Embedment Properties
Corrosion Allowance
Cross Section Design
• Thickness, Corrugation Profile
REDLEAF RESOURCES, INC.
HATCH
35th Oil Shale Symposium, Salt Lake City, UT, October 5-6, 2015
6

Pipe Cross Section Design Criteria

- In-Capsule pipes are exposed to significant construction, thermal, geostatic and subsidence forces.

Indirect Heating using Flue Gas

Ring Deflection
- Pipe deflection should be controlled to minimize impact on process performance (pressure drop)

Ring Stability
- Ring collapse to be avoided to maintain gas tightness and/or process performance

Plastic Rupture
- Pipe design to ensure no plastic rupture to maintain gas tightness (where required), accounting for initial forming strain in corrugated pipes

Creep Rupture
- Pipe design to ensure no creep rupture at high operating temperatures to maintain gas tightness (where required)

Direct Heating

Ring Deflection
- Pipe deflection should be controlled to minimize impact on process performance (pressure drop)

Ring Stability
- Ring collapse to be avoided to maintain process performance

Pipe Cross-Section Model

- Transient finite element analysis (FEA) used to capture complex pipe-shale interaction including:
  - Temperature-dependent elasticity, plasticity, thermal expansion and creep behavior of pipes
  - Temperature and pressure-dependent shale behaviors, including variation of density, Poison's ratio, thermal properties and shrinkage
  - Shale stiffness recovery post pyrolysis by capping hot shale compression after retorting

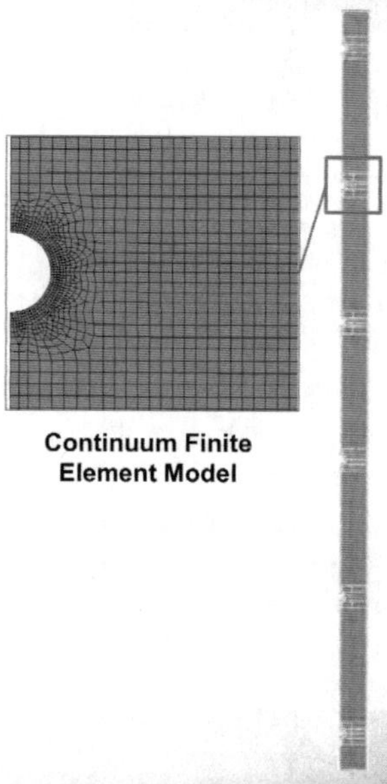

Continuum Finite Element Model

Shale Subsidence

- Retort compression tests (RCT) on shale material showing high sensitivity to shale oil grade, particle size distribution (PSD) and porosity

RCT Equipment Setup

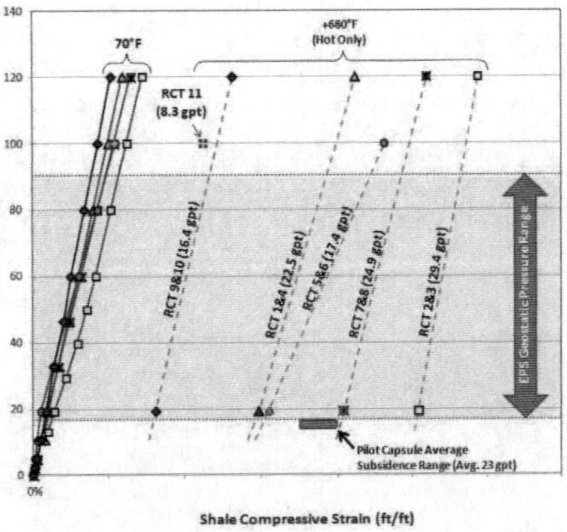

Shale Compression Test Results

Piping Flexibility Analysis

- Significant capsule subsidence during retorting process
- Capsule piping exposed to loads & deformations from soil-pipe interactions

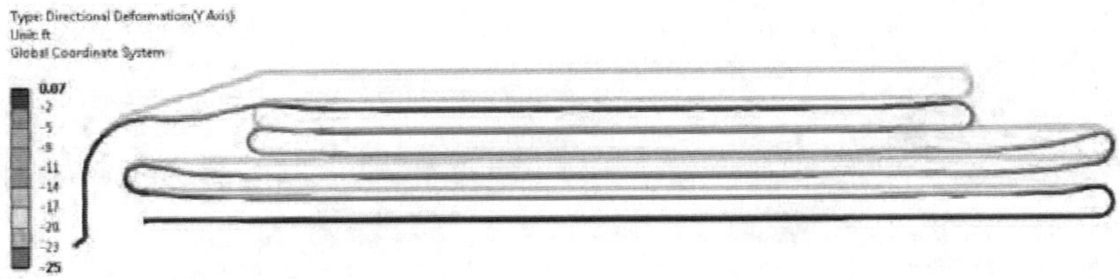

Piping Flexibility Analysis

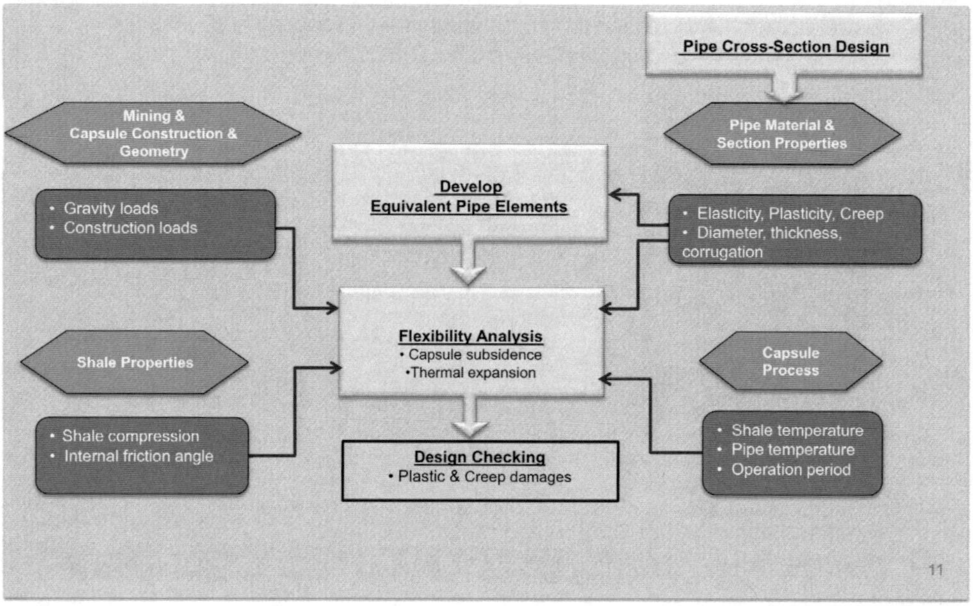

Piping Flexibility Model

Step 1: Equivalent Pipe Element
Replacing corrugated pipe with line element having:

- Equivalent axial stiffness
- Equivalent bending stiffness
- Shear correction factor
- Equivalent creep behavior
- Equivalent plastic behavior

Step 2: 3D Capsule Model

3D Modeling of entire pipe network with equivalent pipe elements under in-plane and out-of-plane forces (e.g., thermal expansion, shale subsidence)

Output

Integrity Checking:

- Pipes
- Manifolds
- Bends
- Downcomers
- Capsule penetration

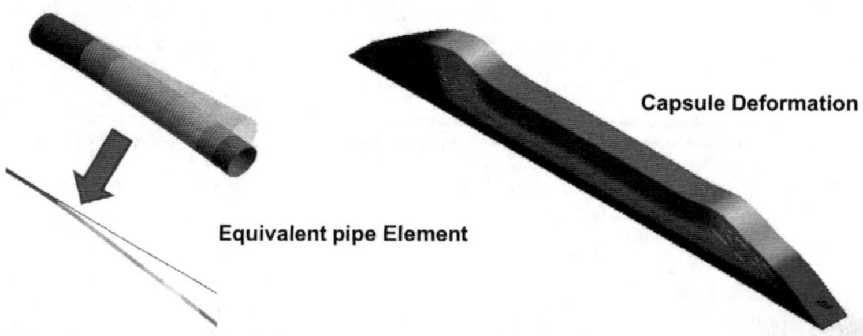

REDLEAF RESOURCES, INC. HATCH 35th Oil Shale Symposium, Salt Lake City, UT, October 5-6, 2015 12

Pipe Component Design

Indirect Heating using Flue Gas

- Multi-segment corrugated bend and downcomer designs increases pipe flexibility under capsule subsidence forces.

Direct Heating

- Slip joints designed to absorb capsule subsidence forces.
- Smaller diameter pipes allow for using off-the-shelf smooth pipes

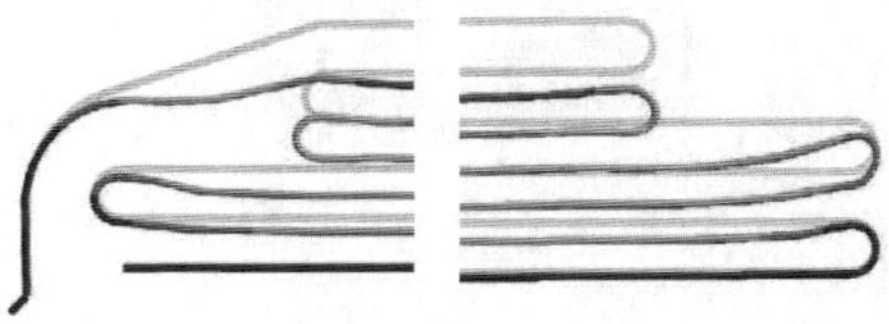

REDLEAF RESOURCES, INC. HATCH 35th Oil Shale Symposium, Salt Lake City, UT, October 5-6, 2015 13

Pipe Embedment Requirements

Indirect Heating Using Flue Gas

- Lower grade shale and controlled particle size distribution (PSD) may be necessary for the pipe embedment region in order to:
  - Reduce deformations and forces on piping system
  - Avoid pipe puncture and loss of gas tightness
  - Enhance capsule heating efficiency

Direct Heating

- Higher ring stiffness of smooth pipes and no gas tightness requirements allow for:
  - Same gpt oil shale materials as entire capsule bed to be used around piping system, avoiding the need for selective mining
  - Avoiding additional crushing and screening of pipe embedment materials

REDLEAF RESOURCES, INC. HATCH 35th Oil Shale Symposium, Salt Lake City, UT, October 5-6, 2015 14

Pipe Testing at Utah State University*

- Large particles in embedment zone resulted in dents and non-uniform deformation (ovality) of corrugated pipes at high geostatic pressures

*Buried Structures Laboratory, Utah State University, Utah, June 2014

Pipe Testing at Utah State University*

- Embedment using controlled shale particle size distribution resulted in adequate support to the corrugated pipes (uniform ovality) without pipe denting

*Buried Structures Laboratory, Utah State University, Utah, June 2014

Pipe Fabrication & Installation Requirements

Indirect Heating using Flue Gas

- Deeply corrugated pipes
- Requiring articulated pipe mill
- Significant lock seam welding
- All welds gas tight
- Significant welding QA/QC
- Limitation on allowable machinery loads
- Time consuming installation

Direct Heating

- Perforation requirements
- Minimized welding
- Minimized QA/QC
- Easier/faster assembly & installation

Conclusions

- Complex FEA models developed to analyze and design buried pipes under significant loads and deformations at elevated temperatures inside Echshale capsule for both indirect and direct capsule heating processes.
- Indirect capsule heating uses large diameter & gas tight corrugated pipes due to process and structural integrity requirements
- Direct capsule heating uses smaller diameter and perforated smooth pipes
- Piping system design for direct capsule heating has numerous advantages:
  - Utilizing off-the-shelf smooth pipes
  - Using much less pipes in capsule
  - Not requiring special pipe embedment materials simplifies mining, material handling and capsule construction activities
  - Not being gas tight simplifies pipe fabrication, welding and QA/QC activities
  - Mitigating risk of flue gas leakage into capsule environment
  - Significantly reducing piping and pipe embedment costs

Thank you

For More Information:

Hamid Ghorbani, PhD. PEng.

Hatch Ltd.
2800 Speakman Dr.
Mississauga, On, Canada

Email: hghorbani@hatch.ca
Phone: +1 416-818-9185

Hydrologic Characterization of the Southeastern Extent of the Birds-Nest Aquifer, Uintah County, Utah

Ryan Clerico
Head of Development and Environment
Enefit American Oil

Enefit

Overview

- History of the Birds-Nest Aquifer (BNA)
  - Early Research
  - Characteristics
  - Location and Extent
- Working Hypothesis and Data Collection in the BNA
  - Theory
  - GW Wells and SW Locations
  - Water Quality Results
- A Unique Situation?
  - Static and Perched Aquifer
  - Implications for Oil Shale Industry

Enefit

The Birds-Nest Aquifer

- History of the BNA
  - Early Research
  - Characteristics
  - Location and Extent
- Working Hypothesis and Data Collection in the BNA
  - Theory
  - GW Wells and SW Locations
  - Water Quality Results
- A Unique Situation?
  - Static and Perched Aquifer
  - Implications for Oil Shale Industry

Enefit

A Brief History of BNA Research

- First set of wells drilled between 1976-1978 by USGS in response to growing interest in oil shale development
- Continuing and focused work was conducted by Holmes (1980) and Holmes and Kimball (1987), focusing on both the BNA and the Douglas Creek aquifers, as features that could be affected by an oil shale industry, specifically from the federal Ua/Ub leasing tracts
- Gap from the late 1980s to late 1990s, as oil shale development interest waned
- BLM permanently closed and abandoned many regional monitoring wells in the early 2000s, but acquired a final WQ dataset in 2002
- Renewed interest in the BNA due to oil shale industry and potential for saline disposal resulted in a new UGS in 2012

Enefit

BNA Characteristics

- Occurs approximately 400 feet above the Mahogany Zone oil shale layer, in the Parachute Creek member of the Green River Formation
- Lower BNA averages 115 feet thick, with areal extent of over 700 square miles
- Isolated and not in hydraulic communication with aquifers stratigraphically above or below

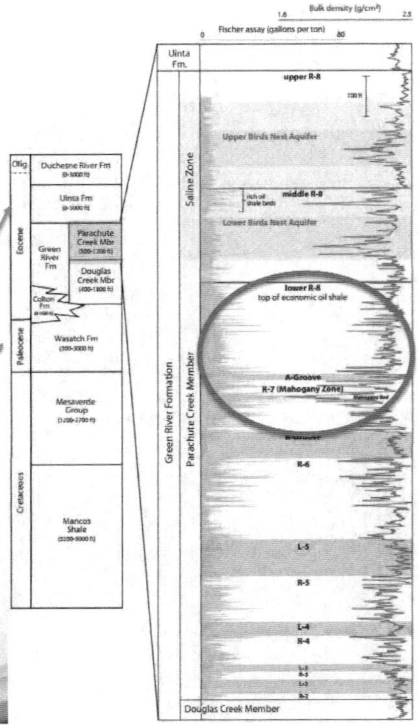

(In Vanden berg et al 2013; after Bradley 1931, Cashion and Donnell 1974, and Hintze 1988)

BNA Characteristics

- Contains cavities of nahcolite ($NaHCO_3$) which, when dissolved away at outcrop, give the appearance of many swallow nests
- *Recharge is primarily from stream infiltration and from downward leakage from the overlying Uinta Formation*
- *Groundwater movement typically follows the slope and direction of major streams (e.g. White River, Bitter Creek, Evacuation Creek)*

Enefit

The Birds-Nest Aquifer…

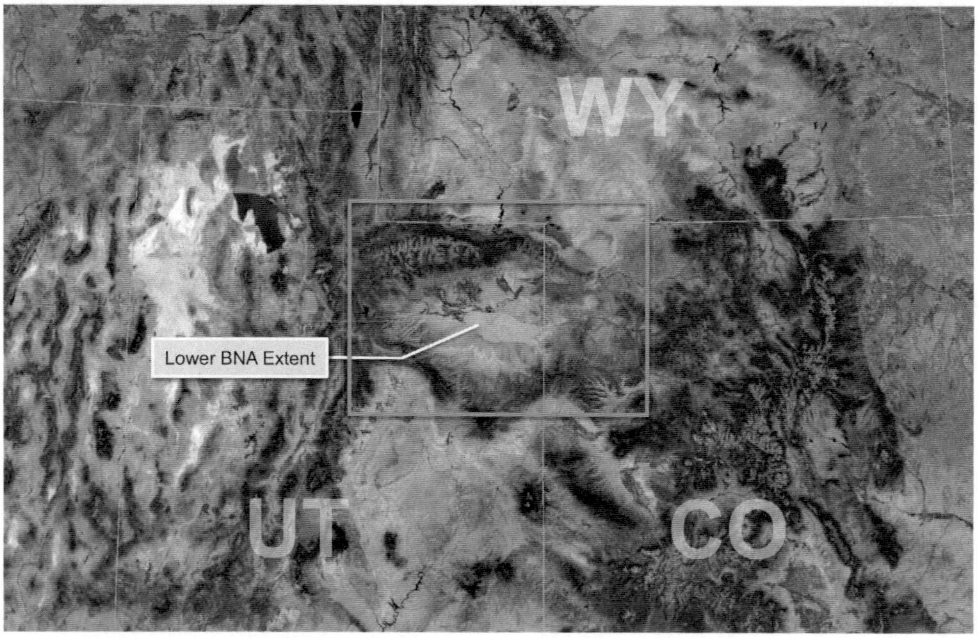

The Birds-Nest Aquifer…

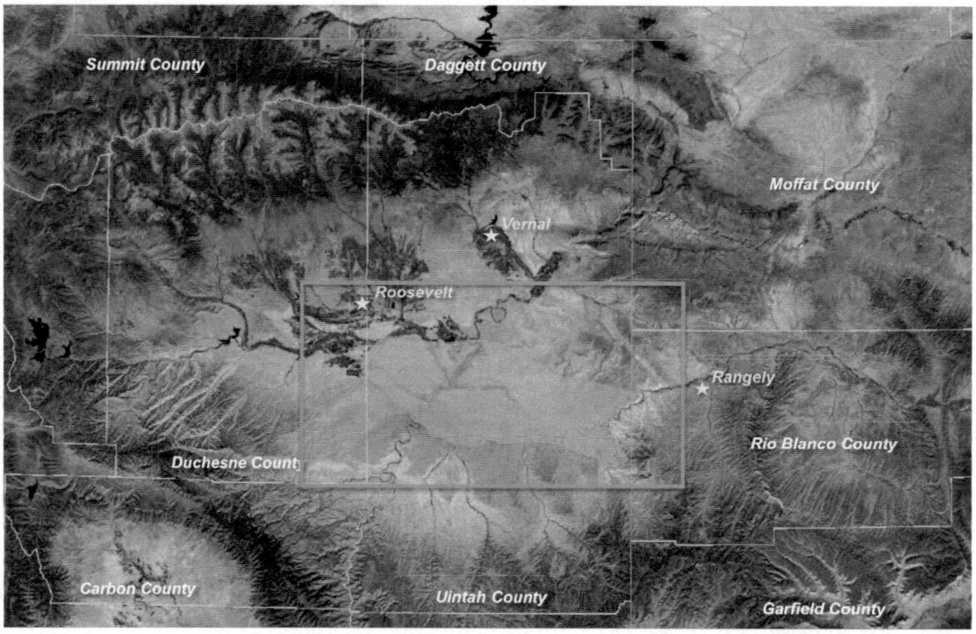

The Birds-Nest Aquifer…

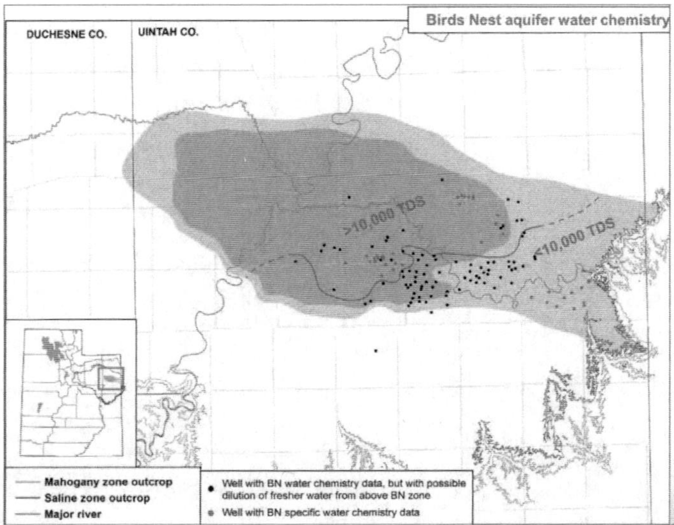

Source: Michael Vanden Berg, Utah Geological Survey

Enefit

The Birds-Nest Aquifer…

Enefit

The Birds-Nest Aquifer…

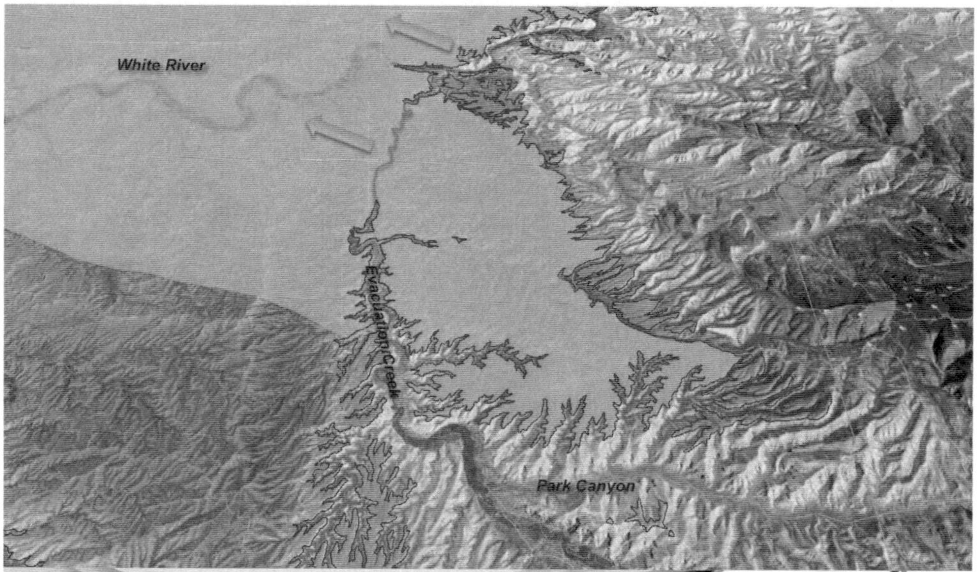

Enefit

The Birds-Nest Aquifer…

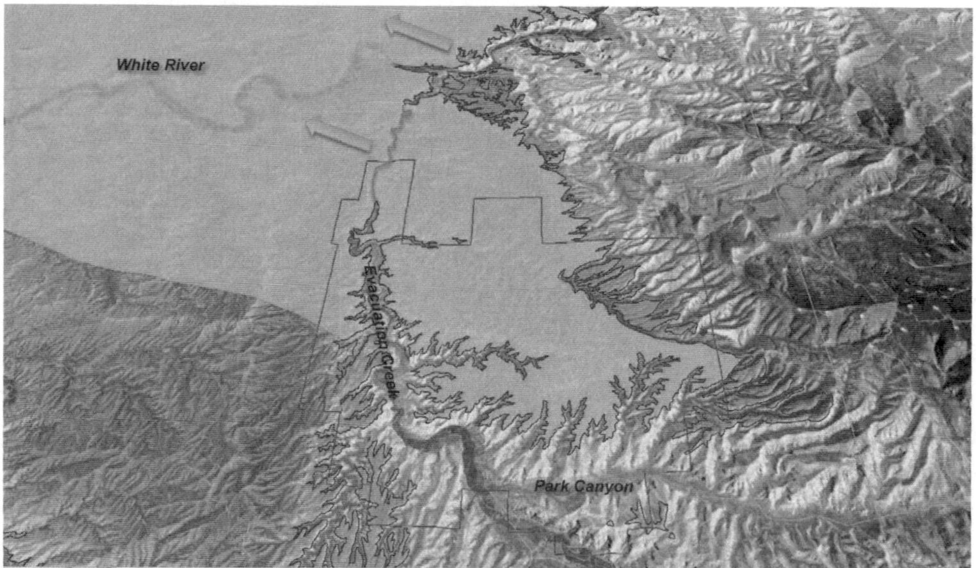

Enefit

The Birds-Nest Aquifer

- History of the Birds-nest Aquifer
  - Early Research
  - Characteristics
  - Location and Extent
- Working Hypothesis and Data Collection in the BNA
  - Theory
  - GW Wells and SW Locations
  - Water Quality Results
- A Unique Situation?
  - Static and Perched Aquifer
  - Implications for Oil Shale Industry

Enefit

Working Hypothesis…

The portion of the Birds-Nest Aquifer that is located east of Evacuation Creek and south of the White River is isolated from the main body of the formation and receives minimal recharge from overlying stream flows and downward infiltration from the overlying Uinta Formation, resulting in unique hydrogeologic conditions.

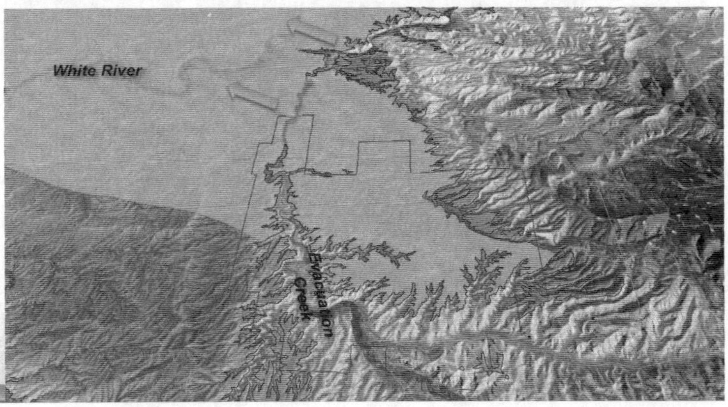

BNA GW Monitoring Locations

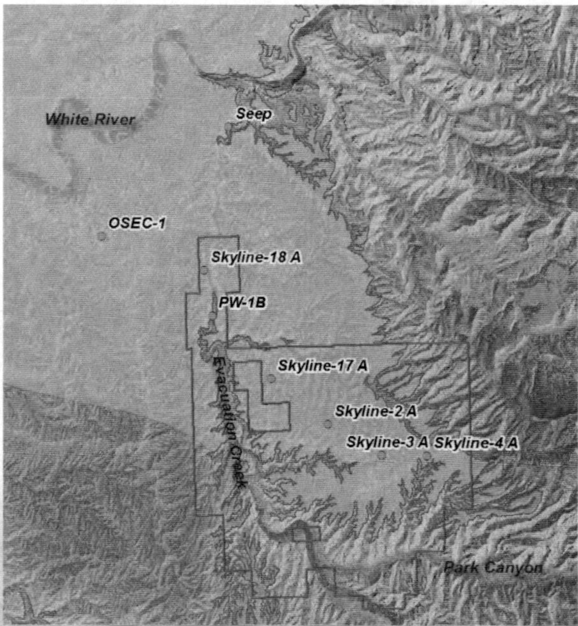

Aquifer	Well	Total Depth (feet below top of casing)
Birds-nest Aquifer	Skyline-2A	305
	Skyline-3A	176
	Skyline-4A	165
	Skyline-17A	161
	Skyline-18A	139
	PW-1B	110
	OSEC-1	690
	Seep	n/a

Enefit

BNA SW Monitoring Locations

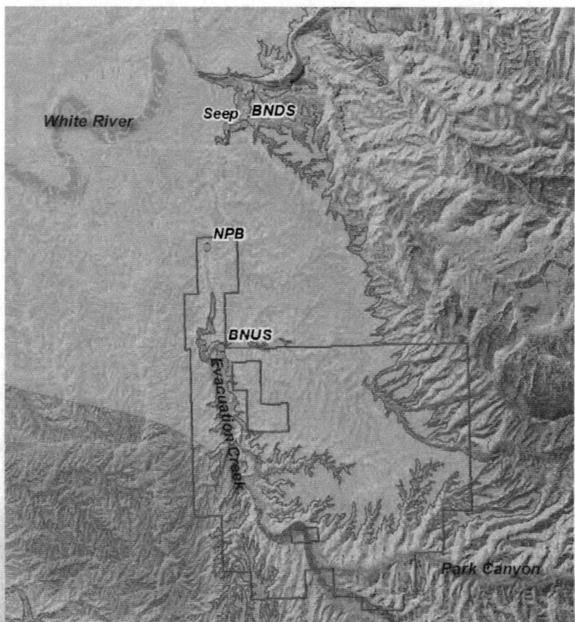

Feature	Location
Evacuation Creek	BNUS
	NPB
	BNDS

Enefit

TDS Results

Total Dissolved Solids, Q3 2013 through Q2 2014 (ppm)				
	Evacuation Creek	Alluvial Aquifer	Birds-nest Aquifer	Seep
Mean	4156	4114	16370	4513
Median	4200	4540	4275	4560
Maximum	7440	5130	115000	4990
Minimum	1650	50	10	3990
n	25	9	34	3

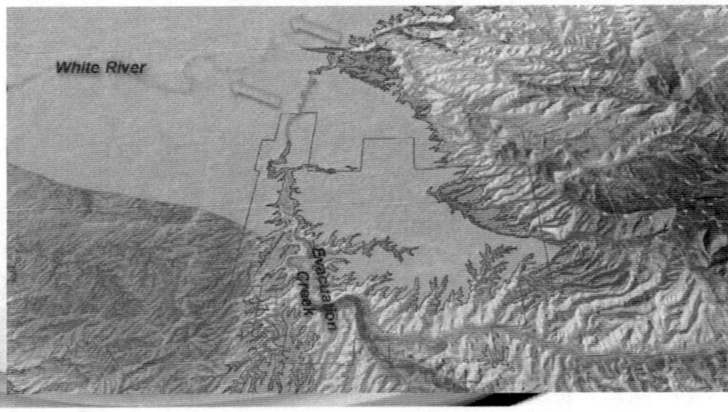

Enefit

TDS Results

Total Dissolved Solids, Q3 2013 through Q2 2014 (ppm)				
	Evacuation Creek	Alluvial Aquifer	Birds-nest Aquifer	Seep
Mean	4156	4114	16370	4513
Median	4200	4540	4275	4560
Maximum	7440	5130	115000	4990
Minimum	1650	50	10	3990
n	25	9	34	3

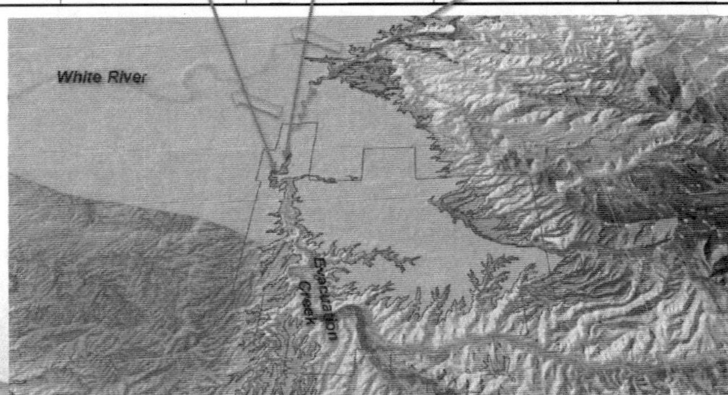

Enefit

TDS Results

Total Dissolved Solids, Q3 2013 through Q2 2014 (ppm)				
	Evacuation Creek	Alluvial Aquifer	Birds-nest Aquifer	Seep
Mean	4156	4114	16370	4513
Median	4200	4540	4275	4560
Maximum	7440	5130	115000	4990
Minimum	1650	50	10	3990
n	25	9	34	3

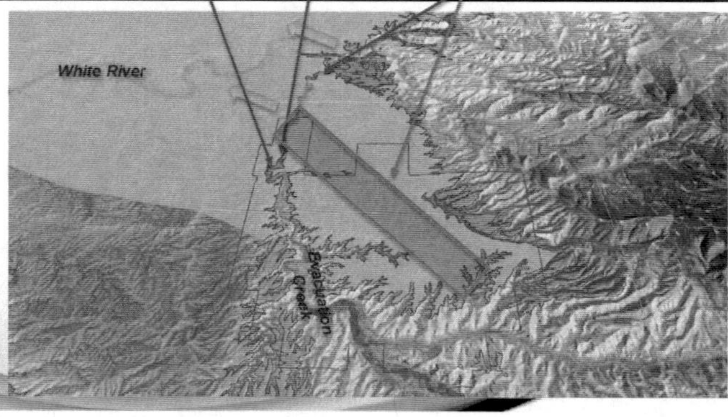

Enefit

TDS Results

Total Dissolved Solids, Q3 2013 through Q2 2014 (ppm)				
	Evacuation Creek	Alluvial Aquifer	Birds-nest Aquifer	Seep
Mean	4156	4114	16370	4513
Median	4200	4540	4275	4560
Maximum	7440	5130	115000	4990
Minimum	1650	50	10	3990
n	25	9	34	3

Enefit

The Birds-Nest Aquifer

- History of the Birds-nest Aquifer
  - Early Research
  - Characteristics
  - Location and Extent
- Working Hypothesis and Data Collection in the BNA
  - Theory
  - GW Wells and SW Locations
  - Water Quality Results
- A Unique Situation?
  - Static and Perched Aquifer
  - Implications for Oil Shale Industry

Enefit

BNA Saline Water Areas

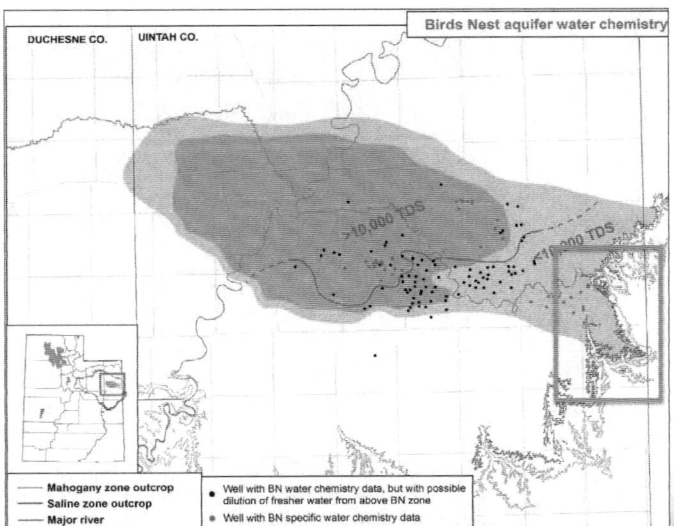

Source: Michael Vanden Berg, Utah Geological Survey

Enefit

"Static and Perched"

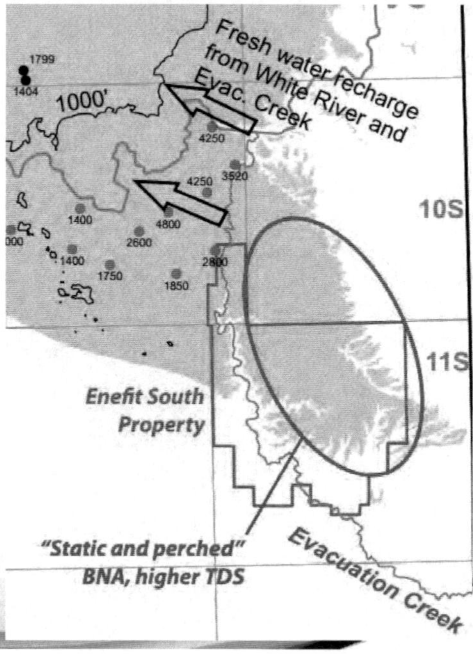

Enefit

Birds-nest Aquifer Age

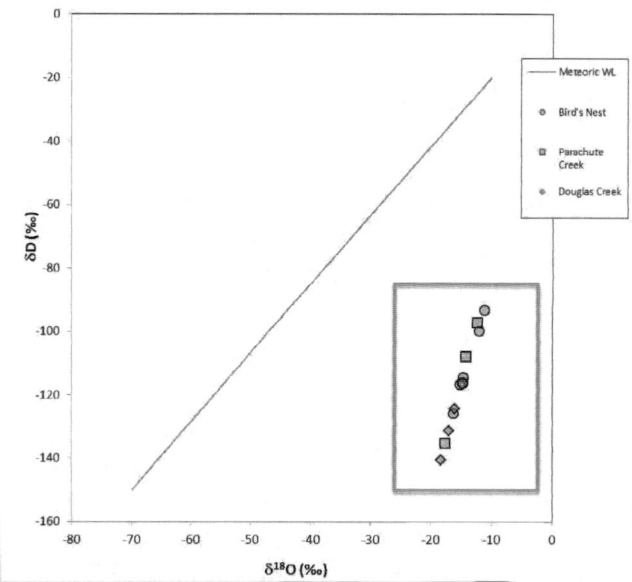

Enefit

Birds-nest Aquifer Age

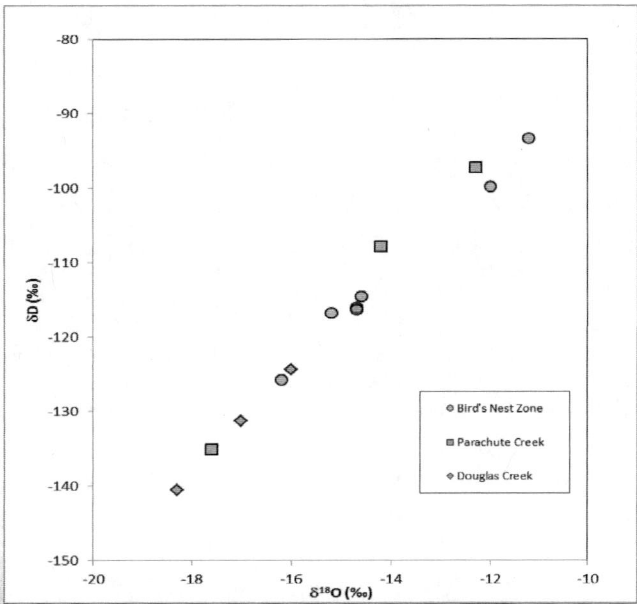

Enefit

Summary and Implications for Oil Shale Industry

- Potentially new, undescribed condition within the Birds-Nest Aquifer that is unique to the area in/around Enefit American Oil's oil shale resource
- Much of the research on the BNA and oil shale still applies
  - The "static and perched" condition appears to be unique to the area upgradient from Evacuation Creek and the White River
  - Possible that this occurs near other outcrop areas depending on distance from fresh water flushing effect
- Recharge and transport conditions for this location need more study
  - Assuming static conditions and "old water", is recharge occurring and from where?
  - Is there horizontal communication from perched aquifer to Evacuation Creek and White River that would affect mining/disposal options?
  - Transmissivity values for BNA in other locations may not apply to this static and perched portion of the formation

Enefit

Ryan Clerico

Head of Development and Environment

Enefit American Oil
307 West 200 South, #4005
Salt Lake City, Utah 84101
801.363.0206
Ryan.Clerico@enefit.com
http://enefitutah.com/

Enefit

USGS
science for a changing world

Evolution of the Lower Tertiary Elko Lake Basin, a Potential Hydrocarbon Source Rock in Northeast Nevada

Ronald C. Johnson and Justin E. Birdwell
U.S. Geological Survey, Denver, CO

U.S. Department of the Interior
U.S. Geological Survey

USGS
science for a changing world

Oil shale of the Eocene Elko Formation was developed as a source of oil in the early part of the 20th century.

Itis thought to be the source for recent oil discoveries in northeast Nevada.

Here we compare the depositional setting and geochemistry of the Eocene Elko Formation to that of the Eocene Green River Formation.

U.S. Department of the Interior
U.S. Geological Survey

Extent of Eocene Elko Formation shown in blue. Extent of lacustrine rocks that may be related to the Elko Formation shown in green.

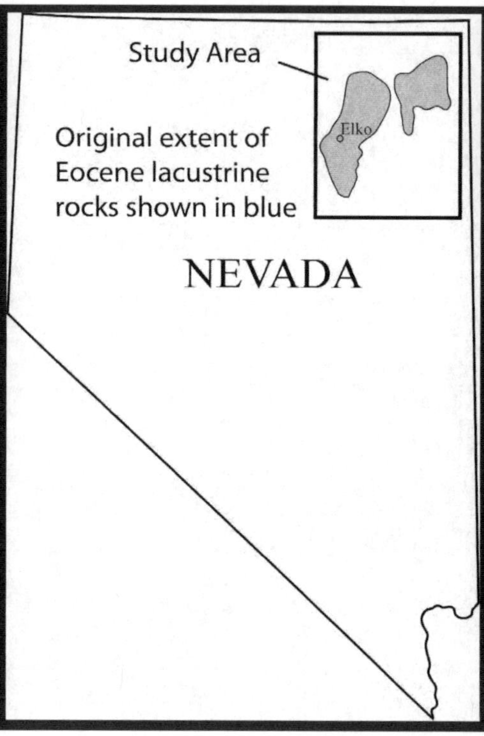

USGS

USGS If you want to work on the Elko Formation bring a shovel. All major outcrops of the Elko were trenched with a backhoe in the past.

This contrasts with the Green River Formation where good outcrops are abundant

USGS

USGS
science for a changing world

Although the Elko lake and Green River lakes are all Eocene in age and deposited kerogen-rich sediments that became oil shale, they don't appear to overlap in time and formed in different tectonic settings.

U.S. Department of the Interior
U.S. Geological Survey

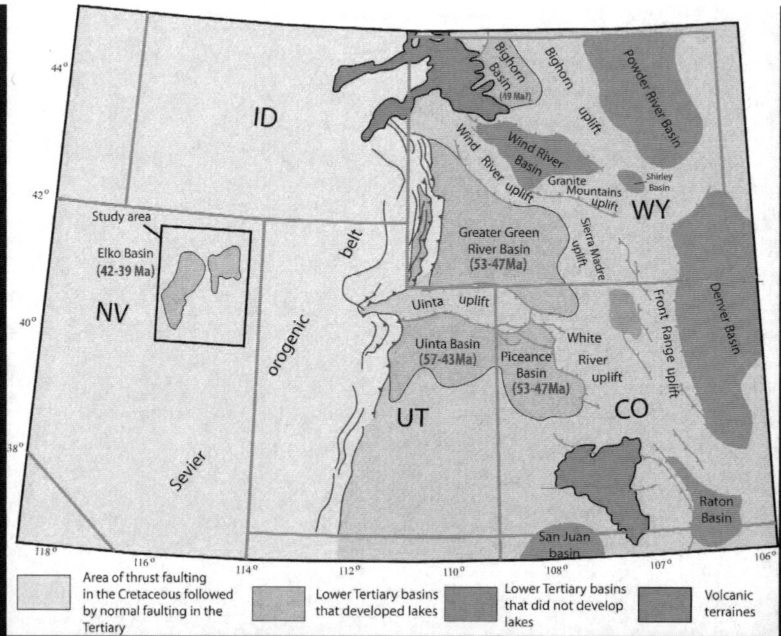

The lacustrine Elko Formation was deposited in sedimentary basin in what was previously part of the Sevier orogenic belt (shown in yellow), an area of thin-skinned thrusting from about 140-50 Ma. The Green River Formation was deposited in Laramide structural and sedimentary basins that formed east of the Sevier orogenic belt.

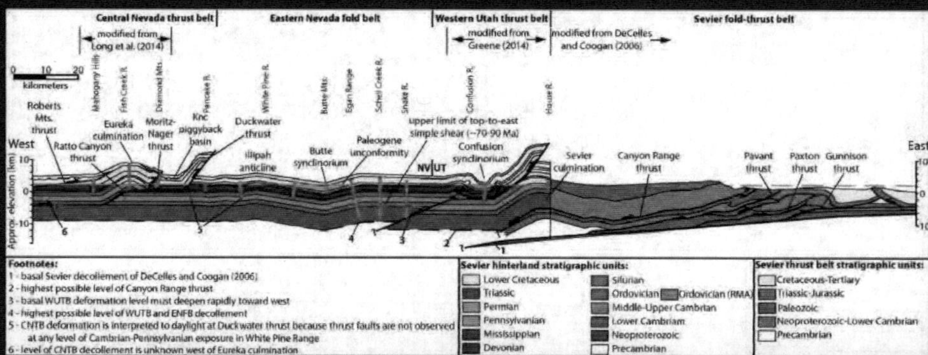

Geological Society of America

The Sevier orogenic belt was an area of thin-skinned thrusting from about 140 to 50 Ma. Most but not all workers (see, for example Ketner and Alpha, 1988) believe that thrusting ended prior to the formation of the Elko Lake basin.

USGS

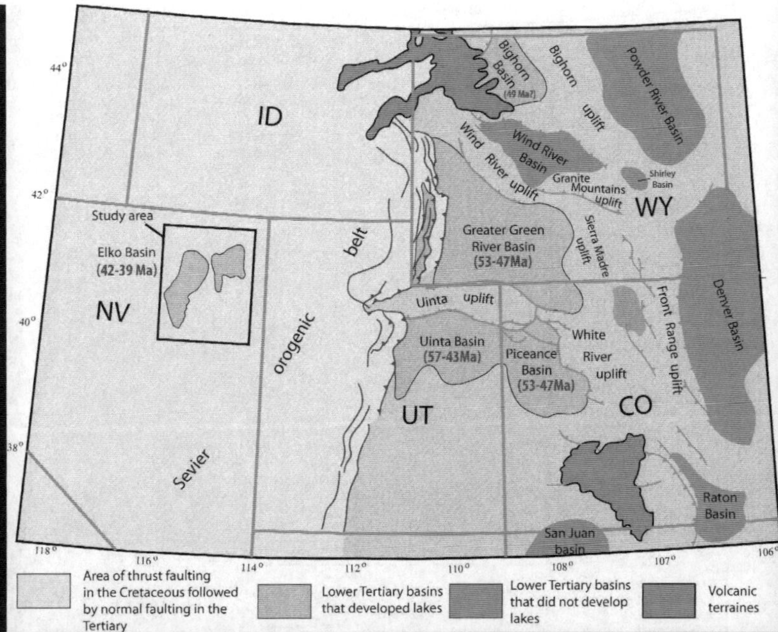

The Green River Formation was deposited in basins formed east of the Sevier orogenic belt during the Laramide Orogeny a period of thrusting involving basement rocks that started 70-80 Ma and ended 35-55 Ma. There is much uncertainty as to when the Laramide orogeny ended, but the youngest lacustrine rocks of the Green River Formation are about 43 Ma (Smith and others, 2007).

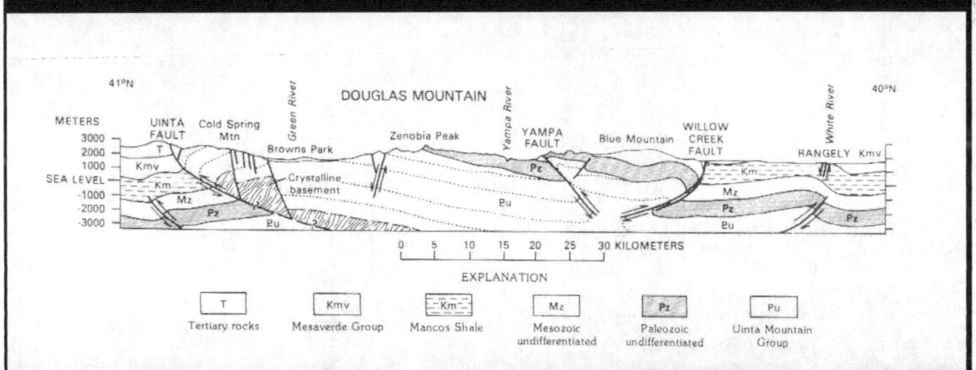

Laramide basins formed during a period of crustal shortening and deep-seated thrusting that involved basement rock. Thrust loading was probably the main cause for basin formation. North-to south cross section of the eastern Uinta Mountains (modified from Hansen, 1986)

USGS

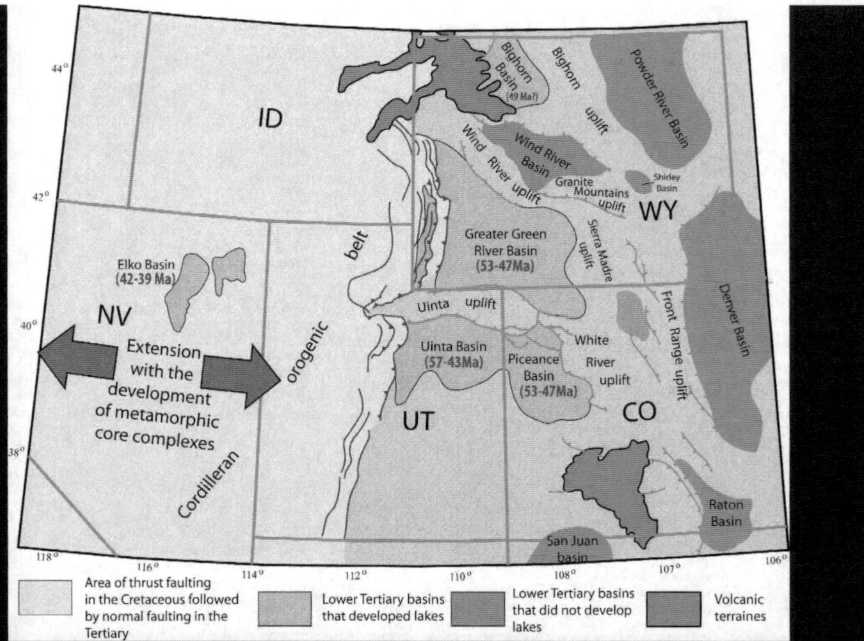

The Elko Lake formed about 42 Ma during an early stage of regional extension that lead to the development of metamorphic core complexes. The lake may overlap in age with the youngest lacustrine rocks of the Green River Formation in the Uinta Basin but is probably entirely younger.

The Elko lake formed west of the Ruby Mountain core complex that formed during an early stage of regional extension that began during the Eocene. Approximate original extent of Eocene-age Elko lacustrine rocks is from Haynes (2003) (blue). Lacustrine rocks also are present northeast of the Elko lake (green). It is not known if this lake is the same age as the Elko lake. Elko uplifts shown in light red.

Recent oil and gas tests targeting the Elko shown in red.

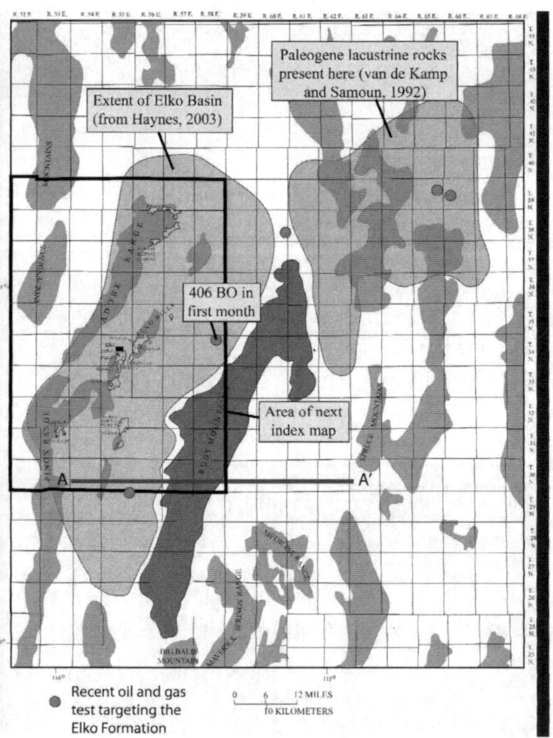

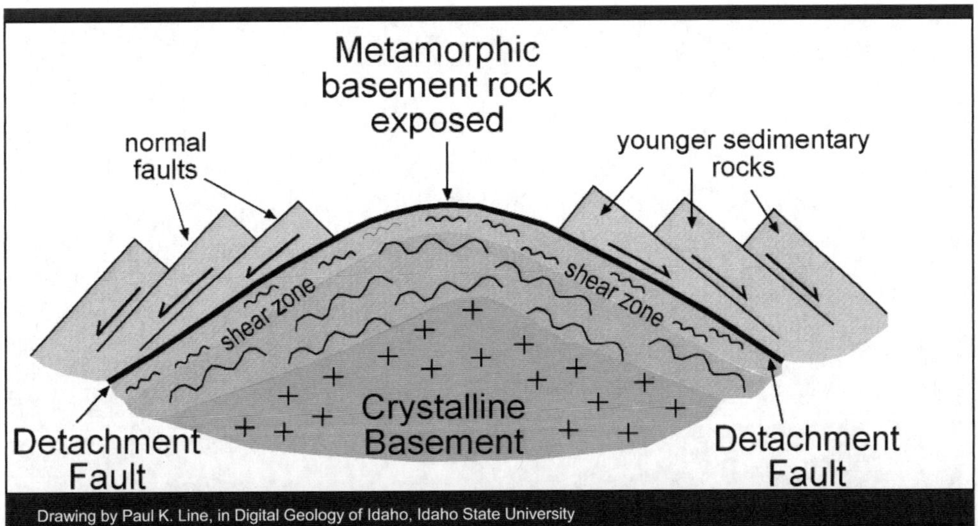

Schematic cross section of a typical metamorphic core complex that brings mid-crustal level basement rocks to the surface in area where the crust is being pulled apart.

USGS

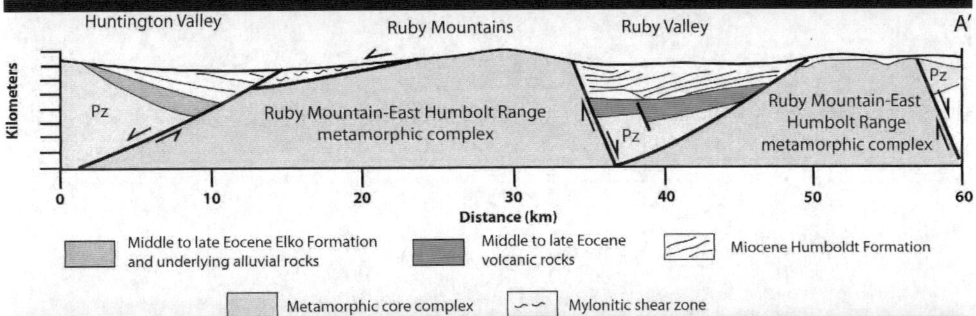

Seismic cross section the Huntington Valley, where the Elko Formation was deposited, the Ruby Mountain metamorphic complex, and the Ruby Valley to the east. Note the mylonitic shear zone. (Modified from Satarugsa and Rohnson, 2000). In this interpretation the two sedimentary basins formed contemporaneous with uplift of the core complexes.

USGS

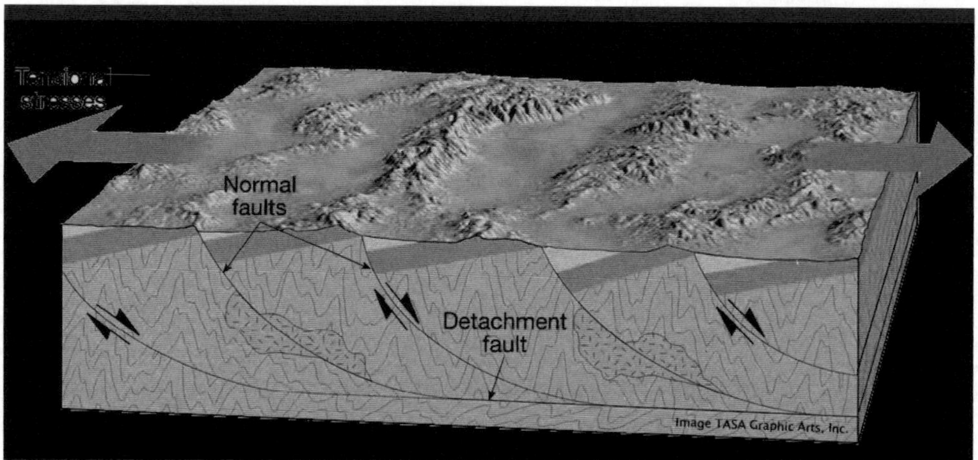

Lacustrine rocks of the Elko Formation were later brought to the surface during a later stage of extension that began about 17 Ma producing the half graben "basin and range" topography seen today.

USGS

Cross section of basin and range faulting. Idaho Museum of Natural History 921 S. 8th Avenue, Stop 8096 Pocatello, ID 83209
Phone: 208.282.3168
Department of Geosciences, ISU
921 S. 8th Avenue, Stop 8072
Pocatello, ID 83209
Phone: 208.282.3235

USGS
science for a changing world

Detailed studies of the Elko Formation

U.S. Department of the Interior
U.S. Geological Survey

The Elko Formation is exposed in widely-spaced outcrops that are difficult to correlate between because of rapid lateral facies changes.

Detailed studies of the Elko Formation were conducted where it crops out in the northern Adobe Range, Elko Hills, and Pinon Range.

In addition, core of the Elko Formation (EOS-1, EOS-6, EOS-3/3A) stored at the Nevada Bureau of Mines core facility was described and extensively sampled for Rock Eval, x-ray, and thin section analysis.

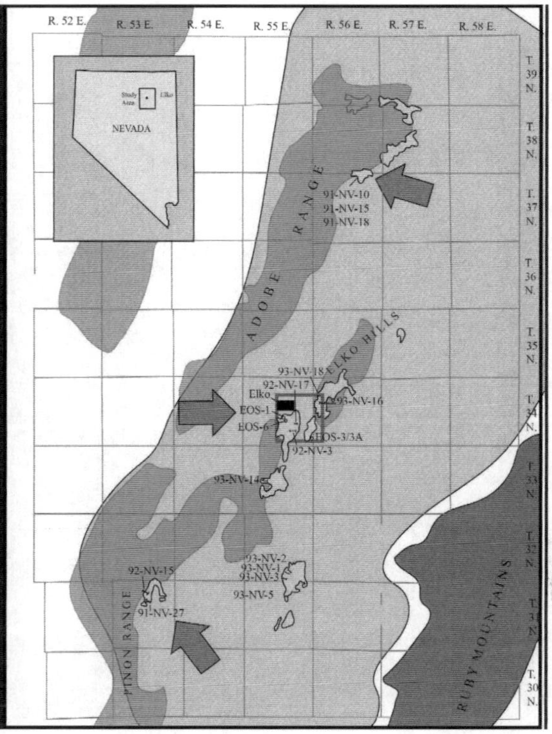

Detailed description of the EOS-3 core will be presented first followed by detailed descriptions of surface sections near Elko, in the Pinon Range, and in the Northern Adobe Range.

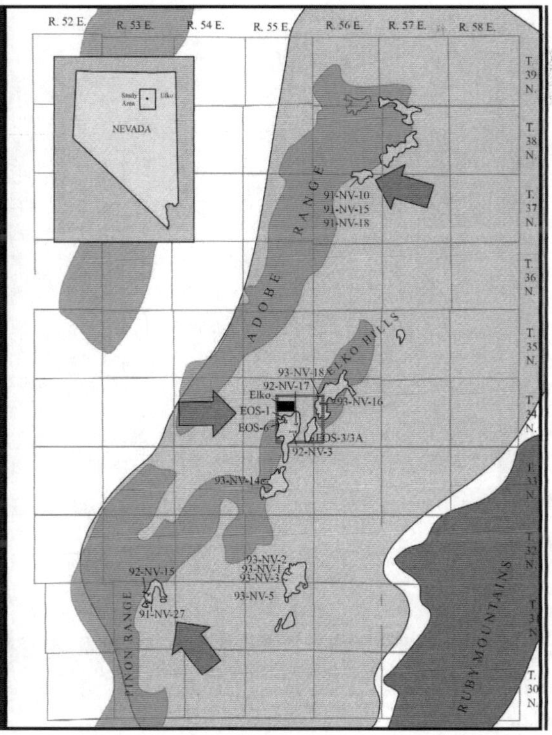

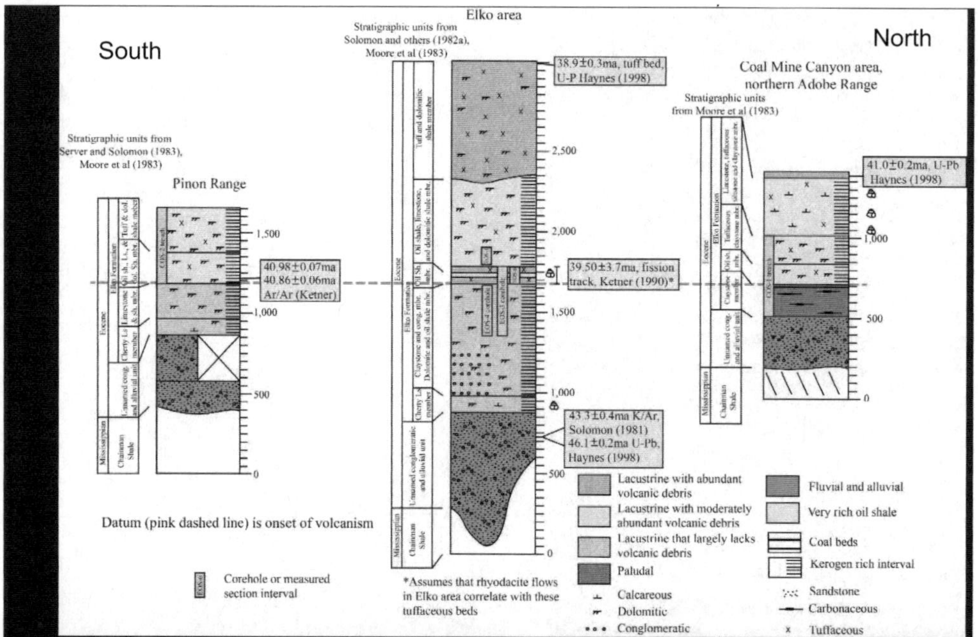

The onset of volcanism at all three locations is interpreted to have occurred at about 40-41 Ma. The lower part of the lacustrine Elko Formation has not been dated, but the lake probably began to form about 42 Ma

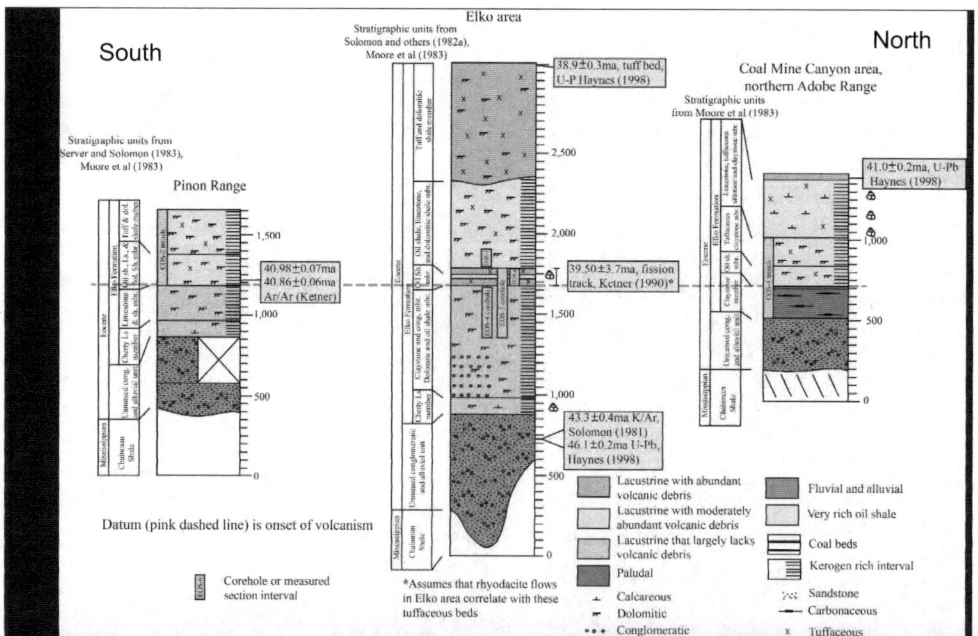

A few hundred feet of alluvial rocks underlies the Elko Formation at all three locations. A tuff in alluvial rocks beneath the Elko Formation has been dated at 46.1 Ma using U-Pb. This thin interval may represent a considerable period of time during which little sedimentation occurred.

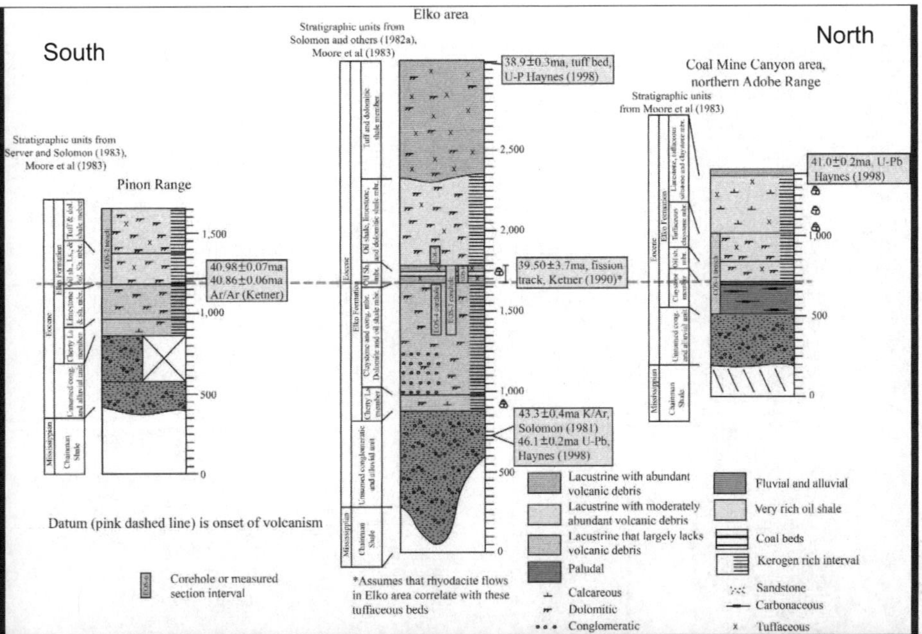

It is unlikely that the overlying 500-1000 ft of overlying lacustrine rock would represent 6 Ma of deposition. Thus, this thin alluvial interval may represent a considerable period of time.

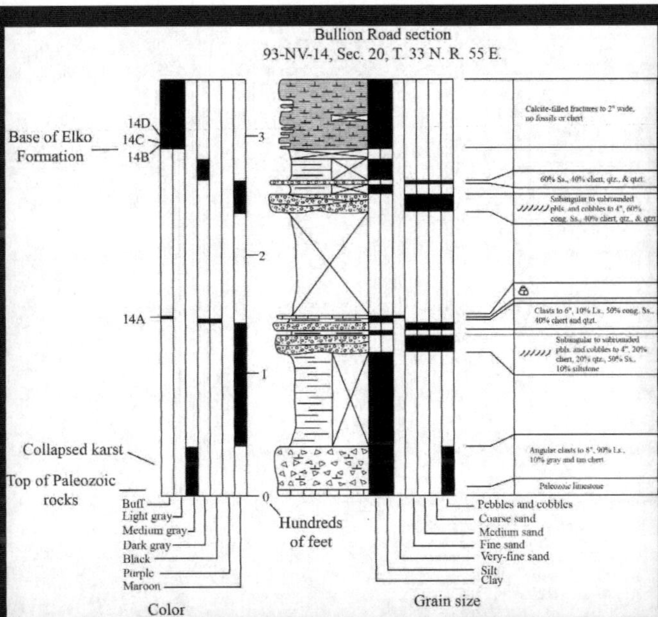

USGS

A collapsed karst is commonly developed above the underlying Paleozoic section

About 425 ft of the Elko Formation was cored in the EOS-3/3A core.

The section can generally be divided into a lower volcanic-free interval with low grade oil shale and an upper volcanic rich interval with very rich oil shale.

<u>The onset of volcanism corresponds to a shift from low-grade to high-grade oil shale</u>

This is the rich interval that was mined during the early part of the 20th century.

USGS

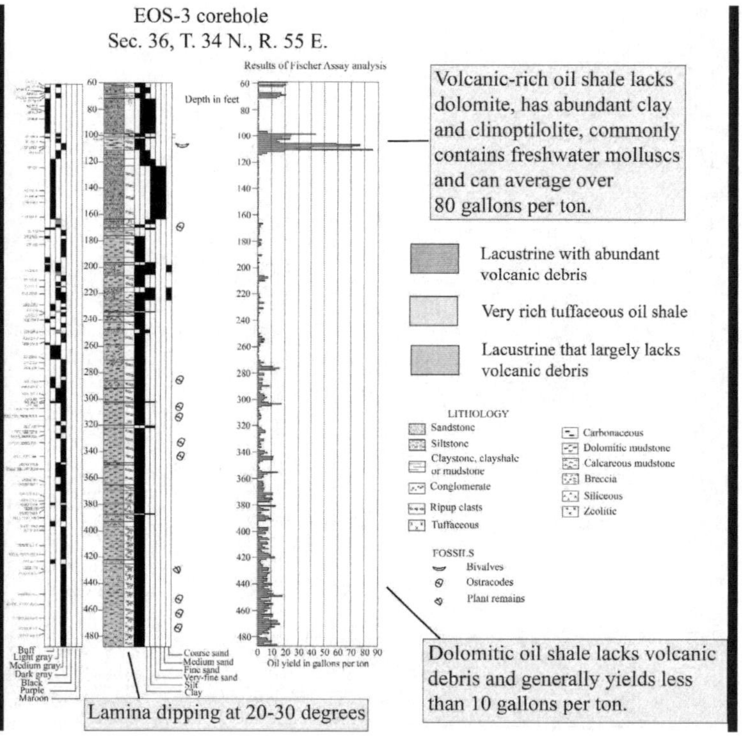

- Rhyolitic rocks are exposed in the Elko Hills near the EOS-3 core. Jager (1987) and Haynes (2003) interpreted the rhyolite as intruded into the Elko Formation after deposition doming up the formation in the Elko area.
- Ketner (1990) considered the rhyolite to be a series of volcanic flows and reworked volcanic rock contemporaneous with the Elko Formation

USGS

Rock-Eval and Fischer assay data indicate that the kerogen in the rich oil shale beds interbedded with these tuffaceous units is thermally immature thus eliminating any possibility that these units are rhyolitic sills injected into the oil shale interval.

In fact, the Elko Formation is thermally immature for oil generation everywhere in the Elko area.

USGS

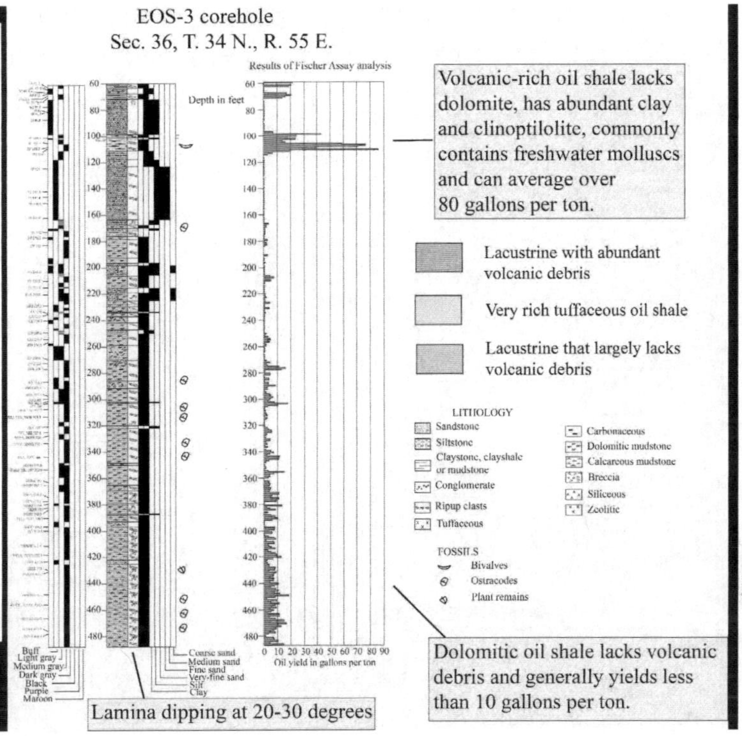

Detailed surface section of the Elko Fm. and underlying Eocene alluvial interval in the Elko Hills near the town of Elko.

The rich oil shale interval is not exposed here. It was exposed nearby and will be presented next.

USGS

Thickness of lacustrine rocks about 900 ft

This section was described about 2 miles west of the EOS-3 corehole. The entire dolomitic low grade oil shale section has graded into conglomerate and mudstone

About 350 ft of fluvial and alluvial rock was deposited on the Paleozoic section prior to the formation of a lake

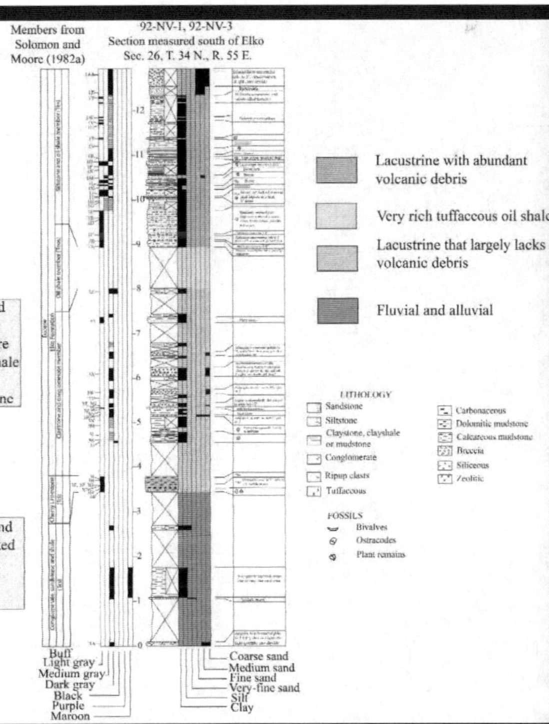

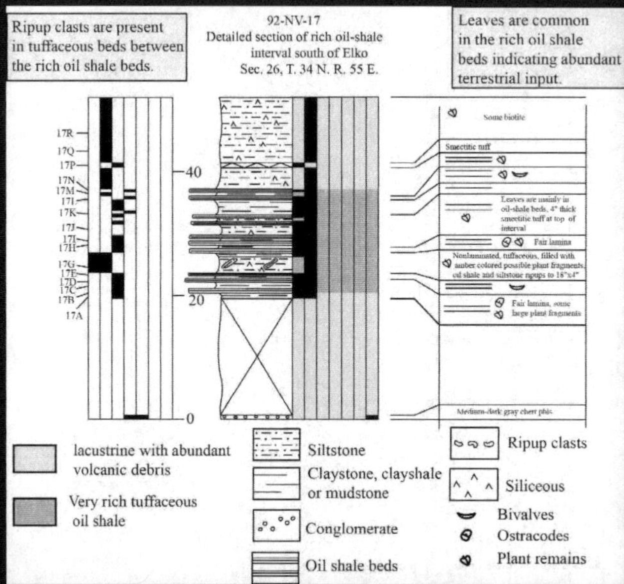

The rich oil shale was exposed nearby and described in detail. The rich oil shale beds are fairly well laminated, contain abundant plant debris and some clam shells. Tuffaceous beds contain large oil shale and siltstone ripups.

Thickness of lacustrine rocks about 900 ft

The Pinon Range section measured 520 ft of Eocene alluvial rock above the Mississippian Chainman Shale. Abundant plant remains are in the rich oil shale interval here as well.

USGS

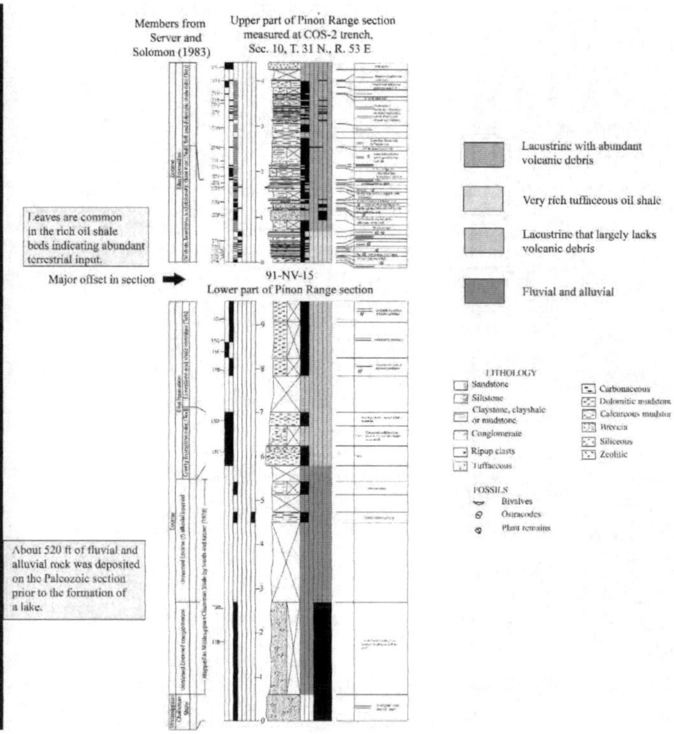

Rich oil shale from Pinon Range (Photo by R.C. Johnson, 1991)

USGS Some of the richest oil shale beds in the Pinon Range section have been affected by mass-movement processes.

Rich oil shale beds in the Green River Formation have commonly been subject to mass-movement processes.

In the Northern Adobe Range, a paludal interval has replaced the pre-volcanic lacustrine phase.

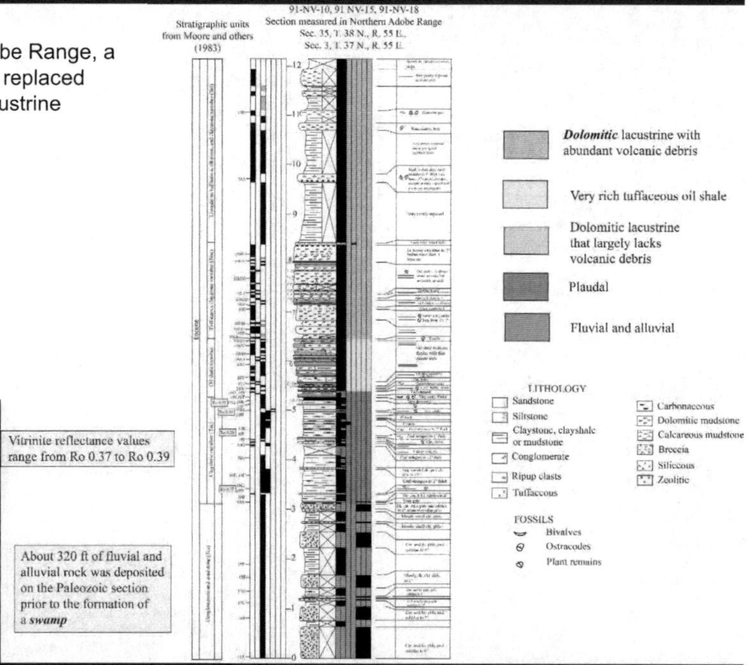

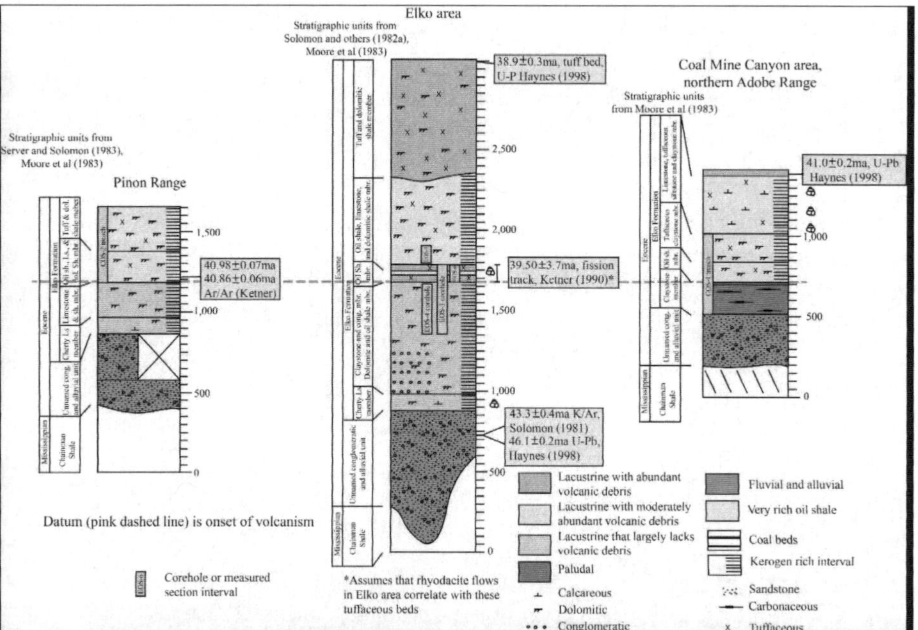

USGS
science for a changing world

History of Elko Oil Shale Development

U.S. Department of the Interior
U.S. Geological Survey

USGS
science for a changing world

Oil shale was first recognized in Nevada in the Elko area in 1875 and has since been recognized in other areas. Between 1917-1924, the Catlin Shale Products Company at Elko produced about 12,000 barrels shale oil, but the enterprise was a commercial failure. The products, gasoline and various lubricants, were of low quality.

(University of Nevada, Reno web site).

U.S. Department of the Interior
U.S. Geological Survey

- On the basis of Fischer-assay determinations of oil yield on continuous core samples from the Elko area, total in-place shale oil has been calculated as 600 million barrels. Of this total, 228 million barrels are from beds averaging at least 15 gal/ton over a 15-feet thickness (Moore et al, 1983).

USGS

- In 1924 five-thousand gallons of lubricating oil were shipped to New York. It was looking good for Catlin's oil company. Five tunnels, 600 feet down stretched four miles.
- Unfortunately, while there was talk of oil, the primary product was paraffin. Oil was secondary because of its high cost of refining. Labeled "Hi-Power Catlin Oil," it sold for $5 a gallon. The lubricant, because of its high paraffin base, worked well in the summer but come winter it congealed and gummed up engines. Durant Motor Company tested the oil in the 1924 Indianapolis 500. E.E. "Monte" Mouton used the lubricant on his Elko to Reno air mail route and claimed his engine ran more than fifty r.p.m. compared to standard oil. (Howard Hickson, January 6, 2003, Great Basin College).

USGS

Photo is of Catlin Oil Shale Plant 1923 (courtesy of Northeastern Nevada Museum document # 46-77)

USGS

Catlin oil shale facility, 2001 (from Elko County Rose Garden)

USGS

Spent shale near Catlin facility , 2001 (from Elko County Rose Garden).

USGS

USGS
science for a changing world

Sign near Catlin oil shale facility:

Remains of a short-lived oil extraction plant, which reached the peak of its productive capacity in the early twenties. Driving of the main shaft began in 1916. Of several tries at extracting oil from shale, this was the only successful operation in Nevada.

Robert M. Catlin, Sr., spent many years experimenting on the extraction of crude oil from these beds before beginning the commercial production of oil. After a production period of less than two years, the plant was closed in the fall of 1924. Hi-power Catlin oil was too expensive to compete with the fossil oils of that day.

Easily 50 years ahead of his time, Catlin did, for a few years, give Elkoans and Nevadans, a dream and the community an oil boom in the "Roaring Twenties."

U.S. Department of the Interior
U.S. Geological Survey

USGS

science for a changing world

Organic geochemistry of Elko oil shale

U.S. Department of the Interior
U.S. Geological Survey

Palmer (1984) performed a study of the hydrocarbon potential and thermal maturity of Elko Formation oil shale outcropping in the Adobe Range. Two organic-rich facies, immature oil shale and siltstone, were identified.

Analyses included XRD, organic petrography, kerogen isolation and elemental analysis, Rock-Eval, TOC, and biomarker analysis of extractable organic matter.

Hydrous pyrolysis experiments were also conducted to assess potential crude oil quality in the Elko.

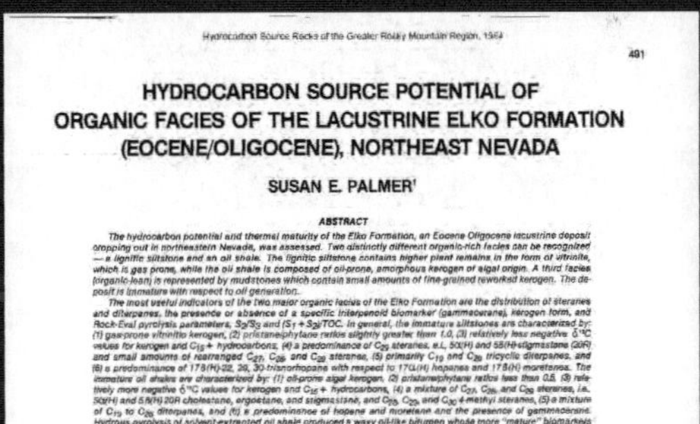
Hydrocarbon Source Rocks of the Greater Rocky Mountain Region, 1984

491

HYDROCARBON SOURCE POTENTIAL OF ORGANIC FACIES OF THE LACUSTRINE ELKO FORMATION (EOCENE/OLIGOCENE), NORTHEAST NEVADA

SUSAN E. PALMER[1]

ABSTRACT

The hydrocarbon potential and thermal maturity of the Elko Formation, an Eocene Oligocene lacustrine deposit cropping out in northeastern Nevada, was assessed. Two distinctly different organic-rich facies can be recognized — a lignitic siltstone and an oil shale. The lignitic siltstone contains higher plant remains in the form of vitrinite, which is gas prone, while the oil shale is composed of oil-prone, amorphous kerogen of algal origin. A third facies (organic-lean) is represented by mudstones which contain small amounts of fine-grained reworked kerogen. The deposit is immature with respect to oil generation.

INTRODUCTION

ACKNOWLEDGMENTS

SAMPLE COLLECTION AND EXPERIMENTAL METHODS

© 2009 Rocky Mountain Association of Geologists

USGS

The USGS examined a set of Elko samples (58) from four cores by Rock-Eval/TOC as part of an examination of the Elko oil shale resource. This is supplemented by a larger number of Fischer assay analyses (640) generated by the U.S. Dept. of Energy.

The Rock-Eval and TOC results showed a wide range of organic carbon and hydrogen contents, similar to what Palmer noted for samples from the Adobe Range.

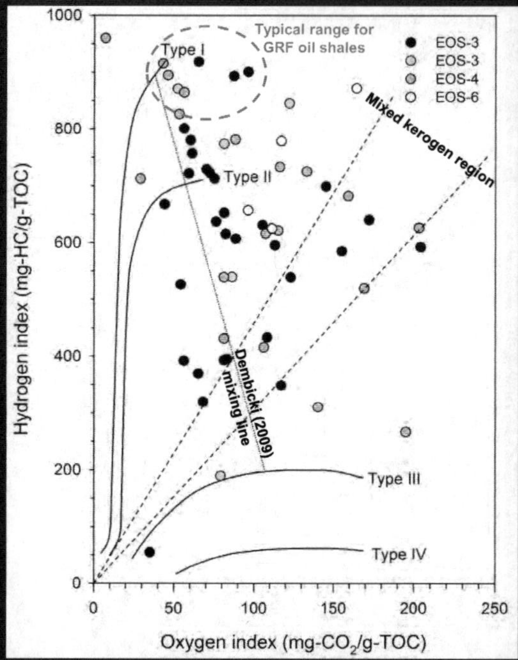

Espitalié et al., 1977

USGS

Analysis of the Rock-Eval results show that most Elko samples examined contain primarily type I kerogen. However, samples containing type III kerogen or mixtures of types I and III were also common. This is consistent with Palmer's geochemical and petrographic results.

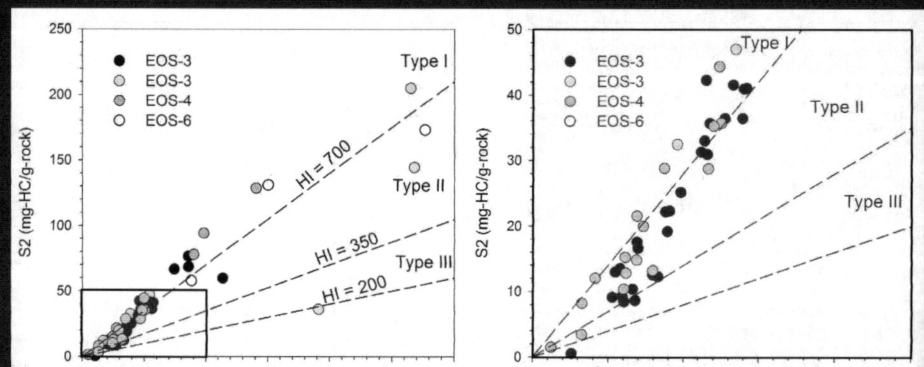

Langford and Blanc-Valleron, 1990

USGS **The organic matter in the Elko samples range in composition with what appears to be varying type I and type III kerogen content.**

The Fischer assay results show that most of the EOS cores samples yielded less than 10 wt. % oil (~27 GPT) and ~2 wt. % gas

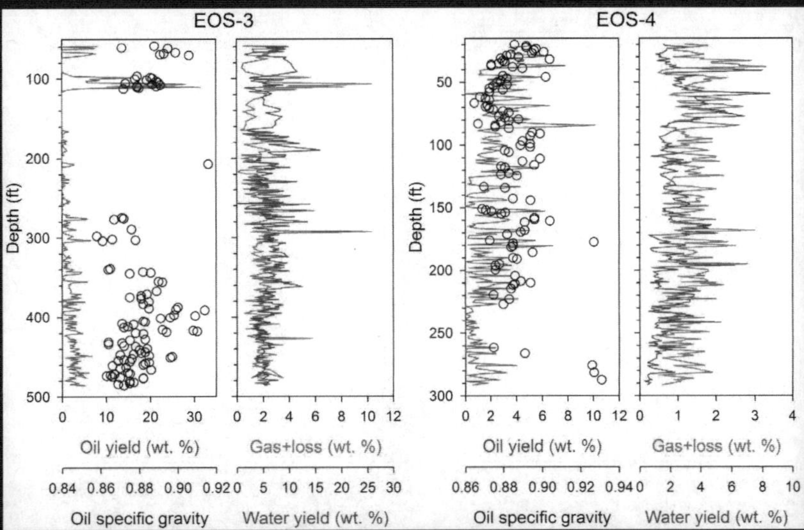

USGS

A few additional oil shales with high yields (10 – 20 wt. %) were present in EOS-6 (not shown) at depths of 120 to 180 ft. EOS-4 does not contain the high yield oil shale interval near the top of EOS-3.

Relative abundance mineralogy data in the EOS-3 (below) shows the Elko contains variably carbonate (calcite and dolomite) and quartz-rich intervals. The EOS-4 core is similar but contains more dolomite and EOS-1 is richer in quartz and clays.

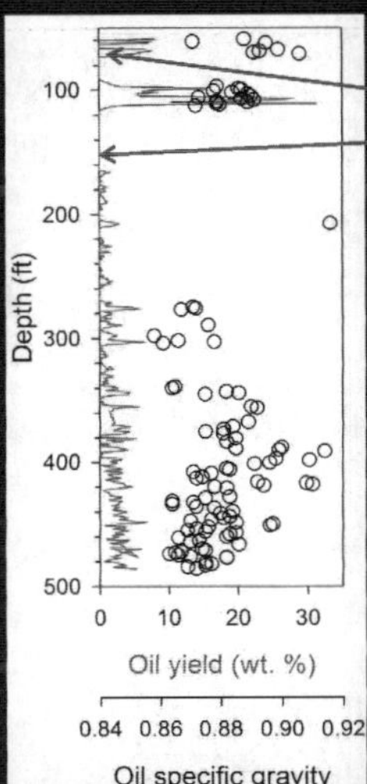

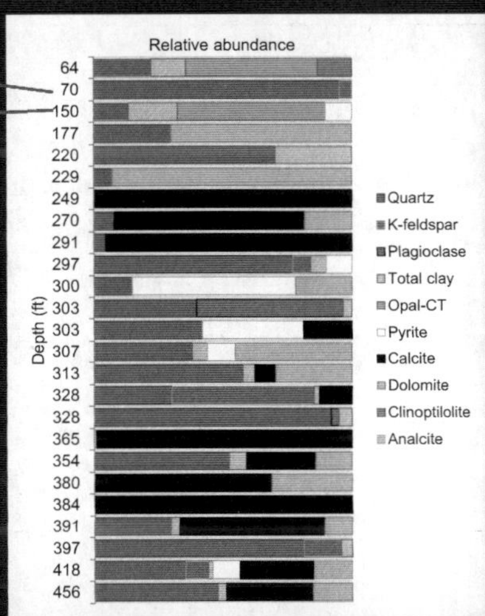

USGS

In contrast, the upper section, where the richest oil shale is found (depth ~100 ft), is bracketed by rocks high in K-spar, clay, and opal.

Elko oil shale **Fischer assay** product distribution (GOR) and oil-generating potential (normalized to TOC) compared to average values for Green River Mahogany zone oil shale (Piceance Basin)

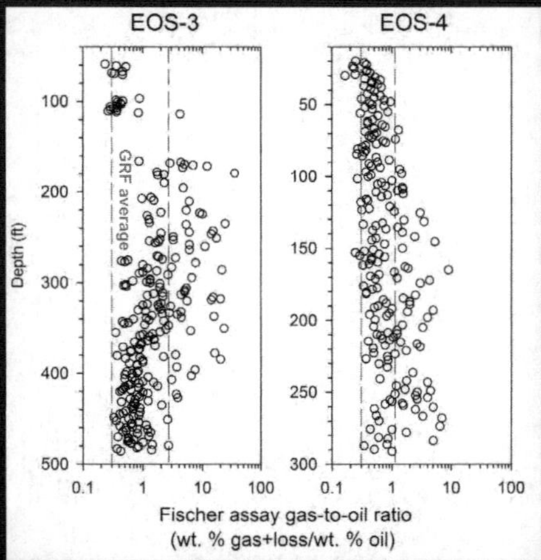

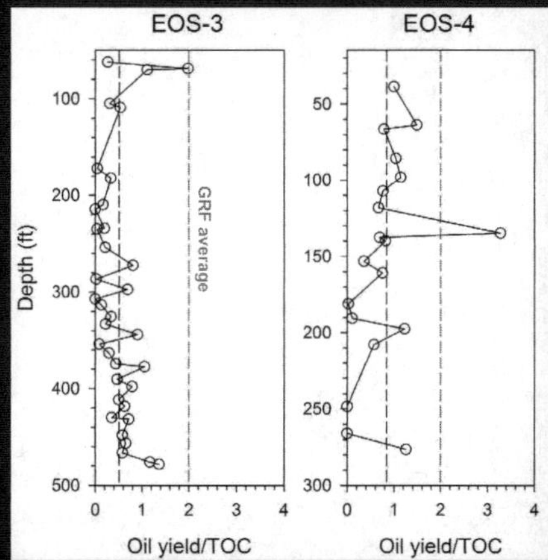

USGS

Although both are Eocene lacustrine oil shales with (mostly) hydrogen-rich kerogen, Elko oil shales generate significantly more gas and less oil than those from the Green River Formation.

Recently, Noble Energy drilled test wells in the Elko to assess its potential as a tight oil play. Target depths for the samples shown here were around 10,000 feet.

Pristane/phytane ratios are generally consistent with mildly reducing and disoxic conditions. The low Pr/C17 and Ph/C18 ratio values and CPI's near 1 indicate that the oils are thermally mature.

Beta carotane was present but not abundant and could indicate that Lake Elko was a brackish to saline lake.

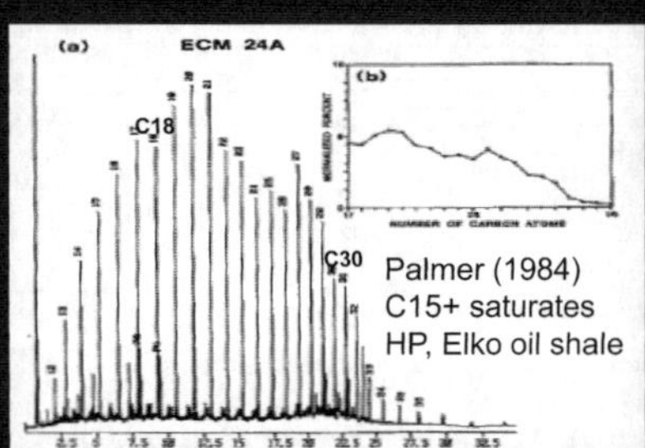

Palmer (1984)
C15+ saturates
HP, Elko oil shale

GC-FID chromatograms for saturate fraction of Elko crude oils

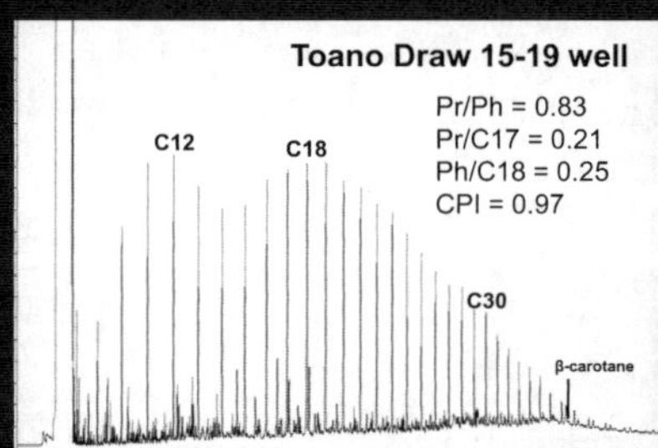

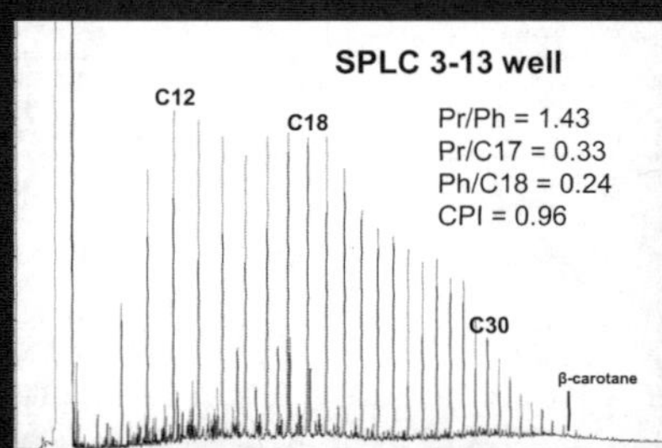

Oil samples provided by Jerry Hansen, Great Basins Exploration

Conclusions

- The Eocene Elko Formation appears to be entirely younger than the Eocene Green River Formation and it is very likely that the Elko Lake(s) did not form until Lake Uinta was completely filled in.
- The Elko Lake probably formed during a period of regional extension, whereas the Green River lakes formed during an earlier period of regional compression (Laramide orogeny).
- The rich oil shale interval in the Elko was deposited immediately after the onset of volcanism in the area, suggesting that the input of volcanic material to the lake may have initiated a period of high organic productivity.
- The organic matter in the Elko is distinctive from that of the Green River oil shale in that it (**1**) is less dominated by type I kerogen in much of the oil shale and therefore (**2**) produces more gas and less oil per unit organic carbon.

USGS

USGS
science for a changing world

Questions?

U.S. Department of the Interior
U.S. Geological Survey

References Cited

- Haynes, S.R., 2003, Development of the Eocene Elko Basin, northeastern Nevada: Implications for paleogeography and regional tectonism: The University of British Columbia MS Thesis, 156p.
- Ketner, K.B., 1990, Geologic map of the Elko Hills, Elko County, Nevada: U.S. Geological Survey Miscellaneous Investigations Map I-2082.
- Ketner, K.B., and Alpha, A.G., 1992, Mesozoic and Tertiary rocks near Elko, Nevada-Evidence for Jurassic to Eocene folding and low-angle faulting: U.S. Geological Survey Bulletin 1988-C, p. C-1 to C-13
- Moore, S.W., Madrid, H.B., and Server, G.T., Jr., 1983, Results of oil shale investigations in northeast Nevada: U.S. Geological Survey Open-File Report 83-586, 56 p.

USGS

References Cited (cont.)

- Palmer, S.E., 1984, Hydrocarbon source potential of organic facies of the lacustrine Elko Formation (Eocene/Oligocene), Northeast Nevada: *in* Woodward, J, Meissner, F.F., and Clayton, J.L., eds., Hydrocarbon Source Rocks of the Greater Rocky Mountain Region; Rocky Mountain Association of Geologists Guidebook, p. 491-511.
- Langford, F.F., and Blanc-Valleron, M.M., 1990, Interpreting Rock-Eval pyrolysis data using graphs of pyrolyzable hydrocarbons vs. total organic carbon, AAPG Bulletin 74, 799-804.
- Satarugsa, Peangta, and Johnson, R.A., 2000, Cenozoic tectonic evolution of the Ruby Mountains metamorphic core complex and adjacent valleys, northeastern Nevada, Rocky Mountain Geology, v. 35, no. 2, p. 205-230.
- Smith, M.E., Carroll, A.R., and Singer, B.S., 2008, Synoptic reconstruction of a major ancient lake system: Eocene Green River Formation, western United States: Geological Society of America Bulletin vol. 120, p. 54-84.

USGS

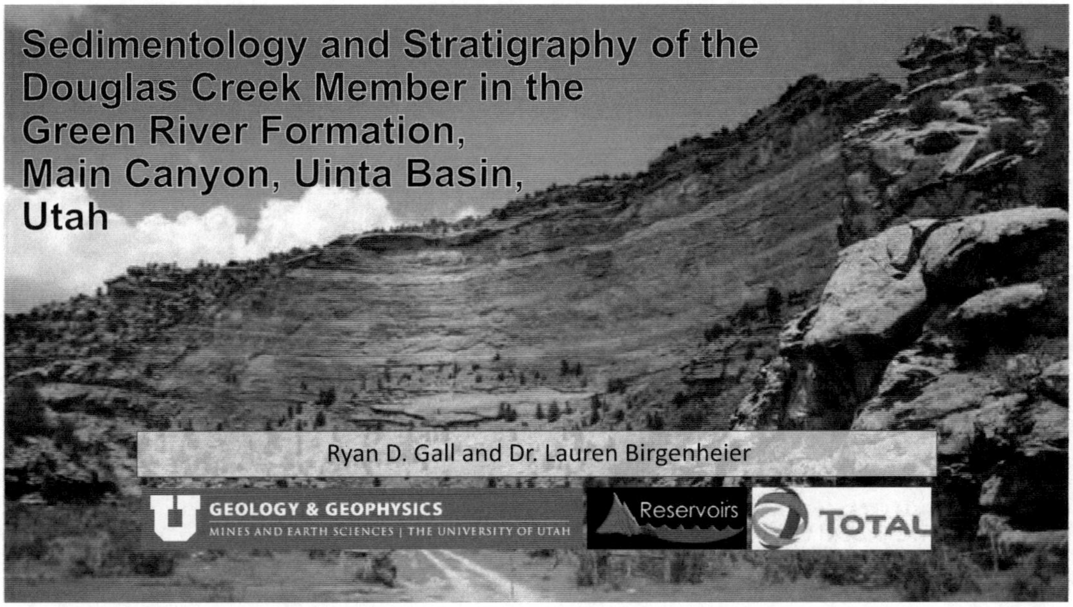
Sedimentology and Stratigraphy of the Douglas Creek Member in the Green River Formation, Main Canyon, Uinta Basin, Utah
Ryan D. Gall and Dr. Lauren Birgenheier
GEOLOGY & GEOPHYSICS
MINES AND EARTH SCIENCES | THE UNIVERSITY OF UTAH
Reservoirs
TOTAL

AGENDA
1. Project Overview
2. Introduction to the Green River Formation
3. Uinta Basin and the Douglas Creek Member
4. Preliminary Facies Interpretations
5. Large-scale Outcrop Interpretations
6. Conclusions
7. Next Steps
8. Acknowledgements

PROJECT OVERVIEW

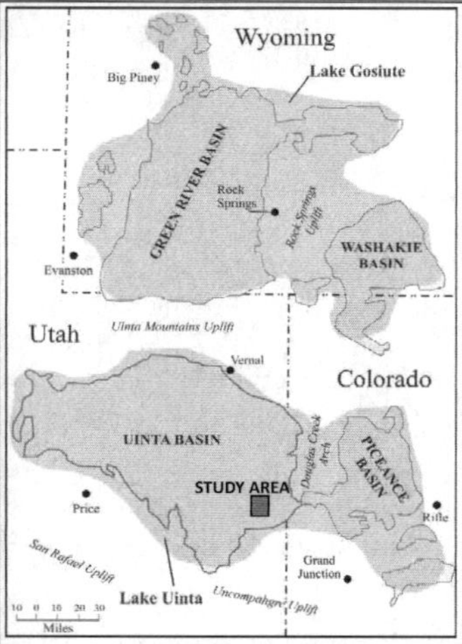

Modified from Birgenheier & Vandenberg (2011)

Location:

- Main Canyon, Uinta Basin, UT
- 20 km transect

Objectives:

1. Assess vertical and lateral variation of Douglas Creek Member in the Green River Formation
2. Determine degree of climate control on deposition

Methods:

1. Vertical and Lateral Variation
   - Measure 6 sections across transect
2. Climate Study
   - Carbon Isotope Analysis

GEOLOGIC BACKGROUND: THE GREEN RIVER FORMATION

- Continental basins formed from Laramide block uplifts
  - Uinta, Piceance, Green River, Washakie basins
- Eocene lacustrine system
  - Lake Gosiute - WY
  - Lake Uinta – UT, CO
  - Lakes isolated or connected as function of accommodation and hydrology

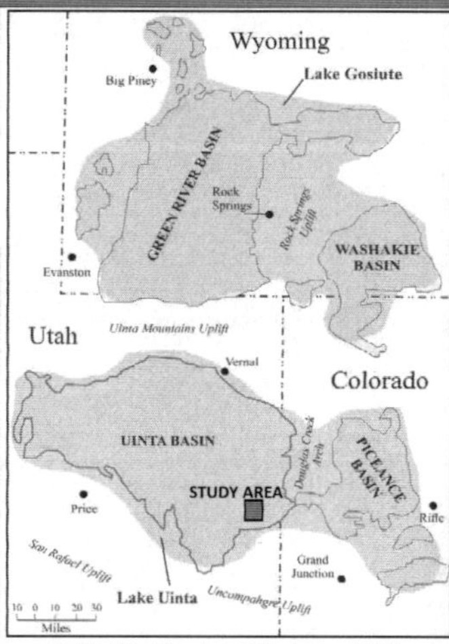

Modified from Birgenheier & Vandenberg (2011)

GEOLOGIC BACKGROUND: THE UINTA BASIN

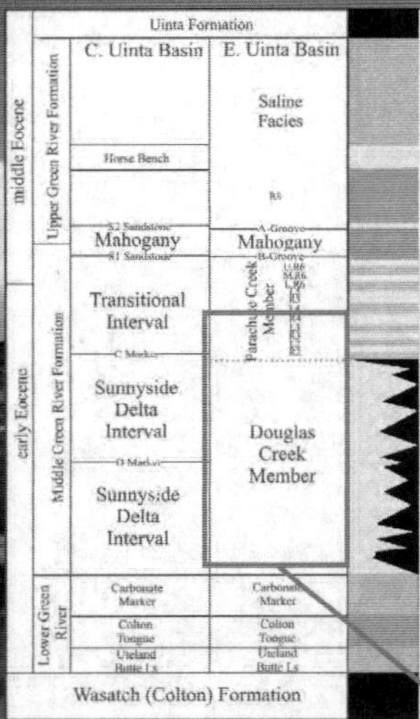

- Lake Uinta Evolution
  - Fluvial-deltaic, to organic-rich and organic-lean profundal mudstone, to saline facies
  - Mixed carbonate and siliciclastic system
- Oil Shale
  - Green River Formation contains largest known oil shale resource (Bartis et al., 2005)
  - Total resource in Uinta Basin: 1.32 trillion barrels in place (Johnson et al., 2010)
- This study focuses on the understudied fluvial-deltaic Douglas Creek Member

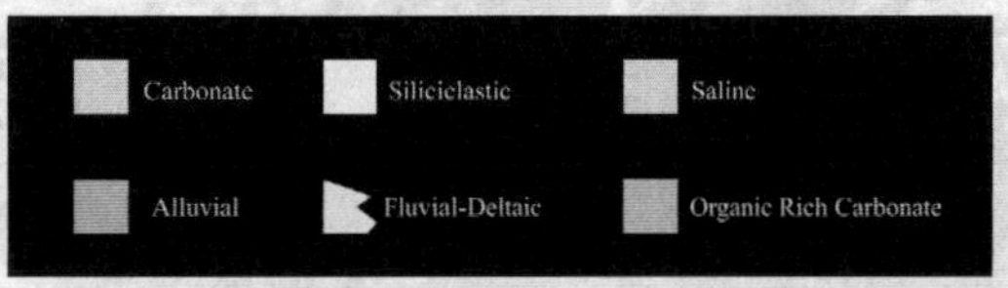

Study Interval

GEOLOGIC BACKGROUND: THE UINTA BASIN

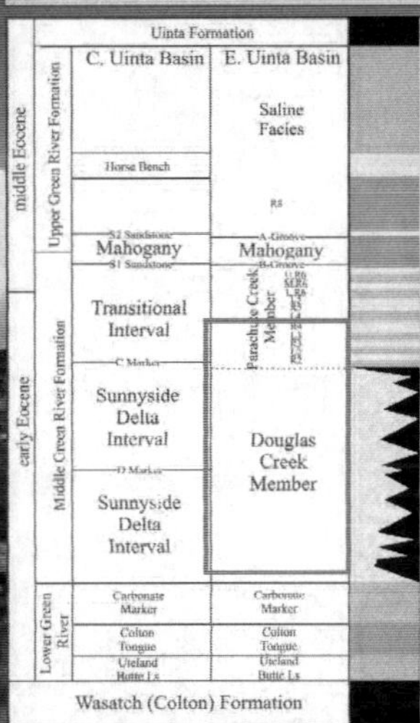

- GRF deposited during major Eocene climatic events
- Are these observable in the lacustrine carbon isotope record?

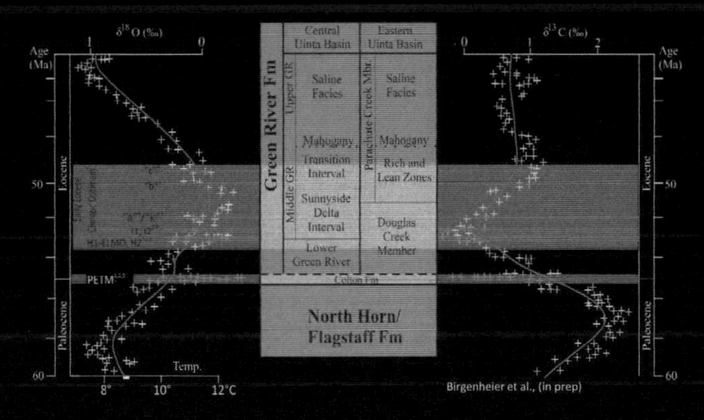

MAIN CANYON STUDY AREA

- ~20 km NW-SE transect
- 6 stratigraphic sections measured to date
  - 1 sampled for carbon isotope analysis
- 1 additional proposed section
- Presenting preliminary results

Main Canyon Fieldwork

- Measured Section
- Sampled Measured Section
- Proposed Section
- "Short" Measured Section

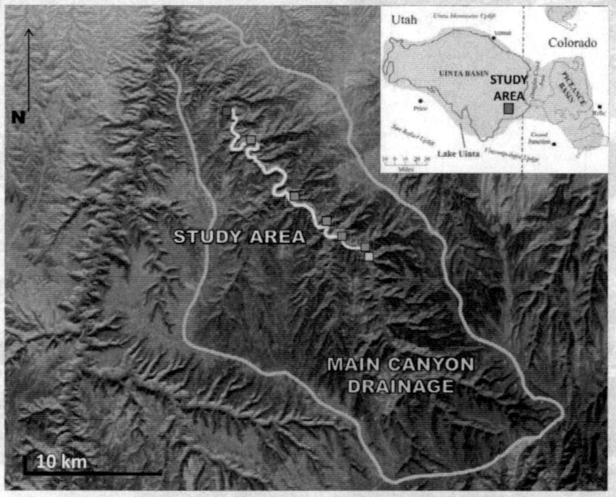

FACIES BOX MODEL

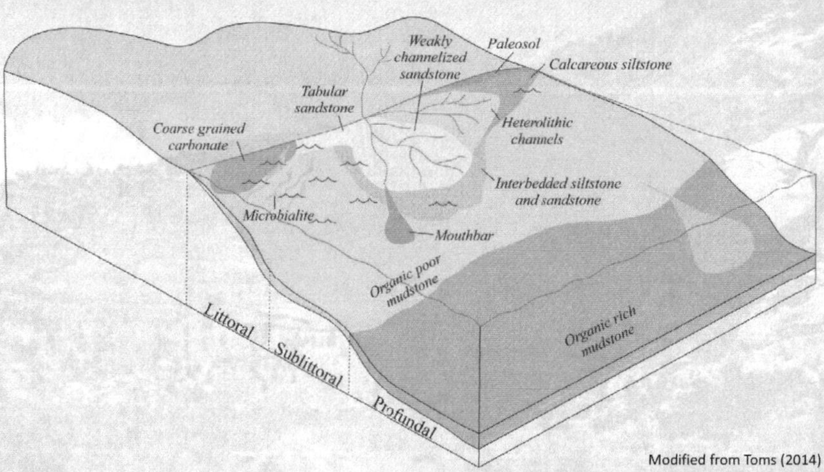

Modified from Toms (2014)

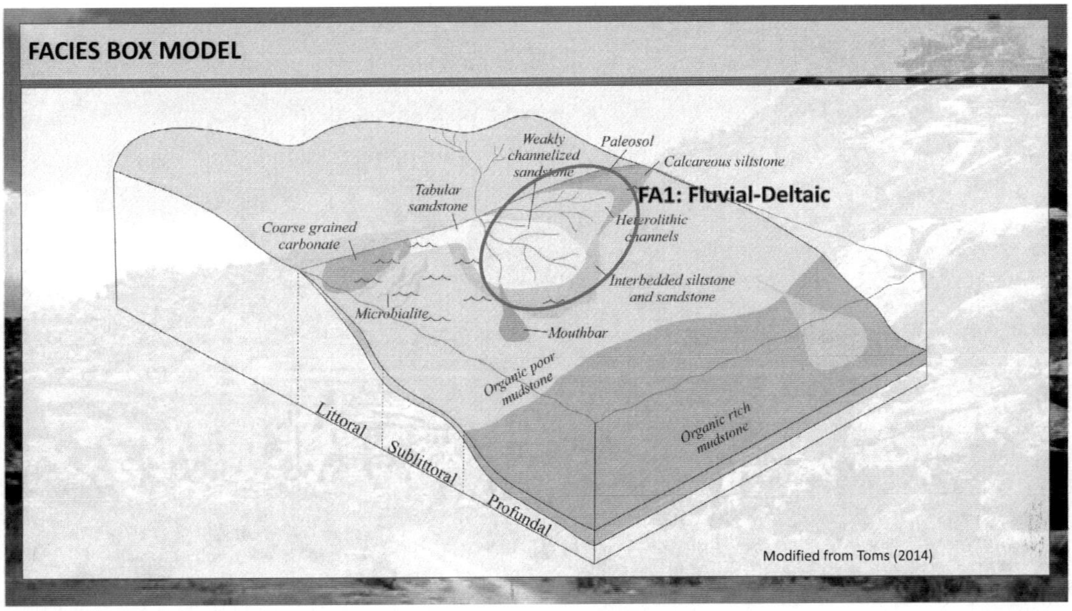
FACIES BOX MODEL
Weakly channelized sandstone
Paleosol
Calcareous siltstone
FA1: Fluvial-Deltaic
Tabular sandstone
Coarse grained carbonate
Heterolithic channels
Interbedded siltstone and sandstone
Microbialite
Mouthbar
Organic poor mudstone
Organic rich mudstone
Littoral
Sublittoral
Profundal
Modified from Toms (2014)

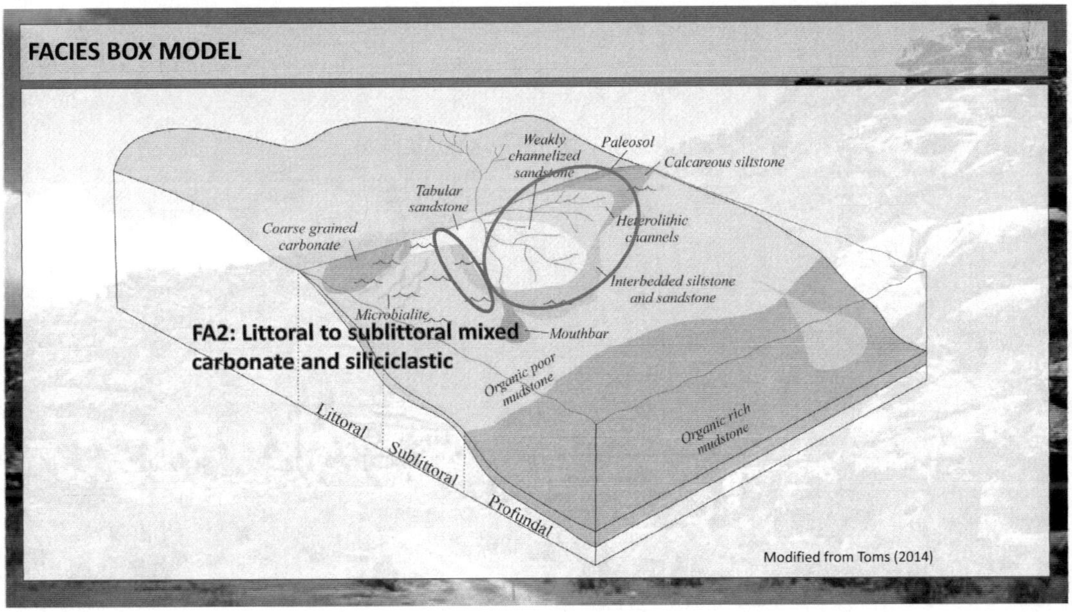
FACIES BOX MODEL
Weakly channelized sandstone
Paleosol
Calcareous siltstone
Tabular sandstone
Coarse grained carbonate
Heterolithic channels
Interbedded siltstone and sandstone
Microbialite
FA2: Littoral to sublittoral mixed carbonate and siliciclastic
Mouthbar
Organic poor mudstone
Organic rich mudstone
Littoral
Sublittoral
Profundal
Modified from Toms (2014)

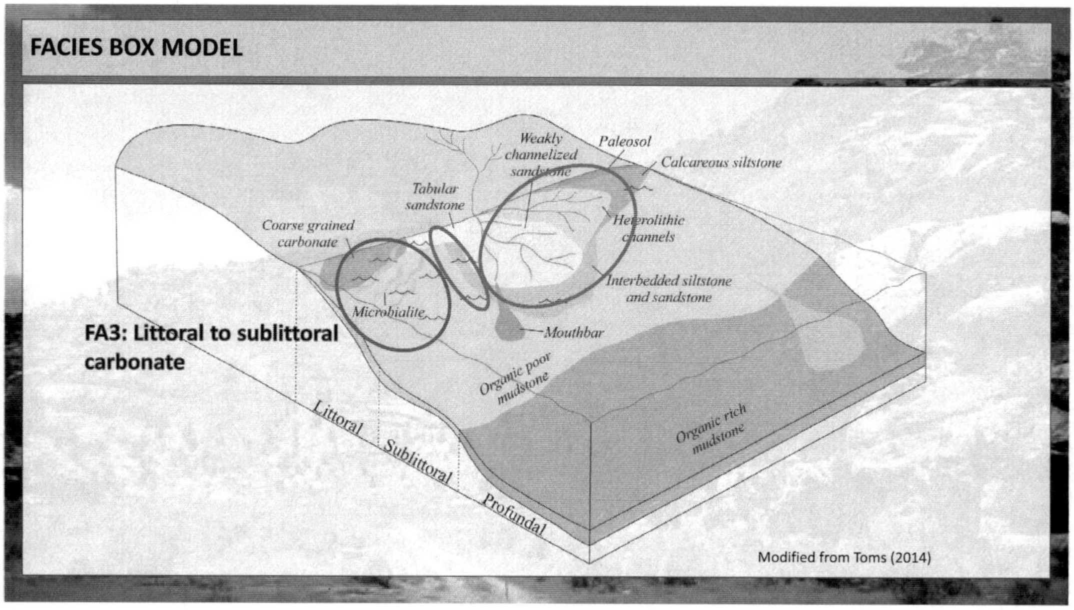
FACIES BOX MODEL
Weakly channelized sandstone
Paleosol
Calcareous siltstone
Tabular sandstone
Coarse grained carbonate
Heterolithic channels
Interbedded siltstone and sandstone
Microbialite
Mouthbar
FA3: Littoral to sublittoral carbonate
Organic poor mudstone
Organic rich mudstone
Littoral
Sublittoral
Profundal
Modified from Toms (2014)

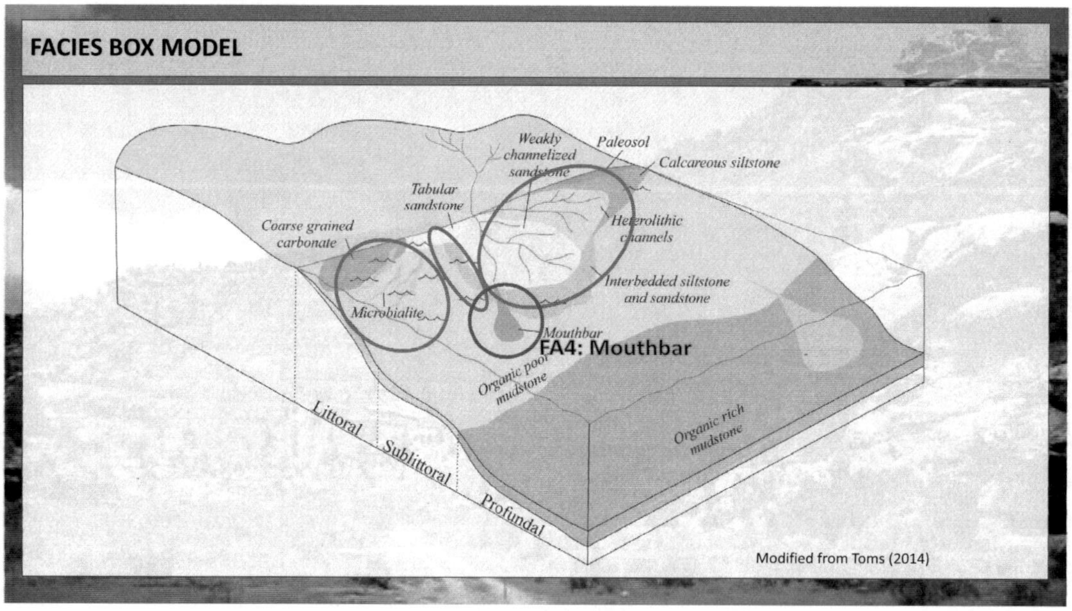
FACIES BOX MODEL
Weakly channelized sandstone
Paleosol
Calcareous siltstone
Tabular sandstone
Coarse grained carbonate
Heterolithic channels
Interbedded siltstone and sandstone
Microbialite
Mouthbar
FA4: Mouthbar
Organic poor mudstone
Organic rich mudstone
Littoral
Sublittoral
Profundal
Modified from Toms (2014)

FACIES BOX MODEL

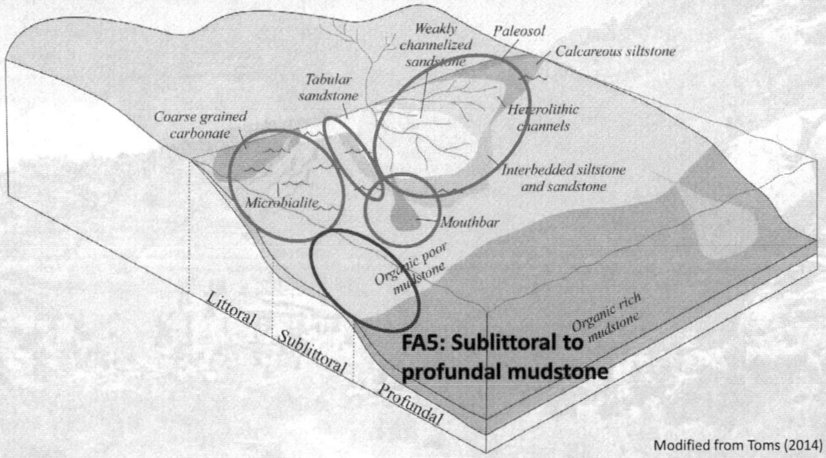

Modified from Toms (2014)

FACIES ASSOCIATION 1: Fluvial-Deltaic Deposits

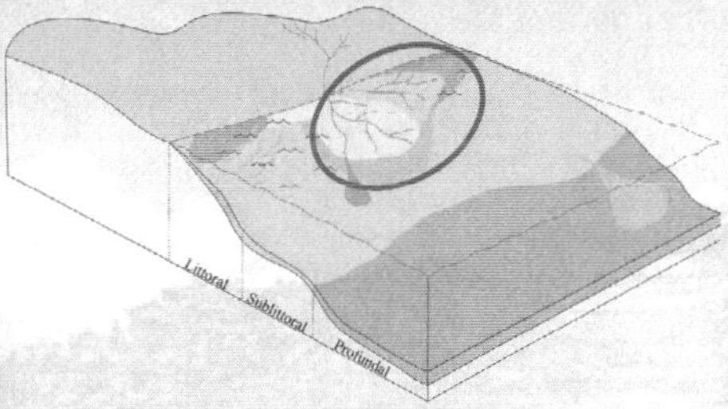

- Weakly channelized sandstone
- Heterolithic siltstone-sandstone channels with significant downcutting

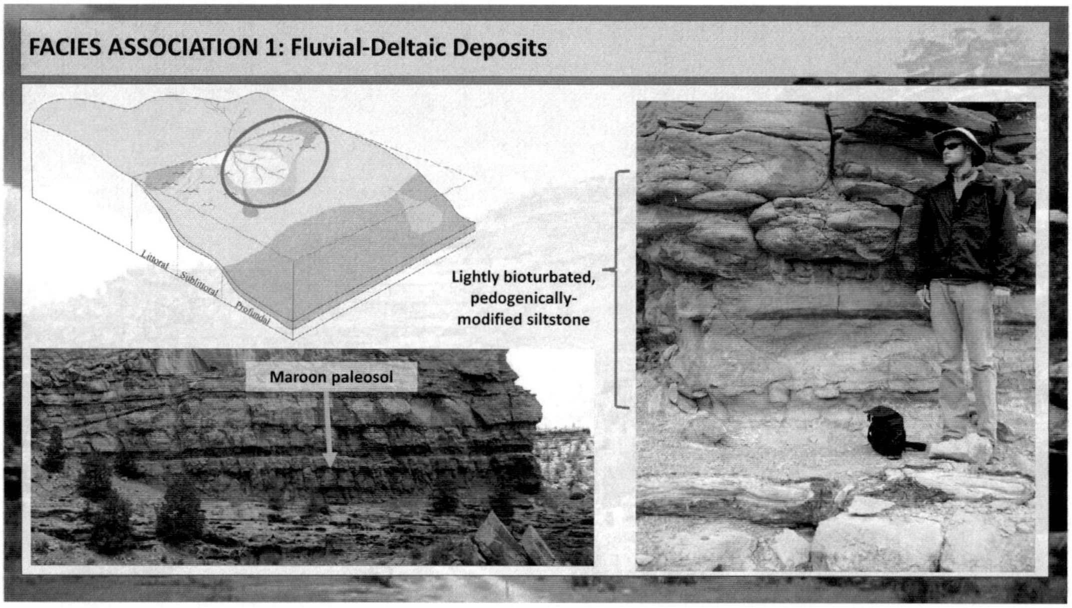
FACIES ASSOCIATION 1: Fluvial-Deltaic Deposits
Littoral
Sublittoral
Profundal
Lightly bioturbated, pedogenically-modified siltstone
Maroon paleosol

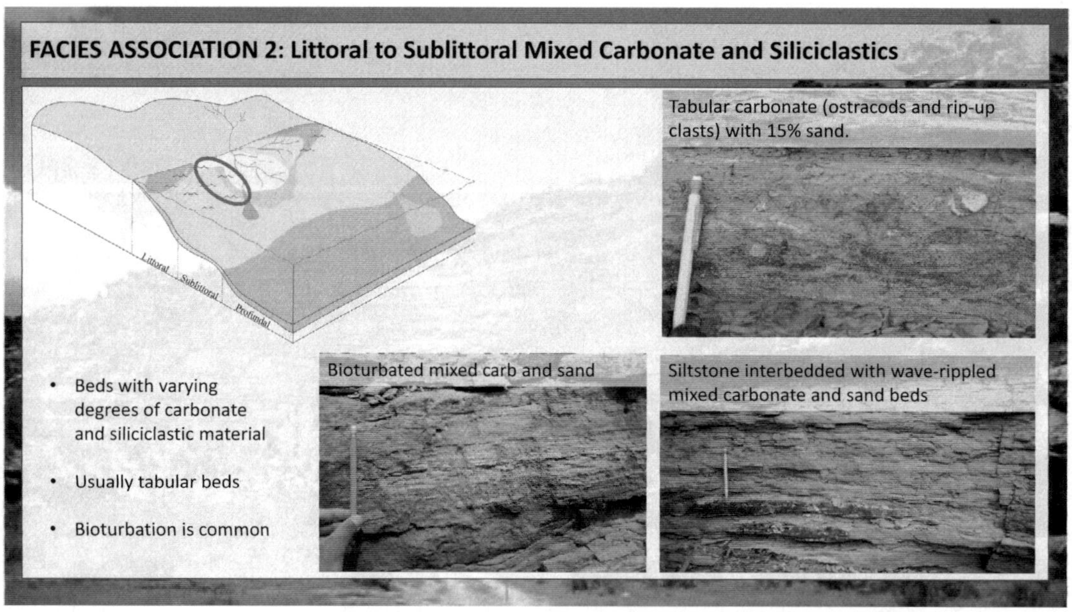
FACIES ASSOCIATION 2: Littoral to Sublittoral Mixed Carbonate and Siliciclastics
Littoral
Sublittoral
Profundal
Tabular carbonate (ostracods and rip-up clasts) with 15% sand.
Bioturbated mixed carb and sand
Siltstone interbedded with wave-rippled mixed carbonate and sand beds
• Beds with varying degrees of carbonate and siliciclastic material
• Usually tabular beds
• Bioturbation is common

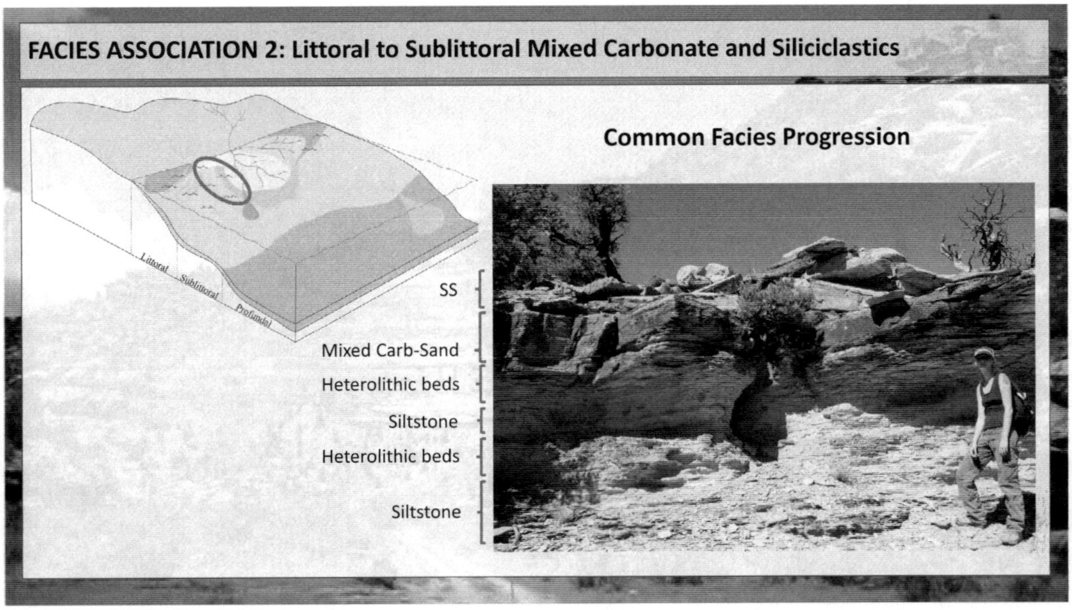
FACIES ASSOCIATION 2: Littoral to Sublittoral Mixed Carbonate and Siliciclastics
Common Facies Progression
Littoral
Sublittoral
Profundal
SS
Mixed Carb-Sand
Heterolithic beds
Siltstone
Heterolithic beds
Siltstone

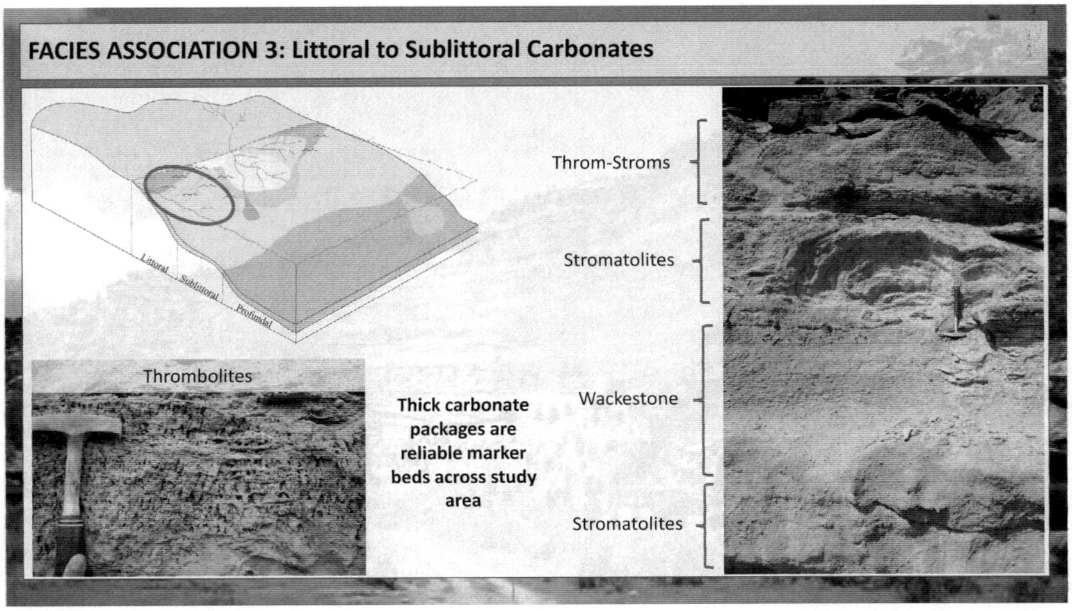
FACIES ASSOCIATION 3: Littoral to Sublittoral Carbonates
Littoral
Sublittoral
Profundal
Thrombolites
Thick carbonate packages are reliable marker beds across study area
Throm-Stroms
Stromatolites
Wackestone
Stromatolites

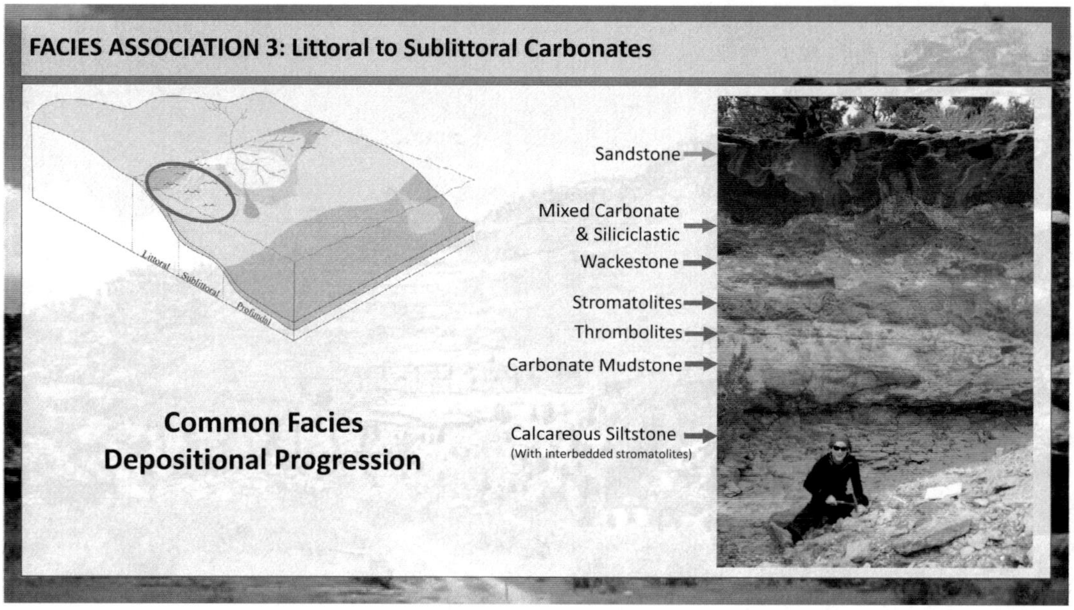
FACIES ASSOCIATION 3: Littoral to Sublittoral Carbonates
Littoral
Sublittoral
Profundal
Sandstone
Mixed Carbonate
& Siliciclastic
Wackestone
Stromatolites
Thrombolites
Carbonate Mudstone
Common Facies
Depositional Progression
Calcareous Siltstone
(With interbedded stromatolites)

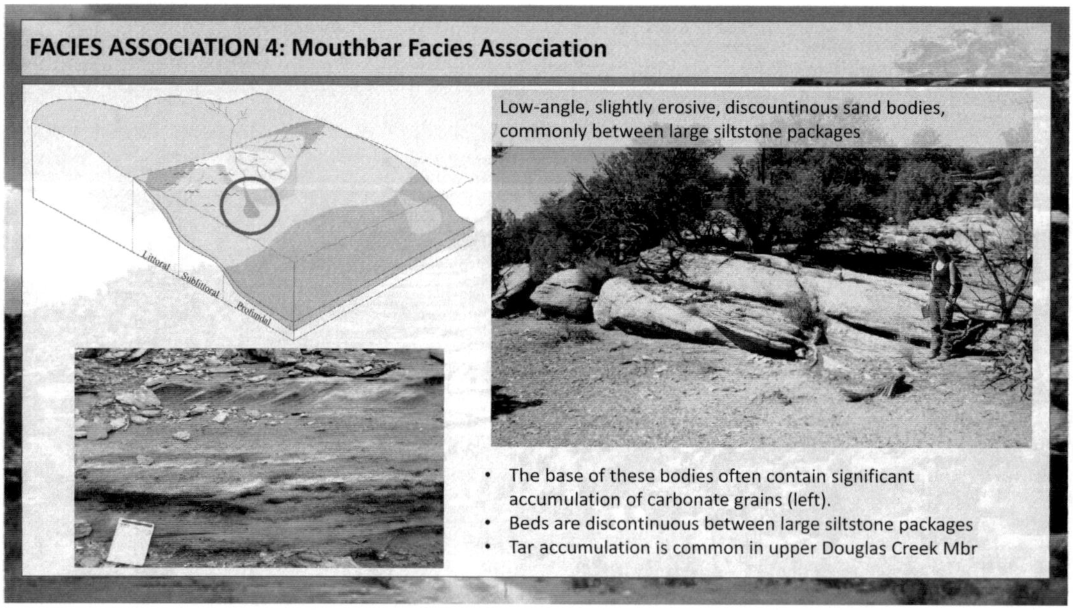
FACIES ASSOCIATION 4: Mouthbar Facies Association
Littoral
Sublittoral
Profundal
Low-angle, slightly erosive, discountinous sand bodies, commonly between large siltstone packages
• The base of these bodies often contain significant accumulation of carbonate grains (left).
• Beds are discontinuous between large siltstone packages
• Tar accumulation is common in upper Douglas Creek Mbr

FACIES ASSOCIATION 5: Sublittoral to Profundal Mudstone

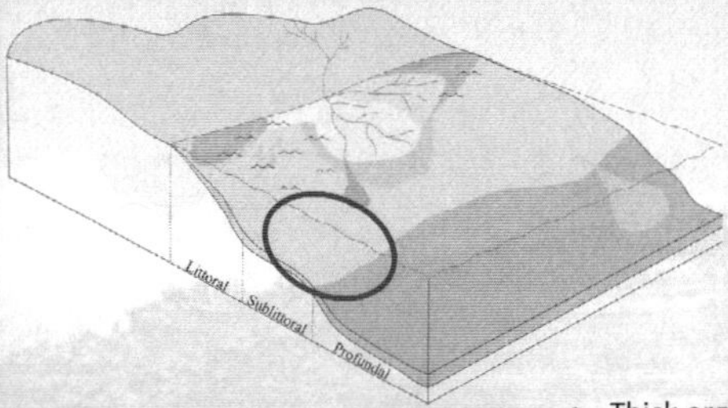

Interbedded siltstone and carbonate mudstone

- Thick organic-lean siltstone packages
- Organic-rich mudstone is rare
- Generally poor exposure, representing Parachute Creek Member

OUTCROP INTERPRETATIONS: VERTICAL VARIATION

FA1: Fluvial-Deltaic Facies

OUTCROP INTERPRETATIONS: VERTICAL VARIATION

Alternating FA2 & FA3:
Littoral – Sublittoral
Mixed Carbonate and Siliciclastic FA
& Pure Carbonate FA

FA1: Fluvial-Deltaic Facies Association

OUTCROP INTERPRETATIONS: VERTICAL VARIATION

FA4: Mouthbar Facies Association

Alternating FA2 & FA3:
Littoral – Sublittoral
Mixed Carbonate and Siliciclastic FA
& Pure Carbonate FA

FA1: Fluvial-Deltaic Facies Association

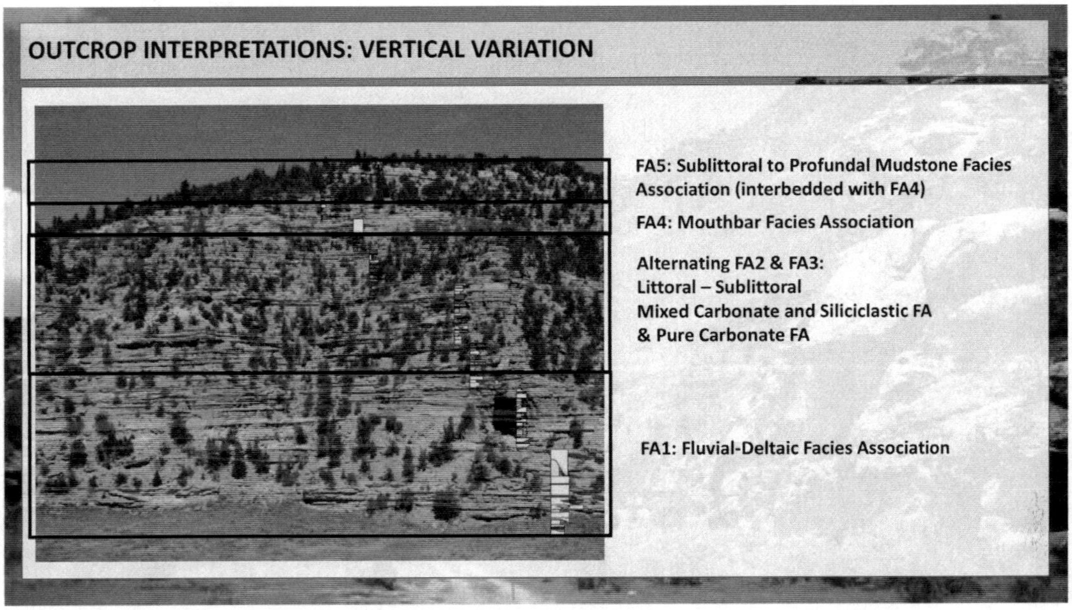
OUTCROP INTERPRETATIONS: VERTICAL VARIATION
FA5: Sublittoral to Profundal Mudstone Facies Association (interbedded with FA4)
FA4: Mouthbar Facies Association
Alternating FA2 & FA3:
Littoral – Sublittoral
Mixed Carbonate and Siliciclastic FA
& Pure Carbonate FA
FA1: Fluvial-Deltaic Facies Association

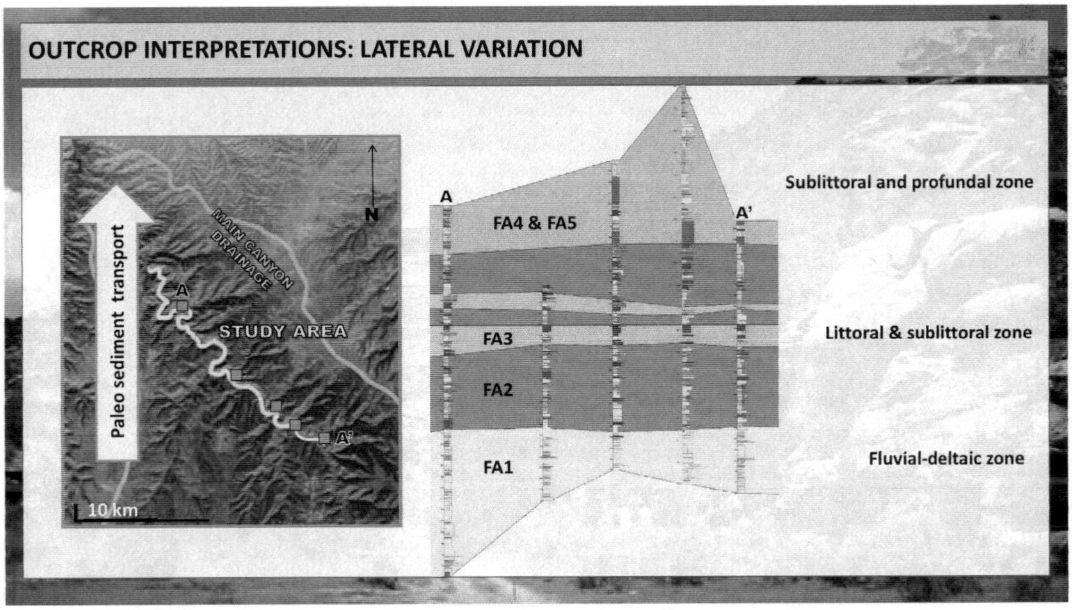
OUTCROP INTERPRETATIONS: LATERAL VARIATION
Paleo sediment transport
N
MAIN CANYON DRAINAGE
A
STUDY AREA
A'
10 km
A
A'
FA4 & FA5
FA3
FA2
FA1
Sublittoral and profundal zone
Littoral & sublittoral zone
Fluvial-deltaic zone

OUTCROP INTERPRETATIONS: LATERAL VARIATION

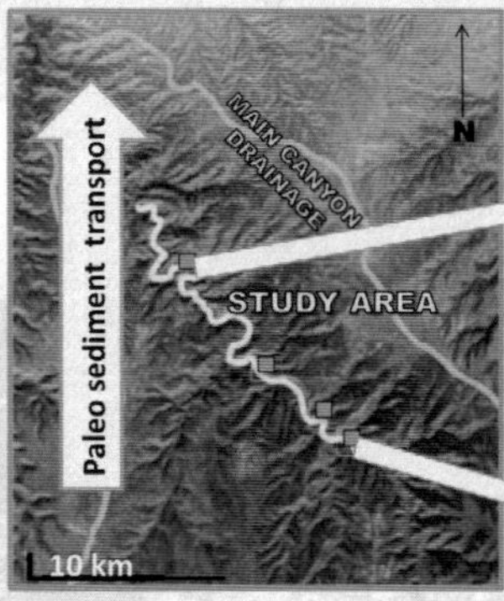

Thicker channelized sandstone and silt-stone packages exist in the northern portion of the study area, whereas packages in the southern portion are smaller and weakly channelized.

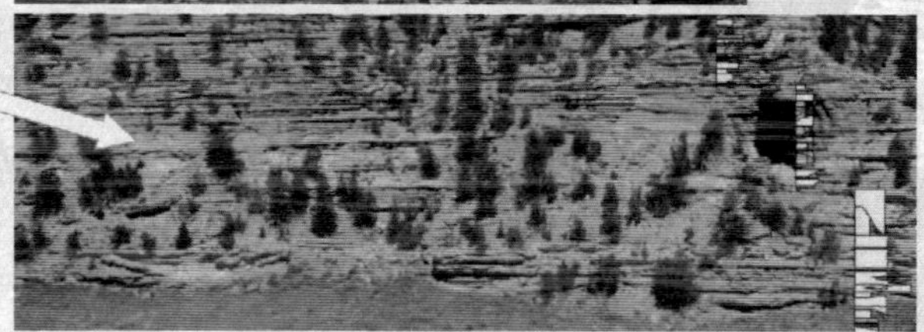

PRELIMINARY CONCLUSIONS

Preliminary study results captures an overall deepening of Lake Uinta in the Douglas Creek Member

- **Vertical variation**
  - Development of a fluvial deltaic system transitioning to deeper sublittoral deposits
  - Continuous interplay between carbonate and siliciclastic systems
- **Lateral variation**
  - Pure carbonate beds are distinct marker beds for correlating field area
  - Increasing amounts of carbonate grains occur in distal study area

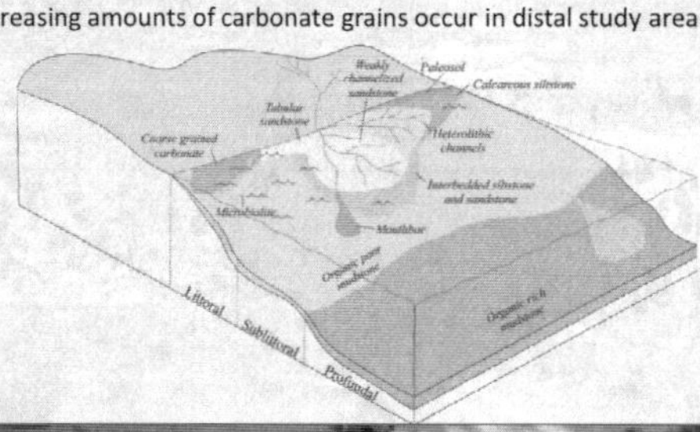

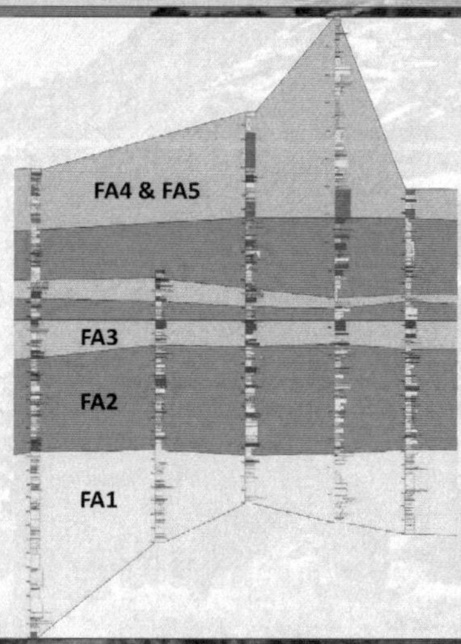

NEXT STEPS

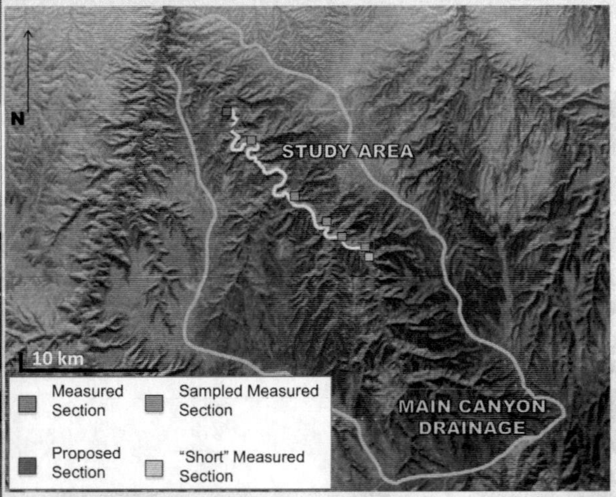

Next Steps:

1. Measure additional section in most distal portion of study area to assess full evolution of the fluvial-deltaic and other facies associations
2. Decipher vertical facies association cycling across the study area to determine high-resolution changes in depositional environments
3. Utilize carbon isotopes from sampled section to determine possible relationship between Eocene hyperthermal events and lacustrine deposition, particularly with the cycling of carbonate and siliciclastic material.

ACKNOWLEDGEMENTS

TOTAL

- **Mike Vanden Berg, UGA**
- **Field assistants: Jennifer Morris, Parker Archer, Shawn Moore**

Analysis of Retorted Water Produced from Partial Combustion of Sultani Oil Shale

Omar Al-Ayed[1] , Surah Al-Harahsheh[2], M. Amer[3] and M.Moutq[4]
[1,4] Al-Balqa Applied University, Department of Chemical Engineering. Amman Jordan
[2] Al Al-Bayt University, Institute of Earth and Environmental Sciences, Mafraq, Jordan
[3] Jordan University, Department of Chemistry, Amman Jordan

[1]omar.alayed@bau.edu.jo

Water and Oil Shale Industry

- Water is necessary for producing shale oil. The amount and quality varies with technology. Typical uses include process cooling, reclamation, dust control and the municipal needs resulting from any population increase due to oil shale workers and supporting businesses.
- Water is produced during the processing (retorting) of oil shale that is not indigenous or tributary to local ground water or streams. It amounts to between 4 and 14% of the shale oil produced. Therefore an average of 10% of the required indigenous water may be saved by treating this water and using it within the project.

35th OSS, Salt Lake City, Oct 5 - 8 . 2015

Partial Combustion of Oil Shale

- In the partial combustion process, the bottom spent shale layer of the fixed or moving bed retort is supplied with oxygen (air) to effect combustion generating heat for the pyrolysis zone.
- In this process, utilizing fixed carbon heat, creating operational problems, lower shale oil quality, improve energy saving, affect retorted water, aggravate ash disposal etc.

Sultani Oil Shale Deposits

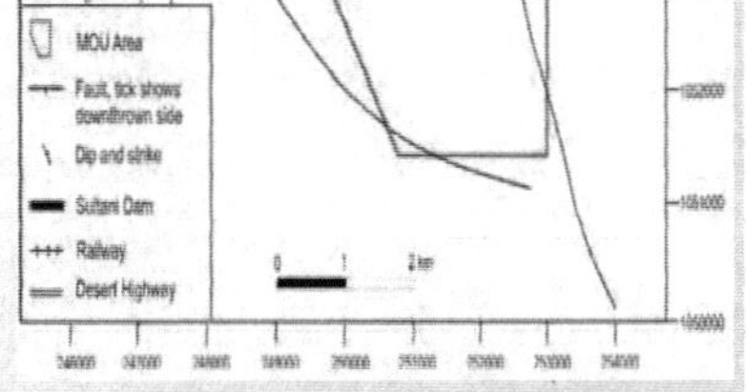

Sultani Oil Shale Deposits

Item	Value	Item	Value
Area, Square kilometer	24	Avg. Oil Content, (wt%)	>7.5
Avg. OS thickness, m	32	Total Organic Material	21.5
Avg. thickness, OB, m	69	Calorific Value, kj/kg	>5,600
Stripping Ratio, m	1.6	$CaCO_3$. wt. %	47
Geological Resv. (Mt)	1130	Sulfur, wt.%	2.4
Open Cast Mining. (Mt) (Oil bbl, M)	942 742.8	Density, (kg/m^3)	1960

35th OSS, Salt Lake City, Oct 5 - 8, 2015

Experimental Setup

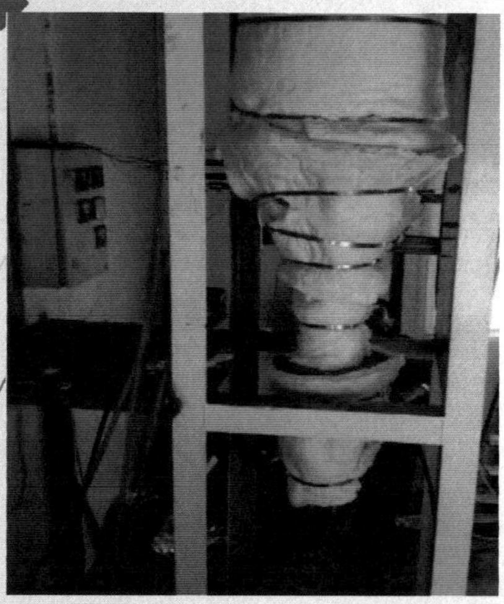

35th OSS, Salt Lake City, Oct 5 - 8, 2015

Experimental Run

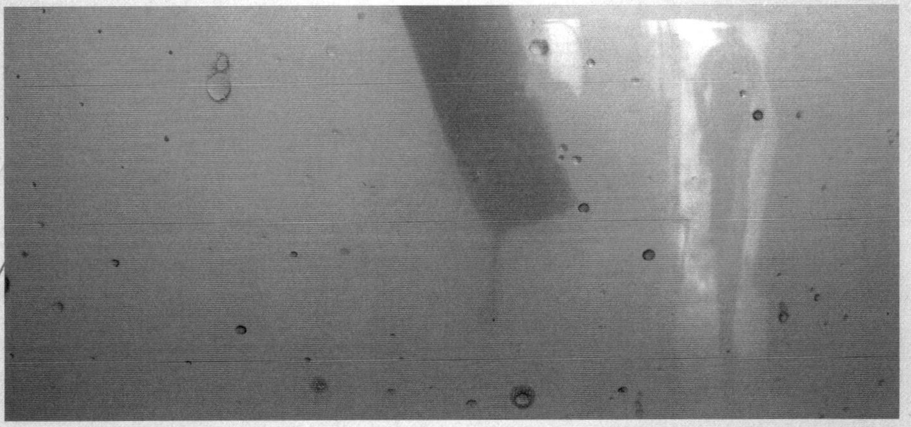

Testing Procedure

- Generated liquids (Water and Oil) are collected, allowed to settle and then separated. Sampling, filtration.
- Produced water from retorting oil shale contains dissolved suspended solids, organic compounds, inorganic materials and hydrogen sulfide etc.

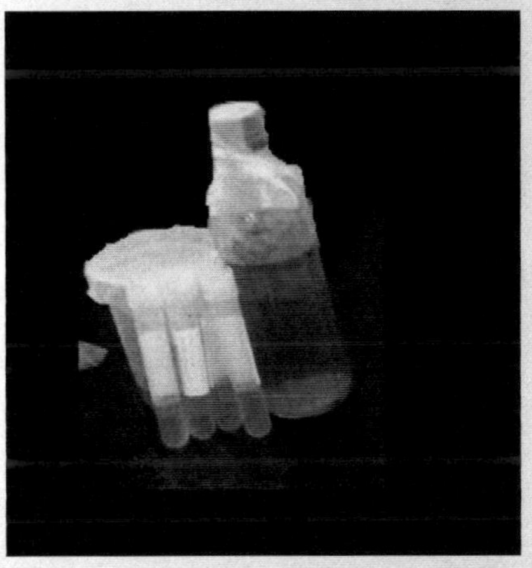

Sample Analysis

➨Total Organic Carbon (TOC)

- TOC = TC-IC , (Disslove CO_2 and bicarbonate)
- 2010 mg/liter in retorted water (this work)
- Al-Jardin reported 3270 mg/L for Ellajjun oil shale.
- Dobson et.al in their study of Rundle oil reported 5000 ppm
- Syamsiah et.al indicated more than 13,000 mg/L
- Venardos et al reported 1800 mg/L in their study of in-situ and normal retorting.

Phenol

➨Phenol is a colorless-to-white solid when pure; however, the commercial product, which contains some water, is a liquid. Phenol has a distinct odor that is sickeningly sweet and lasting. Most people begin to smell phenol in air at about 40 parts of phenol per billion parts of air (ppb), and begin to smell phenol in water at about 1–8 parts of phenol per million parts.

- Test of Sultani retorted water using HPLC indicates 64 mg/L.
- Al-Jaradin reported 29 mg/L for Ellajjun oil shale retorted water
- Syamsiah et.al reported 25 mg/L
- Dobson et al reported 105 mg/L in Rundle oil shale
- Jerry et.al reported 54.3 mg/L in their study on oil shale from Utah.

35th OSS, Salt Lake City, Oct 5 - 8 , 2015

Arsenic Metal

- Long-term exposure to arsenic in drinking water can cause cancer in the skin, lungs, bladder and kidney. Inductively Coupled Plasma Mass Spectrometer is used.
  - In the present work, the amount of Arsenic metal was found to be 0.18 mg/L.
  - Al-Jardain reported 0.23 mg/L arsenic content of the retorted water obtain from Ellajjun oil shale deposits.
  - Matouq et al. have reported 0.15 mg/L for water obtained from Ellajjun oil shale retorting.
  - Venardos et al. have reported a value of 0.133 mg/L.

Phosphate

- It is reported that excessive phosphate consumption leads to a higher mortality in patients with renal disease. Free phosphate, as found in food additives, is entirely absorbed in the gastrointestinal tract, phosphate also seems to damage blood vessels and induce aging processes.
- Stannous chloride test reported 35.7 gm/L in the present work. (water + ammonium heptamolybdate reagent + stannous_chloride reagent → molybdenum_blue) using calorimetry

35th OSS, Salt Lake City, Oct 5 - 8 , 2015

Sulfate

- Sulfate is a substance that occurs naturally in drinking water. Health concerns regarding sulfate in drinking water have been raised because of reports that diarrhea may be associated with the ingestion of water containing high levels of sulfate. Of particular concern are groups within the general population that may be at greater risk when they experience an abrupt change from drinking water with low sulfate to high sulfate concentrations. Ion Chromatography in this work reported 5022 mg/L. Matouq et al reported for Ellajjun 1500 mg/L. Dean et al reported for Rundle oil shale 1916 mg/L. Kekisheva et al. reported 180 mg/L.

Ammonium

- Ammonium sulfate has a low acute toxicity by oral or dermal exposure; after ingestion of high dosages systemic effects as nausea, vomiting, and diarrhea might be possible. Inhalation is considered the major route of exposure. Higher dust concentrations cause discomfort and stress as well as irritating effects on mucous membranes.
- Ion Chromatography is used to measure the water content of NH_4. It is found that retorted water contained 2831 mg/L. El-Jaradin reported 2700 mg/L as NH_3. Colorado oil shale, measured Ammonium as NH_3, and found 1600 mg/L. Kekisheva et al. have reported 1800 mg/L

Nickle

- The most common harmful health effect of nickel in humans is an allergic skin reaction in those who are sensitive to nickel.
- Nickel is the most observed cause of immediate and delayed hypersensitivity.
- All nickel compounds, except for metallic nickel, have been classified as human carcinogens by the International Agency for Research on Cancer (IARC) and the US Department of Health and Human Services. Test using ICP indicated 0.5 mg/L while Al-Jaradin reported 0.8 mg/L for Ellajjun and Matouq et al. found 0.43 mg/L for Ellajjun retorted waters.

Other Elements

Element	Results present of Work	Results of Al-Jaradin	Results of Matouq et al.	Equipment Used
As	0.18	0.23	0.15	ICP
Al	6.7		3	ICP
B	ND		1.8	ICP
Ba	ND		0.07	ICP
Bi	ND		0.2	ICP
Ca	17.5		110	ICP
Cr	0.7			ICP
Cu	ND	0.34	66	ICP
Fe	2238		420	ICP
K	ND		6	ICP
Li	ND		0.01	ICP
Mg	1		0.91	ICP
Mn	31		2.7	ICP
Na	ND		16	ICP
Ni	0.5	0.82	0.43	ICP
Pb	ND	ND	1	ICP
Zn	1	4.74	54	ICP

Conclusions

- Retorted water contains several contaminants that are harmful to human beings.
- Any further use of the retorted water must be subjected to purification and pollutant percentage removal according to end use requirement.
- Depending upon the rate of generated retorted water, the purification process is determined dictated by the end use.

35th OSS, Salt Lake City, Oct 5 - 8 , 2015

Thank you for listening

35th OSS, Salt Lake City, Oct 5 - 8 , 2015

Background

- Jim Patten of Red Leaf Resources contacted Utah Division of Wildlife Resources (UDWR) and expressed an interest in developing a partnership to determine ways that big game habitat can be restored and enhanced following large scale mining operations.
- Two main approaches were discussed.
  - Vegetation reclamation
  - Water development

- A unique cooperative partnership has been formed between UDWR and Red Leaf Resources to implement habitat improvement research
- Seed mix development, procurement, seeding, monitoring, and guzzler construction services are provided by UDWR
- Red Leaf Resources provides funding, equipment, site preparation, site control, and manpower to implement project activities
- UDWR and Red Leaf Resources have worked together closely to identify project locations and activities

Vegetation Reclamation

- Utah Division of Oil, Gas, and Mining (DOGM) regulates mine reclamation.
- DOGM reclamation seed mixes are typically designed to maximize soil stabilization and reduce erosion.
- While DOGM seed mixes do offer considerable benefit to wildlife, UDWR and Red Leaf desire to maximize benefit to mule deer and elk, while still meeting DOGM reclamation standards.

- UDWR has considerable experience performing habitat restoration projects through wildfire reseeding and watershed restoration projects.
- A diverse seed mix was selected based on attributes of elevation, precipitation, soil type, and nutritional value for wintering big game.
- The Red Leaf Seep Ridge Site is located within crucial winter habitat for both mule deer and elk.
- Mule deer and elk need plants with the highest protein to ensure winter survival.

Project Name	Red Leaf Mule Deer Habitat	Acres	Budget	Habitat Treatment	Approximate Date Needed
Database #		3		Drill	

Seed: Common Name	Seed: Scientific Name	Variety	Bulk Pounds per acre	PLS Pounds per acre	Cost per pound	Total Pounds	# Seeds/lb	Average PLS	# of Viable Seeds/sq foot	Total Cost
Annual Sunflower	Helianthus annuus	Any	1	0.92	$15.00	3	58,500	91.52	1.23	$45.00
Big Bluegrass	Poa ampla	Sherman	0.25	0.17	$3.26	0.75	882,000	69.46	3.52	$2.45
Bitterbrush	Purshia tridentata	Any	2	1.66	$24.04	6	15,000	82.81	0.57	$144.24
Blue Grama	Bouteloua gracilis	Alma	0.25	0.17	$10.89	0.75	825,000	66.06	3.13	$8.17
Bluebunch Wheatgrass	Agropyron spicatum	Anatone	0.75	0.57	$7.67	2.25	140,000	76.50	1.84	$17.26
Cicer Milkvetch	Astragalus cicer	Lutana	1	0.86	$5.30	3	145,000	86.07	2.86	$15.90
Crested Wheatgrass	Agropyron cristatum	Douglas	0.35	0.31	$3.03	1.05	175,000	88.12	1.24	$3.18
Fourwing Saltbush	Atriplex canescens	Any	4	1.67	$10.51	12	52,000	41.77	1.99	$126.12
Great Basin Wildrye	Elymus cinereus	Trailhead	0.75	0.65	$10.50	2.25	130,000	87.02	1.95	$23.63
Orchardgrass	Dactylis glomerata	Paiute	0.5	0.42	$2.49	1.5	654,000	83.90	6.30	$3.74
Rocky Mountain Bee Plant	Cleome serrulata	Any	1	0.92	$19.44	3	70,000	91.86	1.48	$58.32
Sainfoin	Onobrychis viciifolia	Eski	2	1.83	$2.81	6	30,000	91.28	1.26	$16.86
Indian Ricegrass	Oryzopsis hymenoides	Nezpar	1	0.91	$6.86	3	141,000	91.32	2.96	$20.58
Small Burnet	Sanguisorba minor	Delar	3	2.73	$2.18	9	55,000	90.84	3.44	$19.62
Thickspike Wheatgrass	Agropyron dasystachyum	Critana	1	0.86	$5.91	3	154,000	86.09	3.04	$17.73
									Total	$522.78

Total Pounds	57
Bulk lbs/acre	18.85
PLS lbs/acre	14.64
Dollars/acre	$ 174.26
Live seed/ft^2	36.81

Project Name	Red Leaf Mule Deer Habitat	Acres	Budget	Habitat Treatment	Approximate Date Needed
Database #		3		Broadcast	

Seed: Common Name	Seed: Scientific Name	Variety	Bulk Pounds per acre	PLS Pounds per acre	Cost per pound	Total Pounds	# Seeds/lb	Average PLS	# of Viable Seeds/sq foot	Total Cost
Alfalfa	Medicago sativa	Ladak	1.5	1.38	$4.37	4.5	225,000	91.95	7.12	$19.67
Alfalfa	Medicago sativa	Spreador 4	1.5	1.38	$4.37	4.5	225,000	91.95	7.12	$19.67
Alfalfa	Medicago sativa	Ranger	1.5	1.38	$4.37	4.5	225,000	91.95	7.12	$19.67
Blue Flax	Linum perenne	Appar	0.75	0.65	$6.22	2.25	293,000	87.15	4.40	$14.00
Forage Kochia - Snowstorm	Kochia prostrata	Snowstorm	0.5	0.27	$28.96	1.5	407,700	53.51	2.50	$43.44
Palmer Penstemon	Penstemon palmeri	Any	0.15	0.12	$24.03	0.45	610,000	80.19	1.68	$10.81
Rocky Mountain Penstemon	Penstemon strictus	Any	0.15	0.13	$10.52	0.45	592,000	88.40	1.80	$4.73
Sagebrush, Wyoming	Artemisia tridentata wyomingensis	Any	1.75	0.34	$18.19	5.25	2,500,000	19.47	19.56	$95.50
Sagebrush, Black	Artemisia nova	Any	0.25	0.08	$18.80	0.75	907,200	32.29	1.68	$14.10
True Mountain Mahogany	Cercocarpus montanus	Any	3	2.32	$24.93	9	59,000	77.38	3.14	$224.37
Western Yarrow	Achillea millefolium	Eagle Mountain	0.05	0.04	$18.13	0.15	3,300,000	89.66	3.40	$2.72
Winterfat	Ceratoides lanata	Any	4	1.18	$13.80	12	56,700	29.38	1.53	$165.60
Yellow Sweetclover	Melilotus officinalis	Any	0.5	0.45	$2.99	1.5	260,000	89.61	2.67	$4.49
									Total	$638.75

Total Pounds	47
Bulk lbs/acre	15.60
PLS lbs/acre	9.72
Dollars/acre	$ 212.92
Live seed/ft^2	63.74

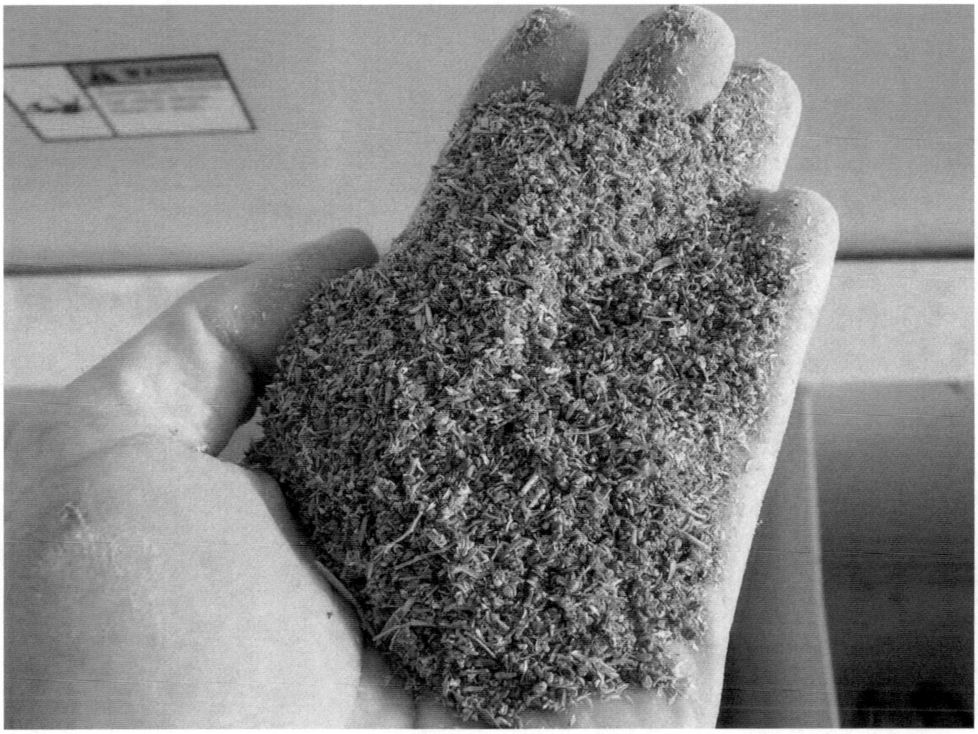

ROUGH RIDER
CAT

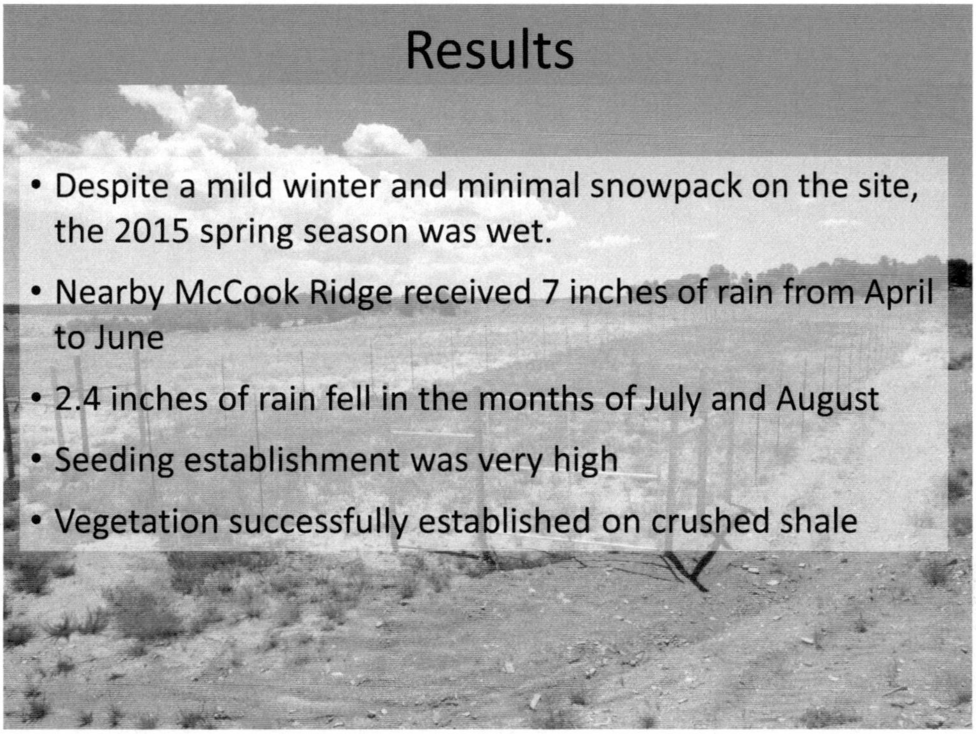
Results
• Despite a mild winter and minimal snowpack on the site, the 2015 spring season was wet.
• Nearby McCook Ridge received 7 inches of rain from April to June
• 2.4 inches of rain fell in the months of July and August
• Seeding establishment was very high
• Vegetation successfully established on crushed shale

• Despite successful establishment of many species, sagebrush establishment was minimal
• Some plant species included in the experimental plot outperformed those species included in the original DOGM seed mix

The site was visited in September to conduct some baseline assessment of vegetation cover
•Many forbs and grasses had already entered dormancy and disappeared from the site until next year
•Browse species remain

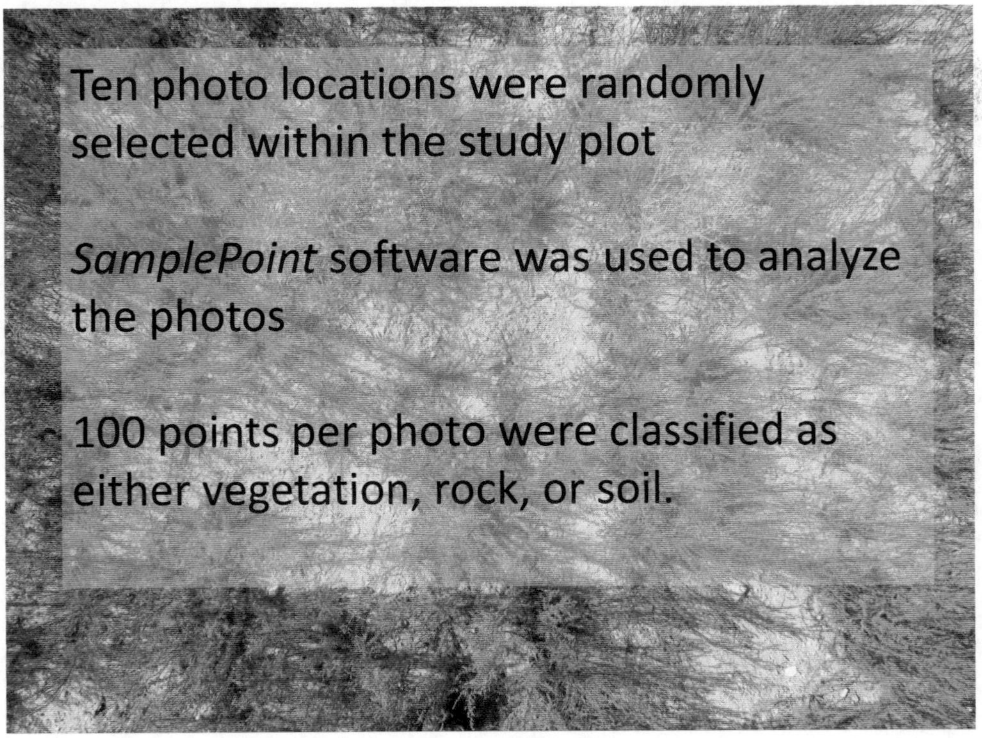
Ten photo locations were randomly selected within the study plot
SamplePoint software was used to analyze the photos
100 points per photo were classified as either vegetation, rock, or soil.

• Results indicate that plants (primarily forage kochia and four-wing salt brush) account for 53.5% vegetation cover
• Rock cover is 6.25%
• Soil cover is 40.25%
• The high level of plant cover is already providing soil stabilization and wildlife forage, even in the first year.

Future monitoring will ultimately reveal which plant materials are best suited for the site
Summer monitoring will provide meaningful data of species composition
Seeding rates and species composition can be modified as this monitoring approach extends into future years

Water Development

- Book Cliffs mule deer are very traditional in their annual migration.
- Every year they begin arriving on winter range around October 1st, despite winter conditions
- Many years, water is absent or severely lacking, concentrating deer on limited water supplies
- In an area generally lacking in perennial water, water catchments known as "guzzlers" can be effective for providing water

What is a guzzler?

- Most guzzlers consist of three parts
- A collection "apron" to catch water and funnel it to a tank
- Storage tank for holding water
- A "drinker" area to allow wildlife to access water
- Guzzlers store water from wet periods and make it available during dry periods

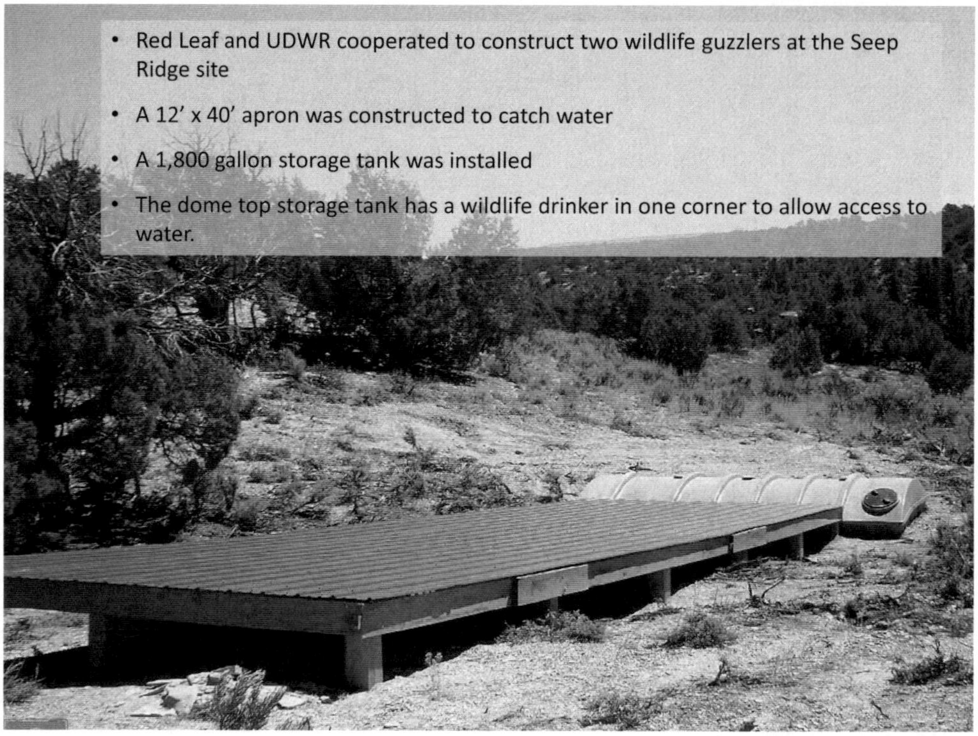
• Red Leaf and UDWR cooperated to construct two wildlife guzzlers at the Seep Ridge site
• A 12' x 40' apron was constructed to catch water
• A 1,800 gallon storage tank was installed
• The dome top storage tank has a wildlife drinker in one corner to allow access to water.

• The two guzzlers constructed at Seep Ridge were completed in August 2015
• By September, enough rain fell to provide some water to wildlife
• Upon completion, camera traps were installed at each guzzler location to document wildlife use.

Red Leaf North Guzzler

Mule deer begin use in October
Camera Name 82°F27°C
09-22-2015 18:51:47

Camera Name 80°F26°C
09-22-2015 18:56:10

Bushnell
Camera Name 80°F26°C
09-22-2015 18:57:09

Bushnell
Camera Name 102°F38°C
09-25-2015 17:37:39

Red Leaf South Guzzler

Captured 6 bear images, 2 red fox images, and 106 elk images
Bushnell
Camera Name 50°F10°C
09-12-2015 03:25:02

Mule deer arrive in October
Bushnell
Camera Name 59°F15°C
09-14-2015 17:52:06

Camera Name 32°F0°C
09-19-2015 05:17:34

Other Mule Deer Guzzler Use
• UDWR has two guzzlers on its ownership 2 miles from the new Red Leaf guzzlers.
• These guzzlers were monitored for one month in March and April 2015
Bushnell
Camera Name 64°F17°C
03-29-2015 18:21:44

434 mule deer images were collected on one guzzler
Bushnell Ⓜ Camera Name 60°F15°C
03-26-2015 10:45:48

954 images of mule deer were collected at the second guzzler
Bushnell Ⓜ Camera Name 60°F15°C
03-24-2015 16:16:51

If the trend holds true, significant deer use is expected to occur on the new guzzlers in the fall 2015 and spring 2016.
Bushnell Camera Name 70°F21°C
04-14-2015 18:15:50

•Additional monitoring of both the reclamation plot and the wildlife guzzlers is needed.
•Monitoring results will guide future decisions and actions regarding reclamation seed mixes and strategies.
Bushnell Camera Name 55°F12°C
08-23-2015 23:22:17

Questions?
Bushnell Ⓜ Camera Name 62°F16°C
08-25-2015 02:28:30

8.3 Some Potential Long-term Effects of Oil and Gas Development on Water and Oil Shale Resources, Piceance Basin, Colorado

Glen Miller

Retired Geologist, USA

The projected BLM (2015) oil and gas development in the Piceance Basin (additional 10,000± wells) can significantly reduce the available resources of oil shale and usable water. The potential for leaky annuli in gas wells will require a "no-mining" safety zone around each well that will "sterilize", basin wide, 10's of billions of barrels of synthetic shale oil. Along with leaked gas, the upward flow of saline "formation water" and resulting degradation of ground water will greatly increase mining costs. A largely unrecognized issue, the future widespread deeper drilling and hydrofracturing for gas in the underlying Mancos Shale, can result in a large "new" aquifer, with a capacity of perhaps 20-40± million acre feet. This aquifer will fill with water. From where? Overlying aquifers via leaky O&G wells is an obvious source. Deeper aquifers may provide recharge. When full, the resulting water quality is likely to be unusable, and will discharge, probably to overlying aquifers and streams. Further such widespread drilling in The Mancos in NW Colorado and NE Utah could potentially create "new" aquifers with total capacities of many hundreds of million acre feet. Such implication require further studies. The above developments, which potentially can amount to a trillion± dollars in loss and costs, can be minimized or prevented by effective planning for well spacing, design, and especially by completing oil and gas wells that do not leak along the annulus, and thus minimize sterilization of oil shale. A scientifically designed and operated ground water monitoring system is a must for multiple resource development in The Basin.

8.4 A New Project in Turkey: Hasanlar (Bolu, Western Anatolia) Bituminous Shale Deposit and its Synthethic Crude Oil Potential

Ilker Senguler[1], Abdurrahman Murat[1], Ferudun Alp Ugur[1], Murat Yukunc[1]

[1]General Directorate of Mineral Research and Exploration of Turkey, Turkey, [2]Turkish Coal Enterprises, Turkey, [3]Turkish Petroleum Corporation, Turkey, [4]General Directorate of Mineral Research and Exploration of Turkey, Turkey

The Hasanlar bituminous shale deposit is located in the Hatildag lacustrine basin at the south-eastern part of Goynuk in Bolu province and the project area licenses belong to the Turkish Coal Enterprises (TKI). Field and laboratory studies have been carried out together with the General Directorate of Mineral Research and Exploration (MTA) and Turkish Petroleum Corporation (TPAO). Bituminous shale occurs in the west of part an E-W oriented syncline of steeply dipping Paleocene-Eocene sediments. The bituminous sedimentary rocks are generally brown and beige coloured. The rock type and content of organic matter change rapidly in the vertical direction. The clastic portion of the bituminous sediment is composed mainly of clay minerals and quartz. The microscopically recognizable organic matter consists of liptinite, bituminite and huminite. It contains around 80% liptinite, 5–10% bituminite and 5–10% huminite. The high liptinitic content shows that the organic matter originated mainly from hydrogen-rich organic remains of algae and pollen. The TOC content of bituminous shale in the basin is high, ranging 0.2 to 28.79 %wt. At the high TOC level are 1.42 S1 (mg HC/g rock), 284.69 S2 (mg HC/g rock), 1.32 S3 (mg CO_2/g rock), Tmax 451 (°C), 989 HI (mg HC/g TOC), 5 OI (mg CO_2/g TOC), RC (quantity of carbon that remains in the sediment following pyrolysis) 4.87 (%) , PC (quantity of carbon that can be pyrolyzed) 23.91 (%) and MinC (mineral carbon the quantity of C contained in minerals) 6.34 (%). The heating value is 3.24 MJ/kg (1 kJ/kg=4.1868 kcal/kg) for the bituminous shale. The percentage Fischer assay yield for the Hasanlar bituminous shale is 90 gallons of synthetic crude oil per ton of shale (GPT) respectively. Oil yields of 10.5 GPT (4%), 26.7 GPT (10.4%), 36.7 GPT (13.8%) and 61.8 GPT (23.6%) classify the shale oil as uneconomical, medium economical, highly economical and very highly economical respectively. Hasanlar field includes shale with a high level of economic potential. Therefore bituminous shale can classified as very highly economical. Exploration work continues at the field. At the end of this study, maps of bituminous shale thickness and richness will be generated with isoresource lines delineating oil yields of 15, 25, 35, and 50 GPT. From these maps, new basin-wide resource numbers will be calculated for each richness grade in the basin.

A Simple Kinetic Model of Oil Generation, Vaporization, and Degradation

35TH OIL SHALE SYMPOSIUM

SALT LAKE CITY, OCT 5-6, 2015

Alan K. Burnham

Objectives of STEMS Project

Develop a comprehensive coupled model for

- Generation and primary migration for natural petroleum source rocks
- Producibility of unexpelled oil and gas
- In-situ and In-capsule oil shale retorts

Chemical and physical processes

- Rock mechanics
- Heat and mass transport
- Chemical kinetics

Purpose of Presentation

Common problems deriving good kinetics are reviewed

- Numerous kinetic studies exist, but many are not very good
- Hopefully you will learn how not to add another bad one

A new, simple model is presented that explains much of the variation in the literature

- Based on an equivalent volatility concept
- Explains the dependence of yield and composition on pyrolysis temperature
- Explains the effects of an external gas purge

Five practices leading to bad kinetic parameters

1. Assuming first-order kinetics for what is really a broad range of consecutive and sequential reactions
2. Performing a single experiment at a single heating rate and fitting it to a kinetic model
3. Not using a wide enough range of temperature or heating rate to properly constrain A and E given random errors
4. Having transient and steady-state temperature errors that distort kinetic parameters even with reproducible data
5. <u>Not considering the dependence of product volatility at different temperatures</u>

A and E values for Green River oil shale are diverse

- Including pre-1975 parameters would give an even larger diversity
- Similar compensation relationships are seen for other materials
- The compensation relation means that rate constants are similar at the measurement conditions
- But predictions diverge upon extrapolation (1-2-3 rule)
- Determinations by the petroleum geochemistry community tend to be in the middle of the cluster

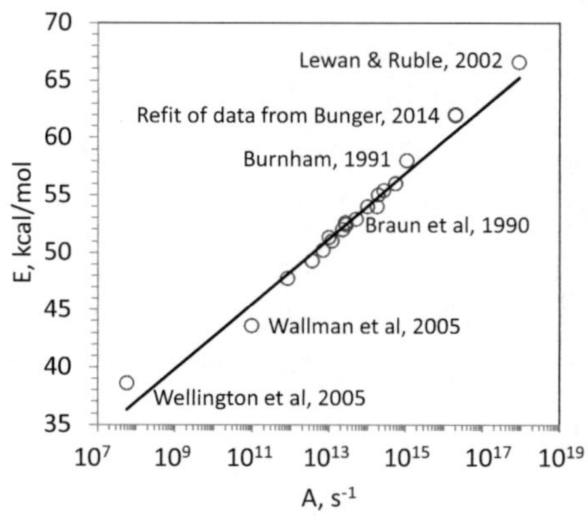

The model presented here is based on kinetics of equivalent volatility

- Most experiments define "oil" as a material that distills out of the reactor
  - "Oil" volatility increases exponentially with temperature
  - "Oil" continuously changes due to progressive reactions, becoming lighter and more volatile if kept in the reactor
- Consequently, experiments at different temperatures and heating rates do not produce the same product
  - This situation basically violates the underlying assumption of the Arrhenius law
- The situation can be corrected by including vapor-liquid equilibria and secondary reactions in the kinetics model
  - Done fairly rigorously in the 1980s but rarely incorporated in process models
  - Addressed in this work by a simple correction term: $E_{eff} = E_{true} + 5*(1-T/500°C)^2$

The difference between oil generation and evolution of "matured" oil is easily calculated

STEMS
Stanford Total Enhanced Modeling of Source rock

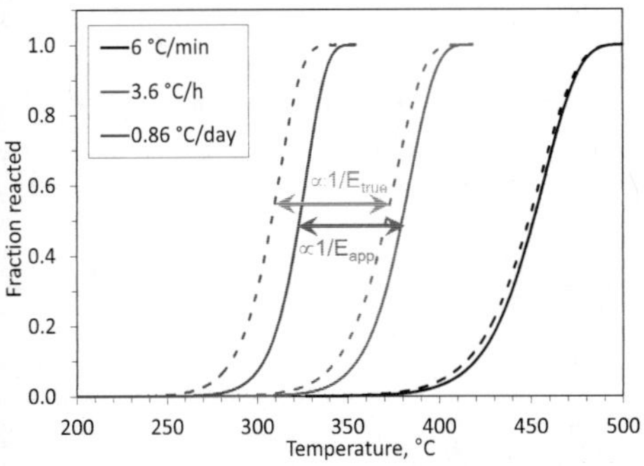

Numerically integrated $d\alpha/dt = k\alpha$ where

$k = Ae^{-E_{eff}/RT}$

and

$E_{eff} = E_{true} + 5(1\text{-}T/500°C)^2$

Also,

$E_{app} = d(\ln k)/d(1/T)$

$= E_{true} + 5(1\text{-}(T/500°C)^2)$

Low heating rate	High heating rate	A_{app}, s^{-1}	E_{app}, kcal/mol
6 °C/min	60 °C/min	1.2×10^{14}	54.6
3. 6 °C/h	36 °C/h	1.8×10^{15}	58.4
0.86 °C/day	8.6 °C/day	1.7×10^{16}	61.2

This simple function explains much of the diverse literature kinetic parameters

STEMS
Stanford Total Enhanced Modeling of Source rock

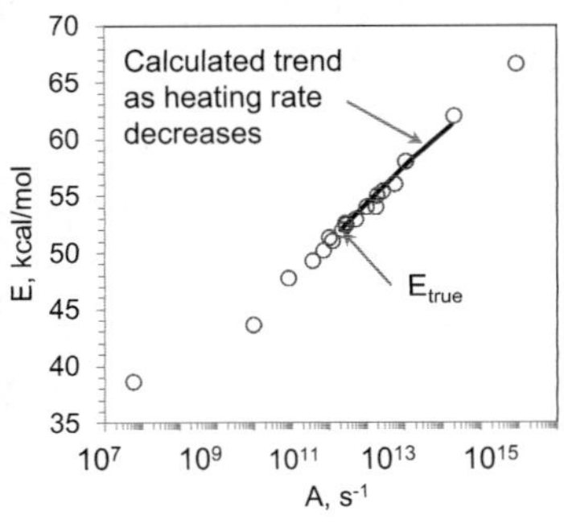

The same two-1st-order reaction model using

$E_{eff} = E_{true} + 5(1\text{-}T/500°C)^2$

with E_{true} = 52 kcal/mol fits both Burnham, 2014, and Burnham, 1991

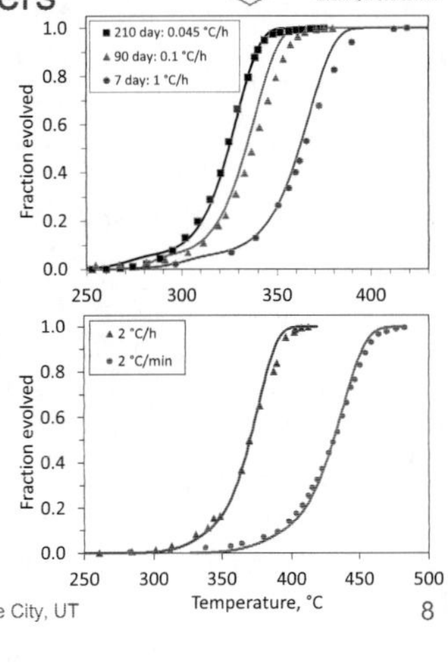

Yields are easily correlated with pyrolysis heating rates and temperatures

STEMS
Stanford Total Enhanced Modeling of Source rock

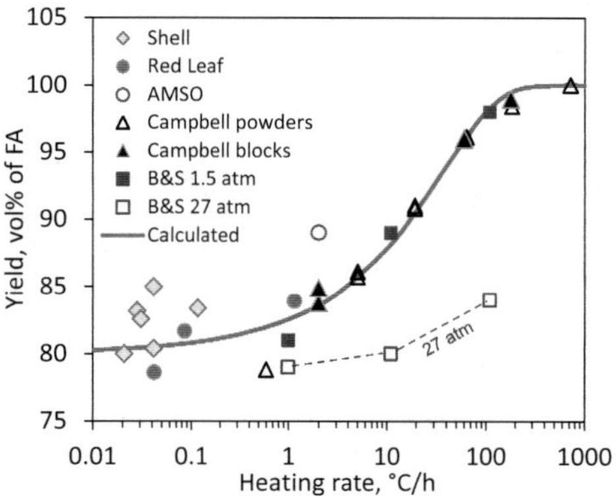

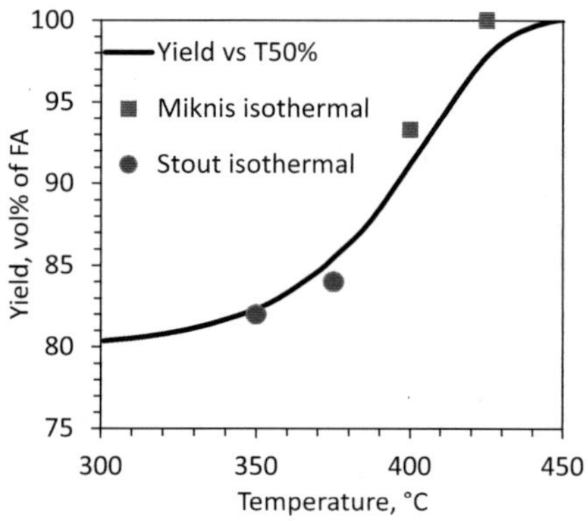

Oil properties are readily correlated with pyrolysis conditions

STEMS
Stanford Total Enhanced Modeling of Source rock

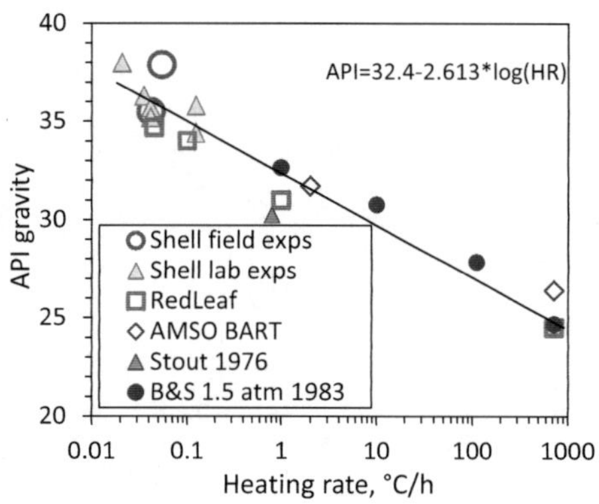

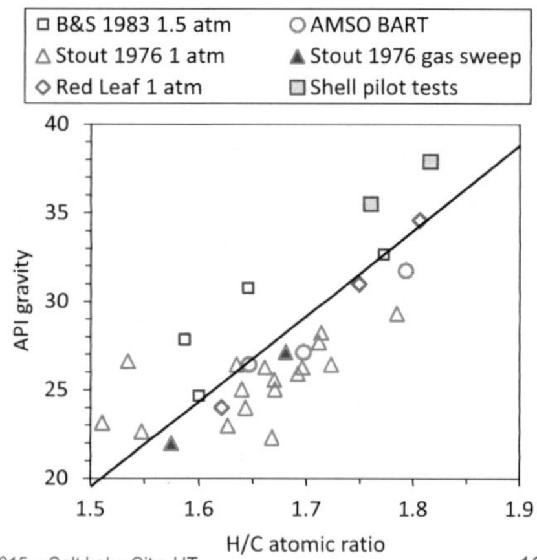

STEMS
Stanford Total Enhanced Modeling of Source rock

Additional oil composition correlations

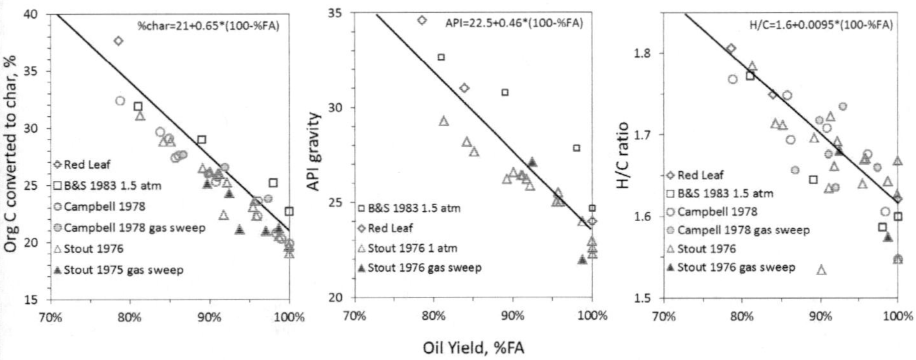

35th Oil Shale Symposium – Oct 5-6, 2015 – Salt Lake City, UT 11

STEMS
Stanford Total Enhanced Modeling of Source rock

Effects of a gas purge can be easily modeled

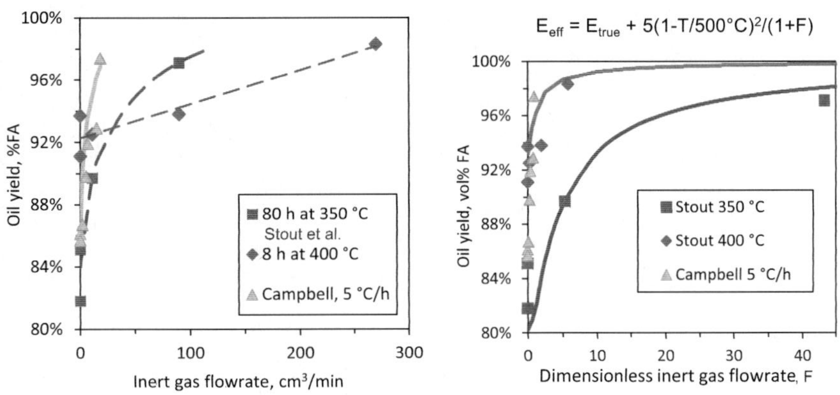

35th Oil Shale Symposium – Oct 5-6, 2015 – Salt Lake City, UT 12

Gas purge and pressure have distinct effects on oil properties

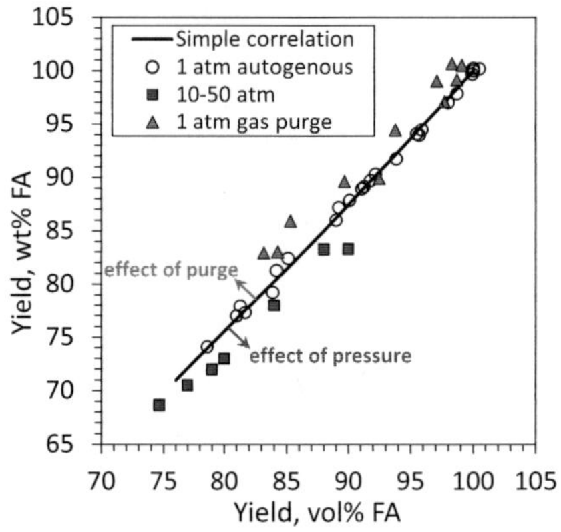

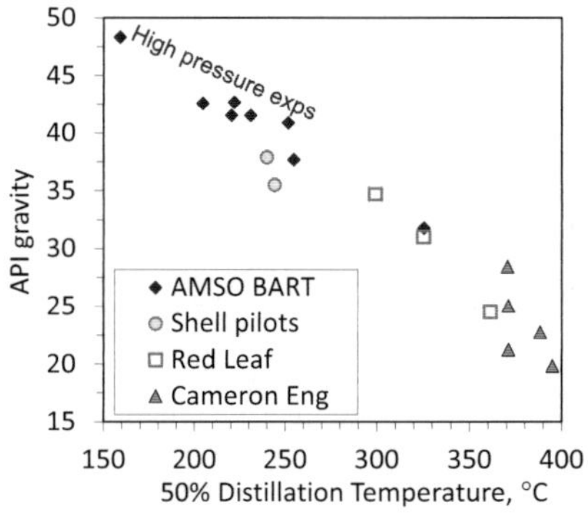

Effects of pressure are easily added

Burnham and McConaghy found that pressure effects could be modeled by $E_{eff} = E_{true} + 67(P_{atm}-1)$

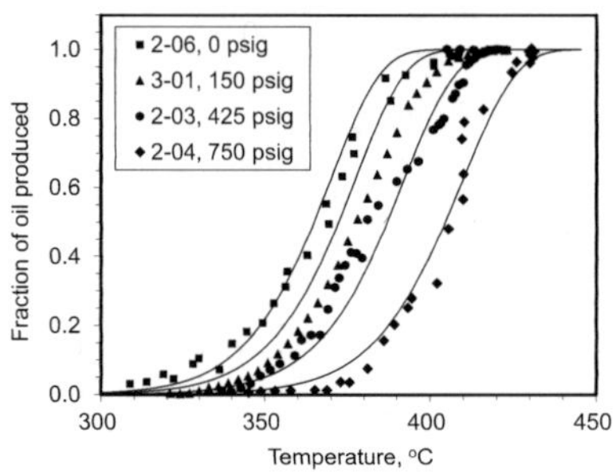

Comparison to Burnham and Singleton (1983) with no adjustable parameters

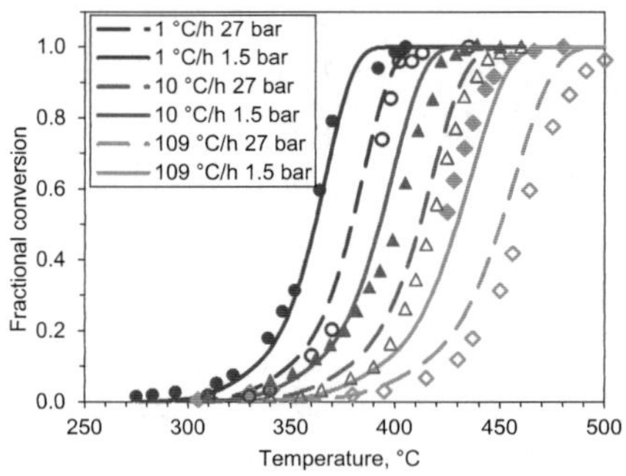

The kinetic parameters also work at high T

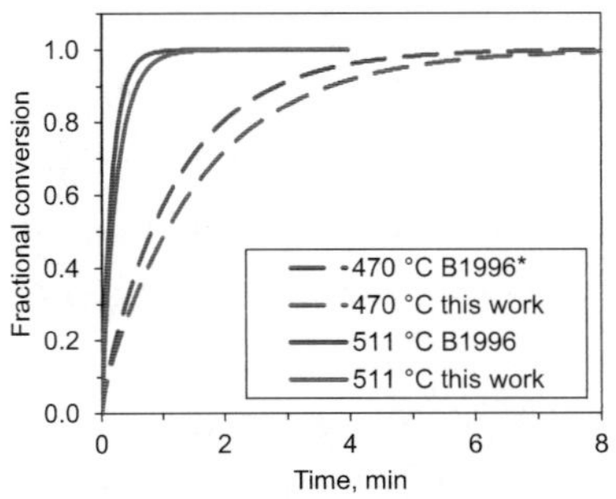

Model validated for:

- Temperatures from 250 to 511 °C
- Time scales of 1 minute to 1 year
- Pressures from 1 to 50 bar
- External gas purge rates

*Burnham et al, Energy&Fuels, 10, 49-59,1996
Data for isothermal pyrolysis in a fluidized bed

The equivalent volatility concept makes modeling diverse pyrolysis conditions easy

- The apparent activation energy becomes a function of pyrolysis temperature and intra-particle gas purge rate
- The greater lag between oil generation and volatilization lower pyrolysis temperatures and heating rates causes oil to coke and crack to a more volatile product
- Resulting oil yields and properties are easily calculated via correlations
- Effects of pressure are easily incorporated

9.2 **Visualization Of Porosity Evolution Of Oil-Shale Pyrolysis Using X-ray Computed Tomography**

Guenther Glatz, Louis Castanier, Anthony R. Kovscek

Stanford University, USA

This paper summarizes results of a successful laboratory investigation to visualize porosity evolution of oil-shale pyrolysis using X-ray computed tomography (CT). Uniform maturation of the sample required the development of a new pyrolysis cell. The cell is designed for cell wall temperatures up to 450 °C at an operating pressure of 5.52 MPa (800 psi). Internal temperatures up to 840 °C and heating rates up to 70 °C/min have been achieved in this new cell. Combining CT imaging techniques with krypton as a pore contrast fluid reveals the porosity distribution of the immature and thermally matured shale sample. Results show dramatic changes in porosity accompanying pyrolysis. Core krypton porosities increase from about 9% to 25% upon pyrolysis. Overall, the samples develop a multimodal distribution of porosity. A key to the success of the CT imaging is the introduction of carbon fiber components.

9.3 Direct Convective Heating with Recycled Pyrolysis Gasses in the Ecoshale Process and Potential Implications for Design, Construction, and Production Efficiencies

Eric Chabal[1], James Patten[2]

[1]*Total, USA,* [2]*Red Leaf Resources, Inc., USA*

One of the most significant evolutions in the Ecoshale technology that is under investigation is the implementation of direct heating by recycled gasses produced by pyrolysis. This involves a change in the heat exchanger architecture from the earlier configuration while retaining convective heating. The concept is to inject externally heated produced gas into the rubble bed through perforated pipes. This process gas then heats the shale bed through convection, as before. Produced gas is extracted from the capsule and cooled to remove condensable hydrocarbon products, but the non-condensable produced gas is externally re-heated and recirculated or re-injected into the capsule to continue convective heating and hydrocarbon production. This approach moves the heat exchanger from inside the capsule to outside the capsule. This could substantially reduce capsule construction cost and complexity, and dramatically reduce the amount of pipe in the capsule and the associated costs. It would allow the elimination of pipe bedding material, which would dramatically reduce materials costs and increase the average grade (kerogen content) in the capsule and also allow the option for internal pipe temperatures to be reduced. It would further allow for a much more robust internal piping system, and would allow the heat exchanger to be re-used rather than left behind in every capsule. The potential danger associated with potential piping leaks would also be eliminated, as no gas entering the capsule would contain any oxygen in this direct heating architecture.

REDLEAF
RESOURCES, INC.

Experimental Apparatus for Investigation of Direct Convective Heating with Recycled Pyrolysis Gasses

Jerry Wiser
Dr. James Patten
Dr. James W. Bunger

RedLeaf Resources Lab

- Can respond quickly and efficiently to answer engineering questions
- Only does work that can't done somewhere else
- Does in-house analysis where necessary

REDLEAF
RESOURCES, INC.

Ore Lab

REDLEAF
RESOURCES, INC.

Retort Lab

REDLEAF
RESOURCES, INC.

Gas Chromatograph Lab

REDLEAF RESOURCES, INC.

Process Control System

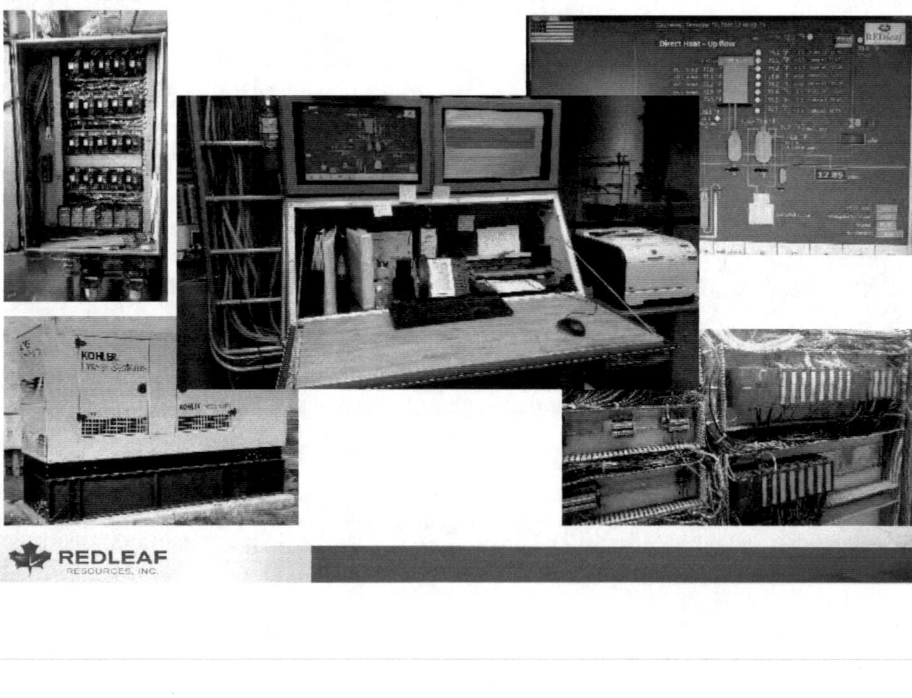

REDLEAF RESOURCES, INC.

Change to Direct Heating

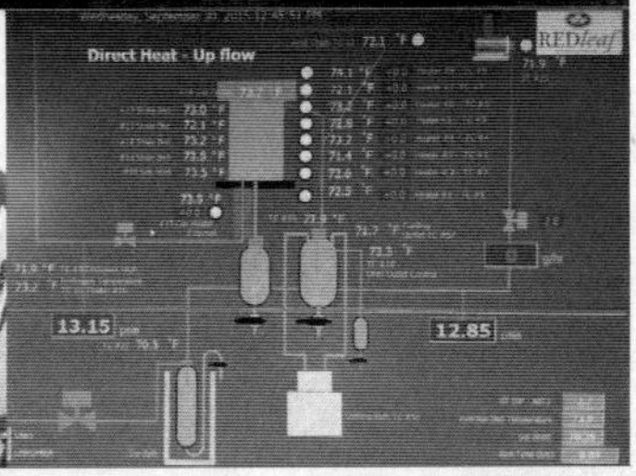

REDLEAF RESOURCES, INC.

Lab Apparatus Design Changes

- Gas inlet to the retort vessel
- Gas recycle system
  - Inlet gas heater
  - Temperature controlled condenser
  - Compressor
  - Flow meter
  - Flow control valve
- Heater control logic changes

REDLEAF RESOURCES, INC.

Direct Heating Changes

Process P&ID

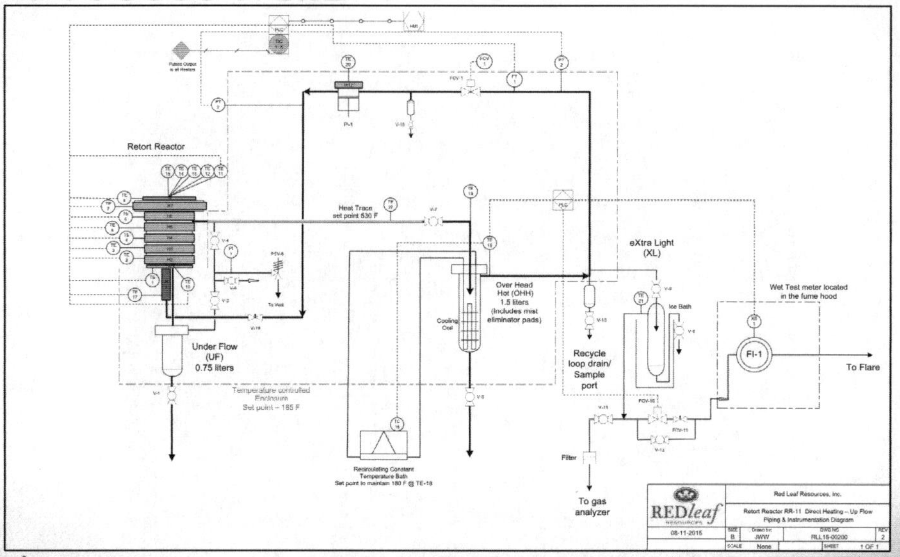

REDLEAF RESOURCES, INC.

Direct Heating Equipment

Direct Heating Test #1

- Objectives:
  - Up flow configuration, 35 % void space
  - Gas flow – 0.0308 lb/hr
  - Gas temperature – 750°F
  - Wall heaters were used to control the heat loss

- The run achieved a 280°F temperature differential
- Actual run time was 17 days because of reactor wall heat loses
- Oil and Gas yield were similar to direct heating
- API gravity lower – 26 rather than 32 typical for indirect heat

REDLEAF RESOURCES, INC.

Direct Heating Test #2

- Objectives:
  - Up flow configuration, 35 % void space
  - Gas flow – 0.0308 lb/hr
  - Gas temperature – 7 day run profile
- Run time was 7 days
- Oil yield 83% of MFA Gas 17% of oil yield
- API gravity – 26.8

REDLEAF RESOURCES, INC.

Direct Heating Test #3

- Objectives:
  - Up flow configuration, 35% void space
  - Gas flow – 0.0694 lb/hr
  - Gas temperature – 750°F
- Run time was 7 days
- Oil yield 84 % of MFA Gas yield 18 % of oil yield
- API gravity – 26

REDLEAF RESOURCES, INC.

Direct Heating Test #4

- Objectives:
  - Down flow configuration, 35% void space
  - Gas flow – 0.0694 lb/hr
  - Gas temperature – 750°F

- Run time was 7 days
- Oil yield 86 % of MFA   Gas yield 18 % of oil yield

REDLEAF
RESOURCES, INC.

Next Direct Heat Tests

- Increase the recycle gas flow rate to reach Ra number
- Condense the recycle stream at 180 F
- Two runs with very different void volumes

REDLEAF
RESOURCES, INC.

Secondary Cracking Experiment

- Recycle gas composition predicted by EPS2 design
- Test sample prepared by combining gas and liquid produced in the lab
- Built test equipment
- Developed an analytical method

REDLEAF
RESOURCES, INC.

Secondary Cracking P&ID

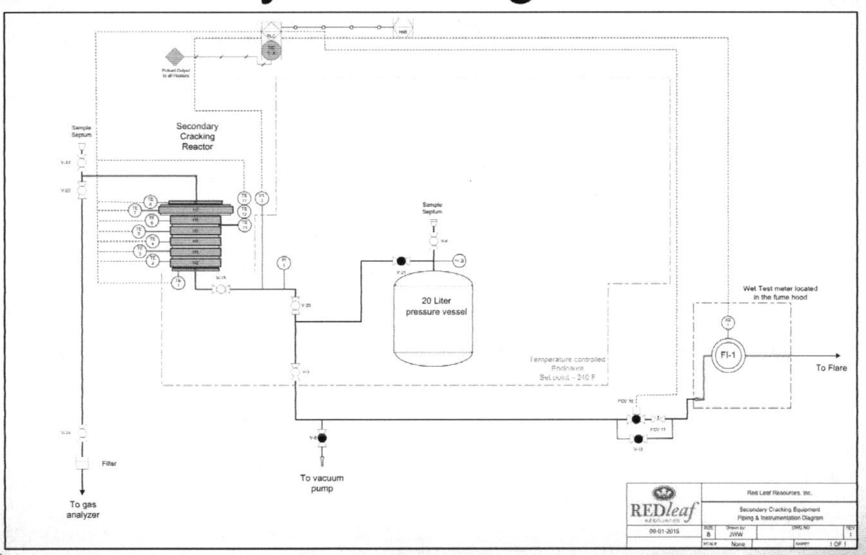

REDLEAF
RESOURCES, INC.

Data Analysis

- Refinery Gas Analyzer (RGA)
  - Non-condensable plus C1 through C5
- Gas Chromatograph with a Mass Spectrometer detector (GCMS)
  - C3 through C13
- Data combined using the following equation:
  - If x, y & z are fractions of H2-C3, C4-C5 and C6 – C13
  - Then the ratios of the integrated areas
    - a = y/x and b = y/z
    - a → RGA
    - b → GCMS
  - Since bz = az we can solve for x, y, z
  - **z = 1/((b/a)+b+1)** **y = bz** **x = y/a**
- Response factors for individual compounds vary but cancel out when looking at differences

REDLEAF RESOURCES, INC.

Future Work

- Build a taller retort reactor vessel that will allow large differential temperatures to develop

REDLEAF RESOURCES, INC.

Some or all of the information contained in the presentation represents views, estimates and assumptions of Red Leaf Resources, Inc., JWBA, Inc., Hatch Ltd. and Total E&P USA Oil Shale, LLC (together, the "Companies"), any of which may change without notice. Neither the Companies nor any of their respective shareholders, affiliates, partners, members, directors, officers, management, employees or representatives makes any representation or warranty, express or implied, as to the accuracy or completeness of any of the information or any other information (whether communicated in written or oral form) transmitted or made available to you.

REDLEAF
RESOURCES, INC.

Regulatory Framework for Oil Shale Development

Challenges and Opportunities

Selma Sierra, Moderator

Kent Hoffman, Deputy Director, BLM Utah

Denise Dragoo, Snell and Wilmer

John Baza, Director of the Utah Division of Oil, Gas and Mining

Map from BLM ROD, 2008

Oil Shale Deposits in the Three-State Area

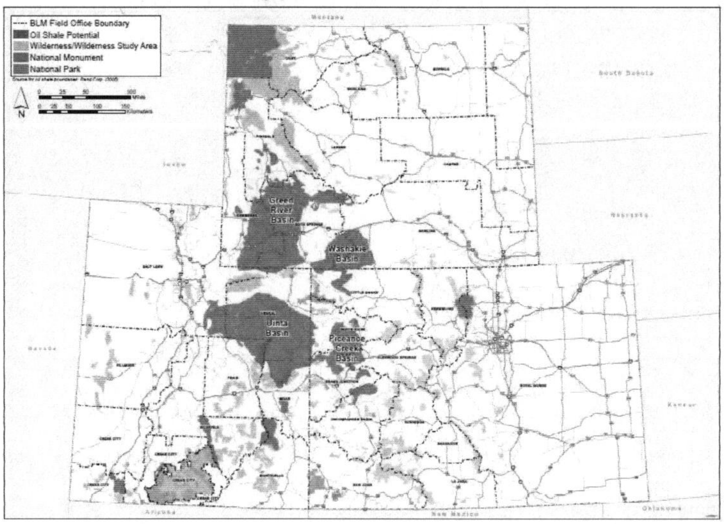

1. NOI - notify the public, Native American tribes, other federal agencies, and state and local governments about its intent to engage in land use planning for a given area. The publication of the NOI in the Federal Register formally initiates the RMP development process.
2. Analysis of Management Situation is an internal examination of the current conditions that should be considered in the development of an RMP, while a scoping report is an external examination of the current conditions. As part of the AMS, the BLM must compile and analyze an inventory of all land resources and other information to characterize the planning area and associated resources. In the scoping process, the public has an opportunity to identify issues to be addressed in the planning process.
3. Draft RMP/EIS: The RMP describes the purpose of and need for the land use plan, the affected environment, available alternatives for managing public lands within the planning area (including the preferred alternative as determined by BLM), the environmental impacts of those alternatives, and the consultation and coordination in which the BLM engaged during development of the plan. The EIS includes a detailed environmental analysis of alternatives identified in the draft RMP and accompanies the draft RMP document.
4. Proposed RMP/Final EIS: The proposed RMP and final EIS build on the draft RMP and draft EIS to include appropriate responses to public comments received on the draft RMP and draft EIS. It also corrects errors that were identified through the public comment process and internal BLM review. Issuance of the proposed RMP/EIS officially occurs with the EPA's publication of a Notice of Action (NOA) for the document in the Federal Register. Individuals and entities have 30 days from the publication of EPA's NOA to file a protest with the BLM Director. The protest period cannot be extended. The BLM must resolve any protests on a proposed RMP/final EIS before issuing a record of decision.
5. The Approved RMP and Record of Decision (ROD) serve as the final planning decision document for land managers and stakeholders. The ROD provides the rationale and basis for decisions in the approved RMP and is issued simultaneously with the RMP. An RMP is officially approved when the State Director or other authorizing official signs a ROD adopting the RMP. The ROD, RMP, and the accompanying EIS comprise several hundred printed pages.

NOTE: RMPS have taken between 5 and 7 years to complete

Resource Management Plan (RMP) Process
BLM

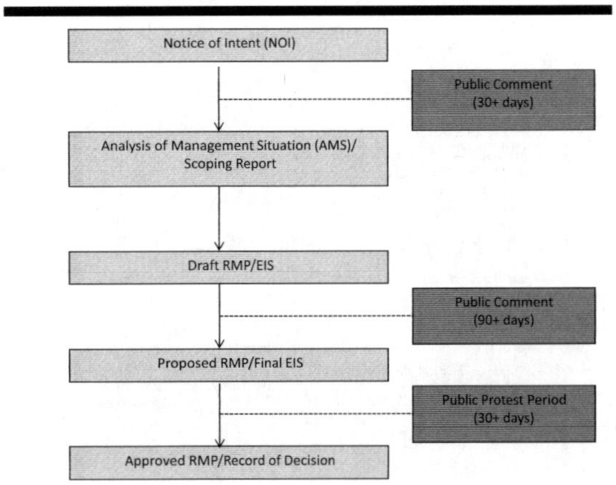

National Environmental Policy Act (NEPA)

NEPA requires federal agencies to conduct environmental analysis in order to identify environmental impacts and alternatives that range from not approving the proposal under review to allowing the proposed action to its fullest extent as proposed. This analysis is done prior to final decisions on the proposal on federal lands.

NEPA is a procedural statute, not an environmental statute. That is, it requires decision makers to consider alternatives and make informed decisions, but agencies are not required to elevate environmental concerns over other valid concerns when making land use decisions. (Park County Resource Council v. USDA 1987 and Robertson v. Methow Valley Citizens Council, 1989)

Programmatic and Site Specific Analysis

A programmatic NEPA analysis is used when evaluating land use planning and new policies. A programmatic analysis addresses a broad geographic area and focuses on the potential cumulative effect that the proposed actions may have on the environment. At the programmatic level, multiple objectives and alternatives are evaluated. Mitigation measures and stipulations to address environmental impacts are generic at this stage. Examples of previously conducted programmatic analysis include the Greater Sage Grouse and oil shale and oil sands.

In contrast to a programmatic analysis, a site-specific analysis is conducted when a specific action is proposed on federal land. For oil and gas, this may be drilling a well or constructing a production facility. Site-specific analyses explore different ways of achieving the specific objective. The geographic area of the analysis is limited, and potential environmental impacts are examined for that specific area. Any mitigation measures are specific and included in the proposal of the action.

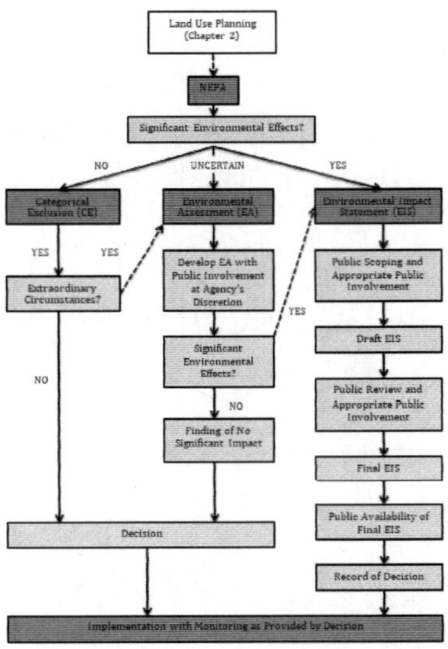

Then and Now

BLM 2008 ROD

- "Designates 1,991,222 acres of land within the most geologically prospective oil shale area as available for application for leasing for commercial oil shale development in accordance with applicable Federal, state, and local regulations and BLM policies;
- Removes the administrative barrier to BLM considering any application to lease oil shale;
- Allows only the use of surface mining technologies in areas in Utah and Wyoming where the overburden is 0 to 500 ft thick;
- Modifies existing decisions related to oil shale leasing in the White River RMP;
- Requires additional NEPA analysis of the environmental, social, and economic effects of reasonably foreseeable development before the issuance of leases for commercial development;
- Requires additional NEPA analysis of the site-specific environmental, social, and economic effects of particular development proposals to consider site-specific and project-specific factors before the approval of project-specific development plans;..."

BLM 2013 ROD

- "Designate 678,700 acres of land within the most geologically prospective oil shale area as available for application for leasing for commercial oil shale development in accordance with applicable Federal, State, and local regulations and BLM policies, including all oil shale resources on split estate lands (Federal minerals, Tribal surface) within the Hill.
- Allow only the use of surface mining technologies in areas in Utah and Wyoming where the overburden is 0 to 500 ft thick;
- Require additional NEPA analysis of the environmental, social, and economic effects of reasonably foreseeable development before the issuance of leases for commercial development;
- Require additional NEPA analysis of the site-specific environmental, social, and economic effects of particular development proposals to consider site-specific and project-specific factors before the approval of project-specific development plans;..."

Discussion Topics
Regulatory Framework for Oil Shale, Pros and Cons
And Modifications for improvement

Snell & Wilmer

OIL SHALE PERMITTING AND DEVELOPMENT ON STATE AND PRIVATE LANDS IN UTAH

Denise Dragoo, Snell & Wilmer, LLP

35th Oil Shale Symposium, Salt Lake City, Utah

October 6, 2015

INTRODUCTION

Constraints imposed by the U.S. Department of the Interior on commercial leasing of oil shale resources on federal lands has pushed development to state and private lands in Utah.

While there is no commercial oil shale leasing on federal lands, at least one oil shale development, Red Leaf Resources, Inc, has received permits for and has initiated activities under Early Production Stage operations, leading to commercial development.

This presentation outlines the steps Utah developers have taken to obtain, defend and maintain permit for their operations in Uintah County, Utah.

1

ACQUISITION OF STATE LEASE LANDS

- The Utah School and Institutional Trust Lands (SITLA) holds oil shale reserves on scattered state lands (4 sections within each Township acquired at statehood) as well as on lands strategically blocked for development through in lieu land selections and state land exchanges.

2

ACQUISITION OF STATE LEASE LANDS(cont.)

- Lands may be acquired for lease on either a competitive basis or a non-competitive, over-the-counter filing. R.850-22-300.
- Lands identified by SITLA with development potential are listed by the State and may be acquired by competitive bid.
- Lands may be proposed for lease by the potential applicant on a non-competitive basis.

3

ACQUISITION OF STATE LEASE LANDS (cont.)

- Leases have a primary term of 10 years, can be held indefinitely held by production or payment of minimum royalty. The lessee pays an annual rental of $1.00 per acre and for the primary term, production royalty of 8% of gross value. After 10 years, the royalty may increase on a sliding scale to 12.5% of gross value, maximum. R 850-22-500.

ACQUISITION OF STATE LEASE LANDS (cont.)

- SITLA relies on the Utah Division of Oil, Gas and Mining (DOGM) to approve the lessee's operation and reclamation plan under the Utah Mined Land Reclamation Act. R.850 -22-700.
- The State does not have a formal NEPA process such as that applicable to federal lands.
- Cultural, paleontological and biological surveys of the lease area required by SITLA are part of the baseline data required for a DOGM mine permit. Id.
- A reclamation bond to address surface damage on the lease, removal of facilities and the cost of reclamation is posted with DOGM. R.850-22-800.

MINE PERMIT

- The DOGM permits and regulates exploration, small mining operations (SMO) and large mining operations on state and private lands. Utah Code 40-8-1 et seq; R 647- 1-1 et seq.
- Large mining operations (LMO) are those exceeding 5 acres of surface disturbance in incorporated areas and 10 acres in unincorporated areas.
- Prior to mining, DOGM must approve a notice of intention to commence mining (NOI) . The NOI application follows the form provided on DOGM's web site.

6

MINE PERMIT (cont.)

- A description and schedule of mine operations, facilities and surface disturbance is required.
- Reclamation activities, including regrading, topsoil coverage and reseeding are set forth in an LMO plan and the operator must post a bond to cover the cost of reclamation.

7

MINE PERMIT (cont.)

- The LMO includes opportunities for public review and comment. Once tentatively approved, the NOI is subject to public review and comment and an informal conference may be requested. Utah Code 40-8-13(6).
- The finally approved NOI may be timely appealed to the Board of Oil, Gas and Mining.
- An NOI is valid for the life of the mining operation. Utah Code 40-8-16.

8

Utah Permits – Indian Country

Uintah and Ouray Reservation

- Portions of SITLA lands located within Duschesne and Uintah Counties are located within Indian Country. 18 U.S. Section 1151
- The boundaries of Indian Country have been contested by the State of Utah and these counties.
- On June 16, 2015, the federal Tenth Circuit Court issued an opinion confirming that certain lands outside the current Uintah and Ouray Reservation of the Ute Tribe are nevertheless "Indian Country" owing to their status as former tribal lands.

9

Utah Permits – Indian Country(cont.)

Federal EPA Region 8 Indian Policy

- Under this policy, unless the Tribe seeks primary regulatory authority, the federal EPA retains responsibility for federal environmental programs within Indian Country.
- The State Department of Environmental Quality retains jurisdiction within Indian Country for those programs which are exclusive to the State.
- The Utah Groundwater Quality Protection Program is unique to the State of Utah and has no federal counterpart.

10

Utah Permits

Groundwater Discharge Permits

- The Division of Water Quality (DWQ) issues groundwater discharge permits "by rule" and individual permits. DWQ also issues construction permits regarding the final design and construction of facilities.
- The DOGM has deferred to the DWQ in determining whether mine permits (NOIs) require groundwater discharge permits, and if so, the terms and conditions of those permits.
- Non-governmental Organizations (NGOs) have challenged DWQ's issuance of permits "by rule" as well as individual permits and associated construction permits.

11

Utah Permits

New Permit Appeal Procedures

- In 2012, as the result of legislative changes to the Utah Environmental Code the ultimate authority to approve permits was vested in the Executive Director.
- Authority for the review of permit appeals was moved from the Boards to administrative law judges.
- This appeal is an "on the record" review of the permit application, public comment and response and the final administrative decision.

Utah Permits

2015 Utah Legislature: SB 282 and SB 173

- DEQ's permit appeal process was further refined in bills which became effective on May 12, 2015.
- Deadlines for appeal proceedings have tightened, page limits are set for briefing and timelines are provided for decisions by the ALJs.
- The standard of review requires the petitioner to demonstrate that a finding of fact or conclusion of law supporting the permit decision was "clearly erroneous."
- New rules to implement these changes were proposed on October 1, 2015. Comments are due by November 5, 2015.

UTAH
DNR
OIL, GAS & MINING
Oil Shale Permitting
John Baza, Director
Utah Division of
Oil, Gas and Mining
Oil Shale Symposium
October 6, 2015

Utah Department of
Natural Resources
UTAH
DNR
NATURAL RESOURCES
Forestry, Fire and Sovereign Lands
Geological Survey
Oil, Gas and Mining
Michael R. Styler
Executive Director
Parks and Recreation
Water Resources
Water Rights
Wildlife Resources

Utah Division of Oil, Gas, and Mining

Oil and Gas

Coal

Administration

Minerals

Abandoned Mine Reclamation

Board of Oil, Gas and Mining

Ruland J. Gill, Chair

Richard K. Borden
Michael R. Brown
Susan S. Davis
Chris D. Hansen
Carl F. Kendell
Gordon L. Moon

Utah Oil and Gas Conservation Act

- Established in 1955
- Utah Code §40-6-2, Definitions (mod. At various times)
- "Oil" means crude oil, condensate, or any mixture of them.
- "Gas" means natural gas, natural gas liquids, other gas, or any mixture of them.
- **"Oil" or "Gas" does not include any gaseous or liquid substance processed from coal, oil shale, or tar sands.**

Utah Mined Land Reclamation Act

- Established in 1975
- Utah Code §40-8-4, Definitions
- "Mining operation" means activities conducted on the surface of the land for the exploration for, development of, or extraction of a mineral deposit, including, but not limited to, surface mining and the surface effects of underground and in situ mining, on-site transportation, concentrating, milling, evaporation, and other primary processing.
- **"Mining operation" does not include: . . . (ii) the extraction of oil and gas as defined in Title 40, Chapter 6, Board and Division of Oil, Gas and Mining.**

Minerals Regulatory Permits

- Exploration – 1 to 2 year activity
  - Drill roads, drill holes & surface sampling
  - Most applications are 6-7 pages in length
  - Two maps are required
  - Bond – standardized unit cost of disturbance
- Small Mines – less than 10 acres (in unincorporated areas)
  - Mining, hauling, and processing
  - Applications are 6-7 pages in length
  - Minimum of two maps are required
  - Bond – standardized unit cost of disturbance
- Large mines – greater than 10 acres
  - Mining, hauling, and processing
  - Application is 50+ pages
  - Bond – detailed calculation based upon specific operations and disturbance

Minerals (cont.)

- Annual permit fees
- Operation and reclamation plans must be maintained and approval sought for modification
- Coordination with other government agencies
  - Land management – BLM, USFS, SITLA, private
  - Other state agencies – DEQ, SHPO

Current Permit Activity

- Active Red Leaf large mine permit in southern Uintah County – Southwest #1
- Active nearby Red Leaf small mine permit – Sweetwater Test Mine
- Proposed TOMCO Energy large mine NOI, also in southern Uintah County – Holliday Block
- No in-situ development has been proposed

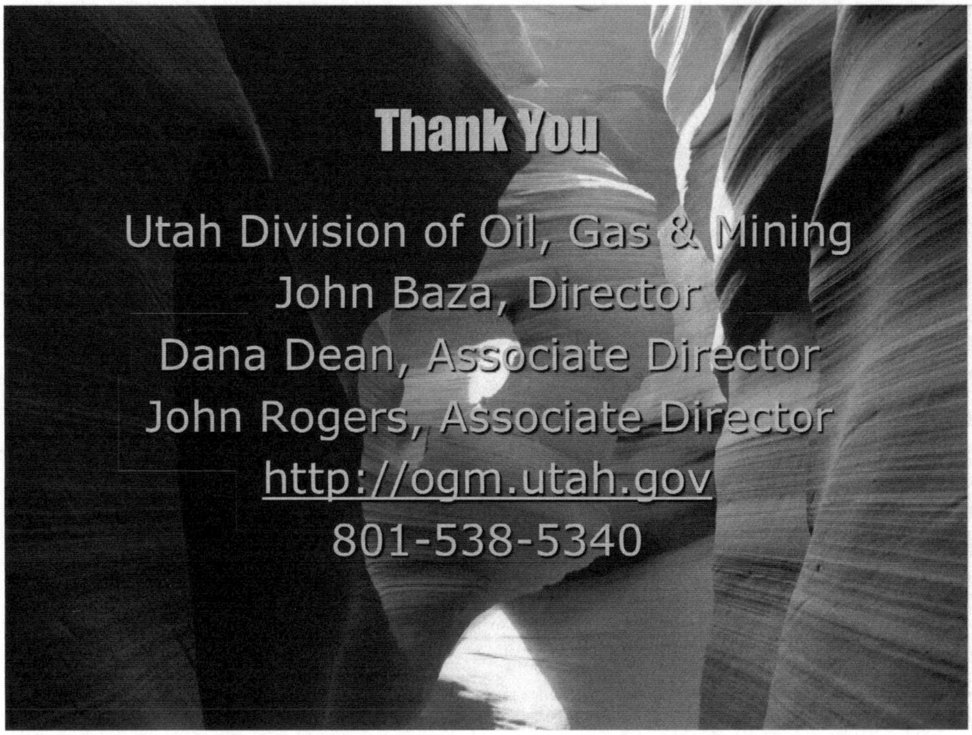
Thank You
Utah Division of Oil, Gas & Mining
John Baza, Director
Dana Dean, Associate Director
John Rogers, Associate Director
http://ogm.utah.gov
801-538-5340

35th Oil Shale Symposium
Salt Lake City, Utah October 6, 2015

Oil Shale Leasing and Development on Public Lands Administered by the Bureau of Land Management

The Regulatory Roadmap:
Policies and Procedures Facing Utah Oil Shale Development

Kent Hoffman
Deputy State Director
Lands and Minerals

Selected Relevant Statutes

- ✓ **Mineral Leasing Act**
- ✓ **National Environmental Policy Act**
- ✓ **Federal Land Policy and Management Act**
- ✓ **Endangered Species Act**
- ✓ **National Historic Preservation Act**
- ✓ **Clean Water Act**
- ✓ **Clean Air Act, et al.**

Regulations and Policies

- ✓ Title 43 Code of Federal Regulations Part 3900 – Oil Shale Management
- ✓ Resource Management Plan (RMP)
- ✓ 2012 Oil Shale and Tar Sands Final Programmatic EIS
- ✓ Research, Development and Demonstration (RD&D) Lease(s)

Some Key Aspects of 43 CFR 3900

- Lands Subject to leasing
- Conformance with land use plan (RMP)
- Compliance with NEPA & other pertinent law
- Who may hold a lease
- Bidding and sale procedures
- RD&D conversion to commercial lease
- Lease size and duration
- Diligent development, rentals & royalties
- Bonding and long-term water treatment trust fund

Some Key Aspects of the 2102 Oil Shale & Tar Sand Programmatic EIS

- **Identifies specific areas available for leasing**
  - 678,700 acres in "most geologically prospective" areas
  - Amended Resource Management Plans
- **Requires surface mining only in areas with less than 500' overburden** (Utah and Wyoming)
- **Requires site specific NEPA analysis**
- **Leasing constraints**
  - "...areas allocated as open for future oil shale leasing, are at this time, only to [RD&D] leases."

Research, Development and Demonstration Leasing

- "First Round" RD&D leases = 160 acres
  - ✓ Preference right lease area (PRLA) = 4960 acres
- "Second Round" RD&D leases = 160 acres
  - ✓ PRLA = 480 acres
- Conversion from RD&D to Commercial (PRL)
  - ✓ Documentation of commercial production from the lease
  - ✓ Bid payment of "fair market value" (>$1000/ac)
  - ✓ Consultation with state and local officials regarding impacts
  - ✓ Post adequate bond

Thank You

35th
Oil Shale
Symposium

2012 USGS in Place Oil Shale Reserves

- 1.52 Trillion Barrels Piceance Basin, Colorado
- 1.44 Trillion Barrels Green River Basin, Wy.
- 1.32 Trillion Barrels Uintah Basin, Utah
- 4.28 Trillion Total Barrels in Green River Formation

Draft OSTS PEIS 1-8

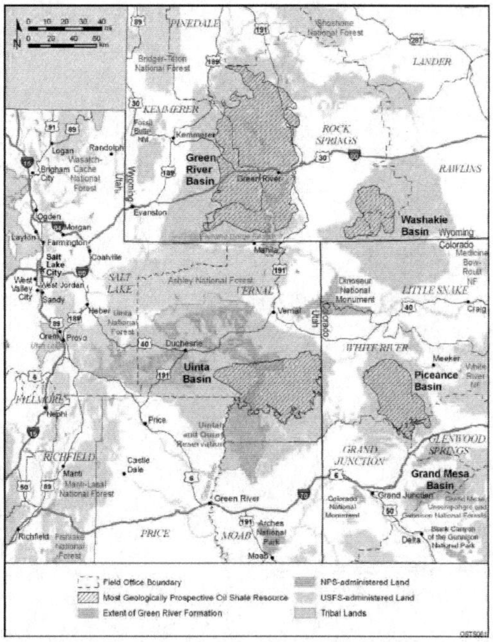

FIGURE 1.2-1 Most Geologically Prospective Oil Shale Resources within the Green River Formation Basins in Colorado, Utah, and Wyoming

Draft OSTS PEIS 2-22

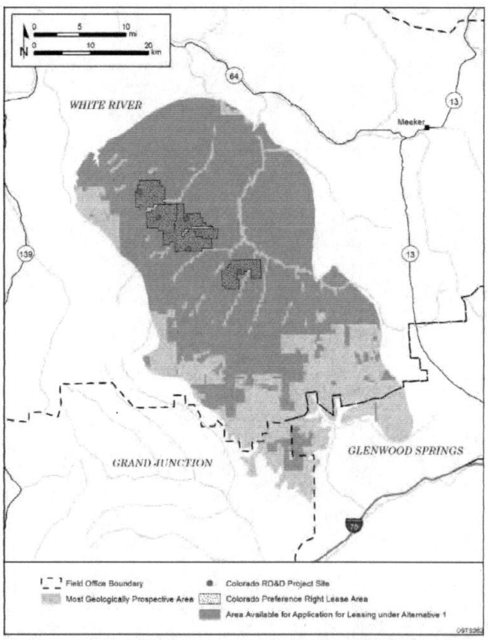

FIGURE 2.3.2-1 Lands Available for Application for Leasing under Alternative 1 in Colorado

Draft OSTS PEIS 2-40

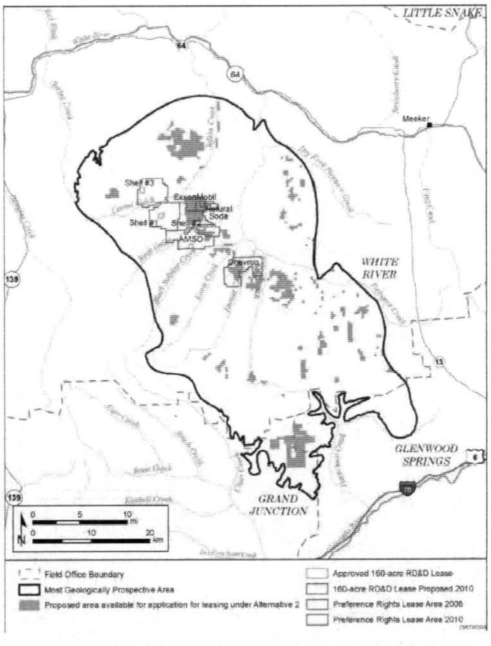

FIGURE 2.3.3-4 Lands Available for Application for Oil Shale Leasing under Alternative 2 in Colorado

Draft OSTS PEIS 2-24

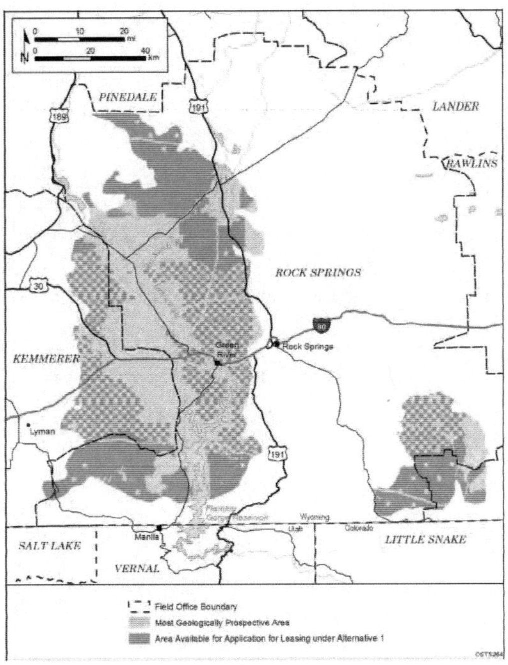

FIGURE 2.3.2-3 Lands Available for Application for Leasing under Alternative 1 in Wyoming

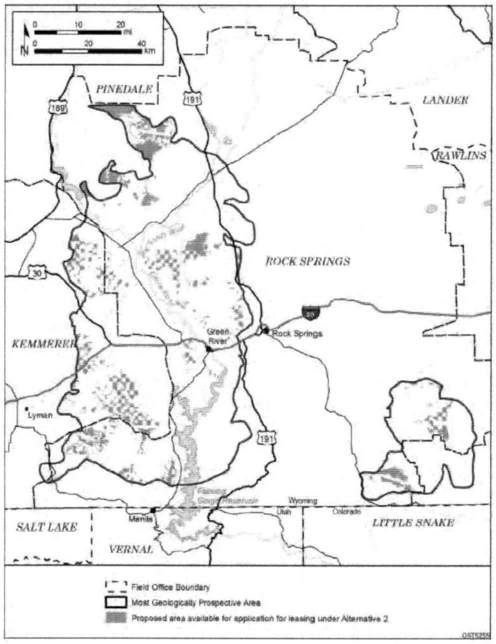

FIGURE 2.3.3-6 Lands Available for Application for Oil Shale Leasing under Alternative 2 in Wyoming

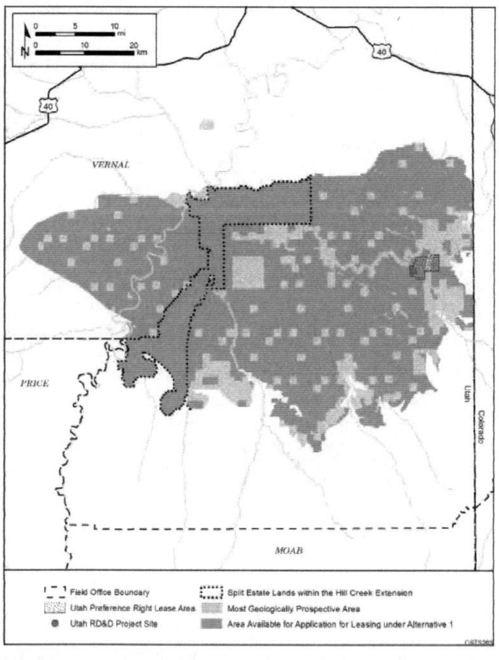

FIGURE 2.3.2-2 Lands Available for Application for Leasing under Alternative 1 in Utah

Draft OSTS PEIS 2-41

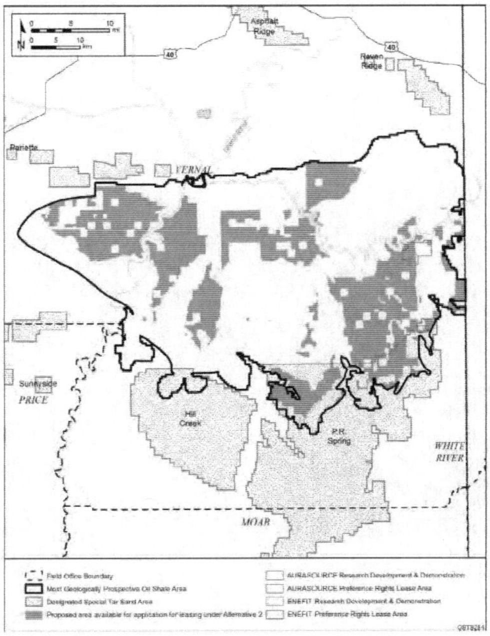

FIGURE 2.3.3-5 Lands Available for Application for Oil Shale Leasing under Alternative 2 in Utah

Penstemon Agreement Plan

- Prevented listing of two plant species
- Provides regulatory certainty for the next 15 years
- Allows for development of natural resources
- Protects current uses on private property
- The agreement will provide protection and mitigation on non-federal lands that would not occur under ESA
- Approximately 35 % reduction in total affected land
- Will protect 4 USFWS-defined White River Beardtongue core areas on non-federal lands

White River Beardtongue in shale habitat

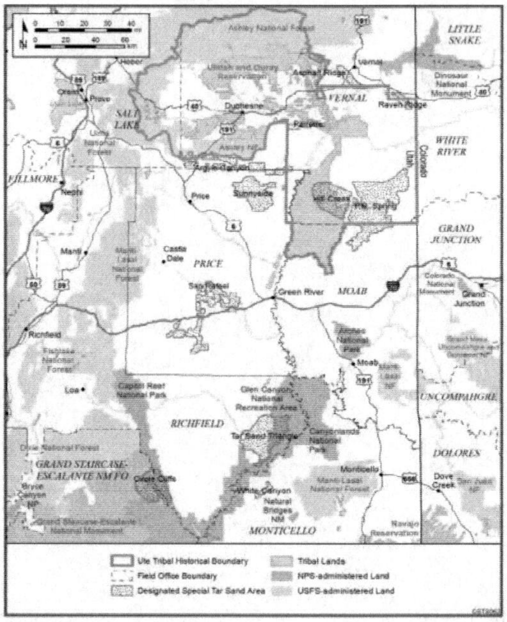

Figure 14: Special Tar Sand Areas in Utah

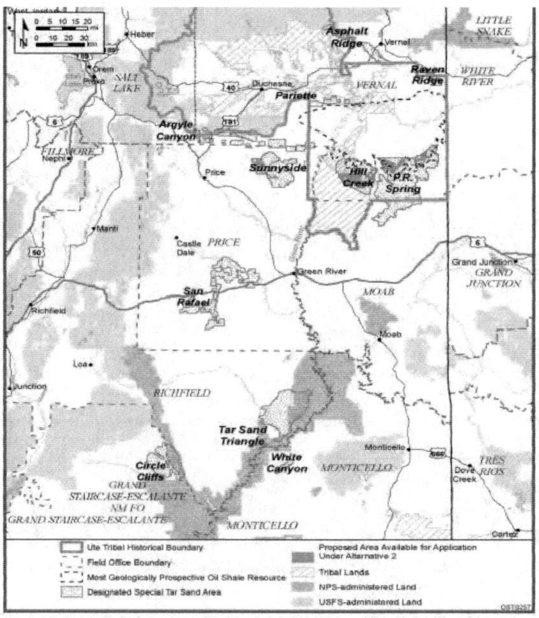

Figure 17: Lands Available for Application for Tar Sands Leasing within the STSAs in Utah

Ozone Non-attainment Areas

Major Source Thresholds:

- Marginal or Moderate > 100 tons per year
- Serious > 50 tons per year
- Severe > 25 tons per year
- Extreme > 10 tons per year

no, we do not have oil here...

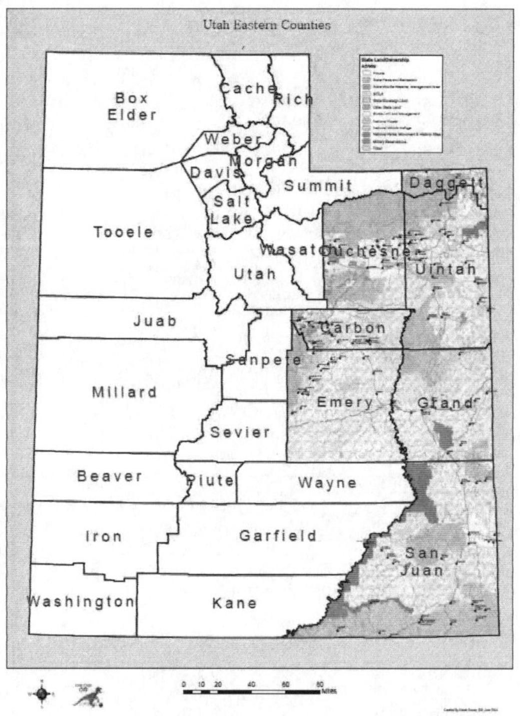
Utah Eastern Counties
Box Elder
Cache
Rich
Weber
Morgan
Davis
Summit
Daggett
Salt Lake
Tooele
Wasatch
Duchesne
Uintah
Utah
Juab
Carbon
Sanpete
Millard
Emery
Grand
Sevier
Beaver
Piute
Wayne
Iron
Garfield
San Juan
Washington
Kane

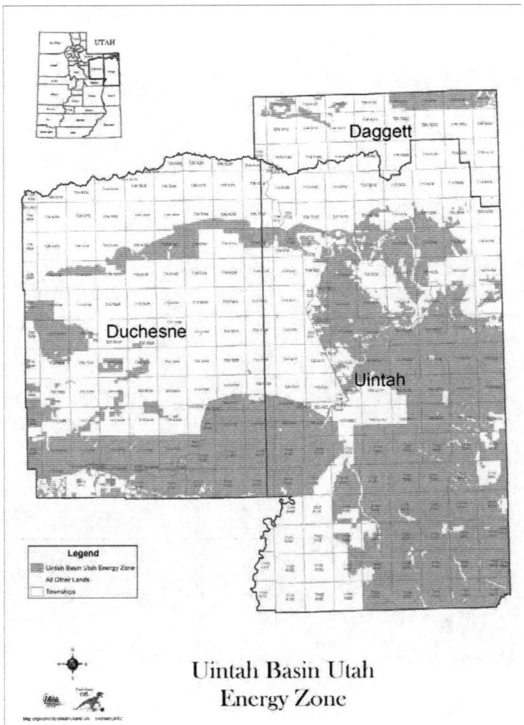
UTAH
Daggett
Duchesne
Uintah
Legend
Uintah Basin Utah Energy Zone
All Other Lands
Townships
Uintah Basin Utah
Energy Zone

Colorado School of Mines 35th Oil Shale Symposium

Tom Faddies
State of Utah
SITLA

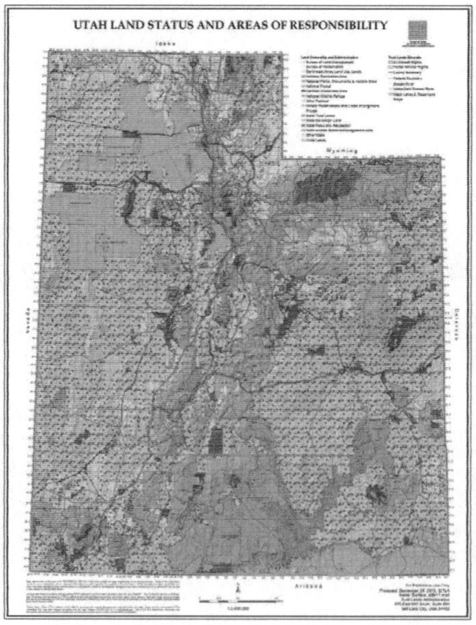

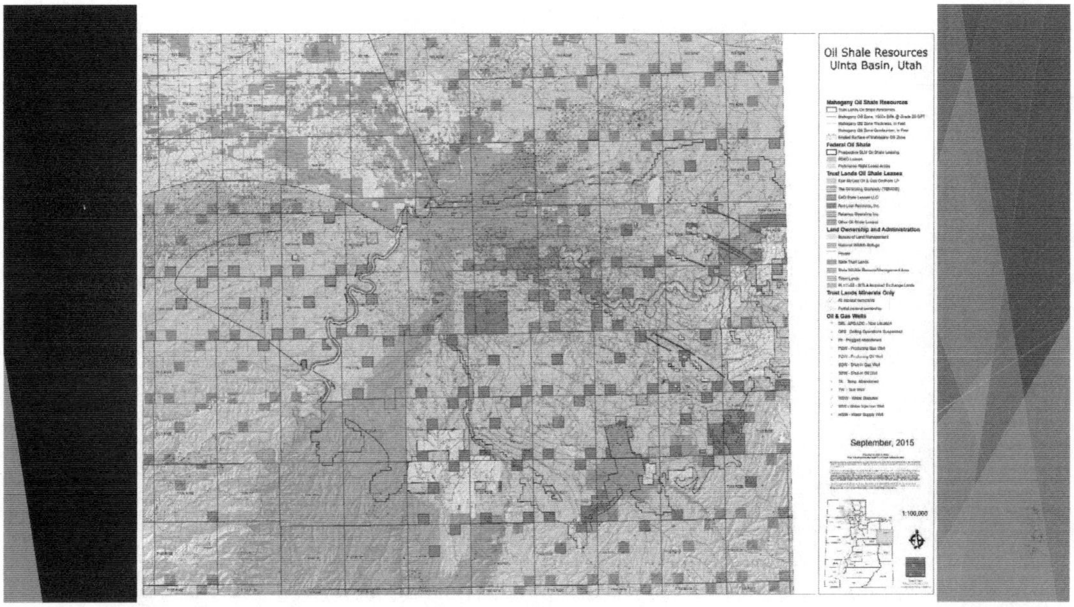
Oil Shale Resources
Uinta Basin, Utah
Mahogany Oil Shale Resources
Federal Oil Shale
Trust Lands Oil Shale Leases
Land Ownership and Administration
Trust Lands Minerals Only
Oil & Gas Wells
September, 2015
1:100,000

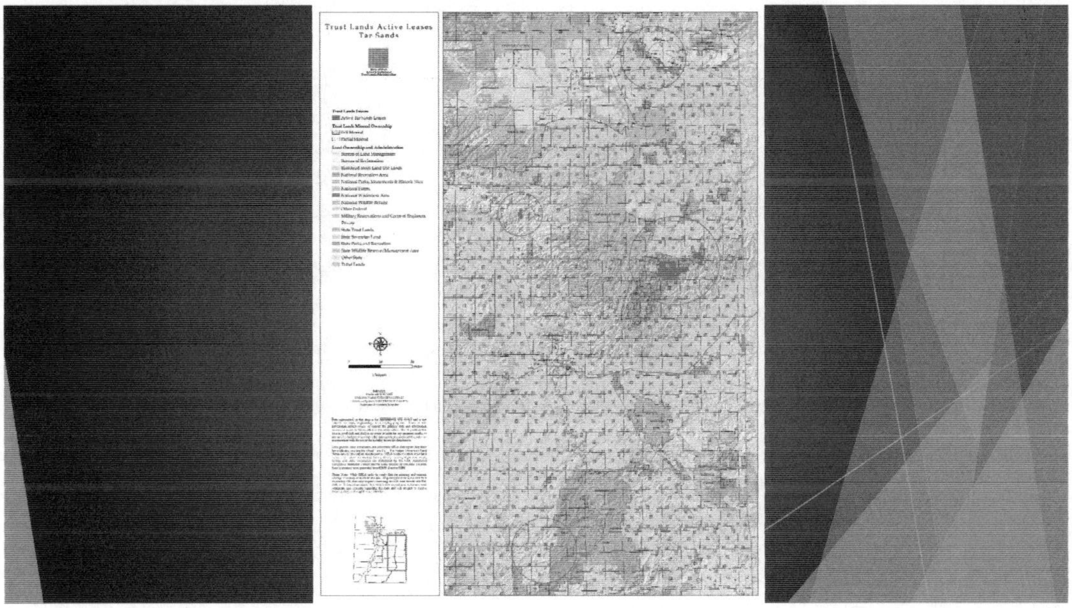
Trust Lands Active Leases
Tar Sands

35th Oil Shale Symposium

October 2015 Salt Lake City

Gladstone New Fuels Project
Update of Status and Progress

John Parsons
QER PTY LTD

QER
Creating new fuels for Australia

About QER

- A brief recap
- Who we are
- Our progress so far
- Achievements
- Next Steps

Oil Shale Tenements

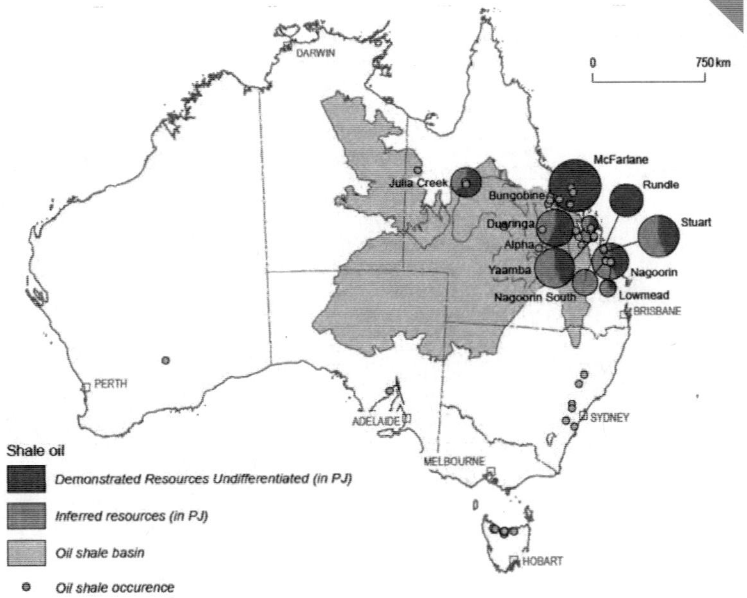

Source: Australian Energy Resource Assessment 2010 Figure 3.20 p.58

The Project Concept

(**GNF2** – Gladstone New Fuel Project - Stage 2)

- A modest scale commercial shale to liquids facility
- Yields an attractive return for investors
- Removes scale up and technical risks
- Does not compromise the remainder of the Stuart deposit
- Confirms and indeed strengthens our social license to operate

GNF2

Feed Preparation	Retort & Oil Recovery	Oil Upgrading	Gas/water treatment & utilities
✓ Simple mining operation ✓ Conventional processes ✓ Extensive equipment tests at Stage 1 with vendors	✓ Extensive optimisation during Stage 1 operations	✓ Conventional 'off the shelf' technology ✓ Tier 1 licensors	✓ Conventional 'off the shelf' technology ✓ Tier 1 vendors

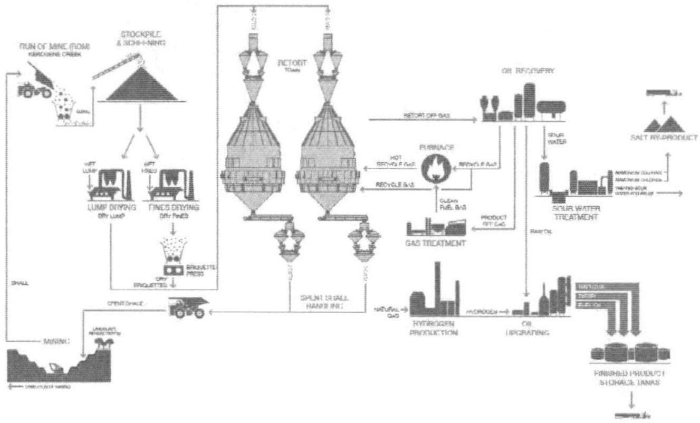

Engineering

- Two programs - Colorado pilot plant and Australian TDP

Engineering

- Continue the Capital efficiency improvements

Mining

- Evaluated several mining methods

Surface Miner

Conventional Truck & Shovel

Process Technology

- Vertical shaft retort remains attractive
- Have adopted a split-feed oil upgrading process (hydrotreater/hydrocracker) for maximum diesel to suit local Australian (QLD) niche
- Evaluating Scope 1 - GHG reductions (solar thermal etc.)

Approvals and Stakeholders

- Have completed the field studies and impact assessment for the GNF2 project
- QER still enjoys high levels of community and local government (county) support

Products-Evaluation Marketing

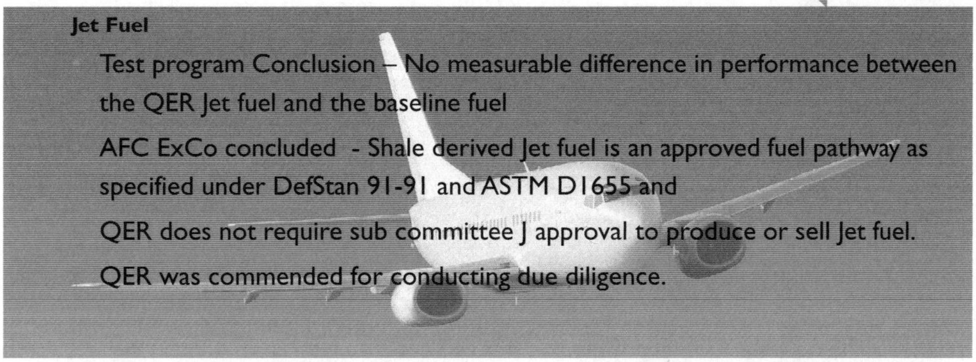

Jet Fuel

- Test program Conclusion – No measurable difference in performance between the QER Jet fuel and the baseline fuel
- AFC ExCo concluded - Shale derived Jet fuel is an approved fuel pathway as specified under DefStan 91-91 and ASTM D1655 and
- QER does not require sub committee J approval to produce or sell Jet fuel.
- QER was commended for conducting due diligence.

Elastomer Testing

High Reynolds Number Thermal Stability Tester

Mobile Emissions Laboratory

Aviation Fuels Thermal Stability Test Unit

Test Rig

Products-Evaluation Marketing

Diesel

- Conclusion - QER diesel, either neat or blended with retail diesel has comparable performance to conventional automotive diesel with regard to power, emissions and fuel consumption.

Performance and Emissions Testing of "Conventional" and Shale Derived Diesel

Investor Search
- what could our project partners look like?

- **Strategic** (where GNF2 is an entrée to a greater prize)
- **Off takers** (Objective is to secure access to reliable supplies of transport fuels at reasonable price)
- **Financial** (looking for attractive project returns with minimal risk)

Investor Search – GNF2

- Met with 35 different companies
- 11 agreed to sign NDA's and access the data room
- 5 companies set up internal deal teams

External Forces

- During 2013-14 the project was being buffeted by external forces
- Persistently high project execution costs (Gladstone LNG projects)
- Erosion of benchmark oil prices in mid 2014
- Generally depressed commodity prices - "*the end of the current mining boom*" (Australia is highly exposed in this regard)

Next Steps

- Expect – "Lower market price for longer"
- Continue to optimise the project flow sheet
- Look for ways to increase capital efficiency
- Reduce project execution times (FID to full production)
- Monitor project costs in Australia and in overseas module yards
- Continue to engage with government and other stakeholders (maintain our social license to operate)
- Monitor and engage with potential investors as market rebounds

Summary

- At around $80/bbl. the project was very attractive
- The project fundamentals are sound
- During this "lower for longer" oil price dip we have an opportunity to make both process and capital efficiency gains
- QER will be in a strong position to progress when we emerge from the current "winter"

For more information

PHONE +61 7 3222 0600
EMAIL info@qer.com.au

LEVEL 10, 200 MARY STREET
GPO BOX 5214
BRISBANE QLD 4001
AUSTRALIA
www.qer.com.au

QER
Creating new fuels for Australia

11.2 Start-Up and Operating Experience of the 3 Petroter Oil Shale Processing Plants in 2010-2015

Jaanus Purga

VKG, Estonia

Global shale oil production growth in 2000-2015 has been mainly caused by added capacity in two countries – China and Estonia. Rest of the world has been in development stage during that period of time. In China majority of the growth has been achieved in Fushun Mining Corporation, in Estonia in Viru Keemia Grupp AS. These two companies are competing for the title – World Largest Shale Oil Producer.

In 2015, Petroter technology is most advanced commercially proven ex-situ oil shale processing technology in the world. It is based on well-known solid heat carrier process, but includes modern state-of-the-art technological solutions, what are now commercially proved by operation experience from Petroter I (in operation since 2010) and Petroter II (in operation since 2014) oil shale processing plant. All changes and modifications from Petroter I and Petroter II were included into design of Petroter III plant, commissioned in summer 2015.

This paper presents real operational data from Petroter I and Petroter II plant and latest information about construction and start-up of world newest oil shale processing plant – Petroter III. Evaluation of the economics of shale oil production anno 2015 will be provided as well as producer opinion of the perspectives of new "green field" oil shale projects in current economic situation. Finally – latest shale oil selling prices statistics and comparison with conventional oil products will provided.

New Developments in Chinese Oil Shale Activities

Yue Ma, Shuyuan Li

China University of Petroleum-Beijing

OUTLINE

1. Oil shale distribution in China
2. Shale oil production in China
3. New Technology
4. New Research work

1. Oil shale distribution in china

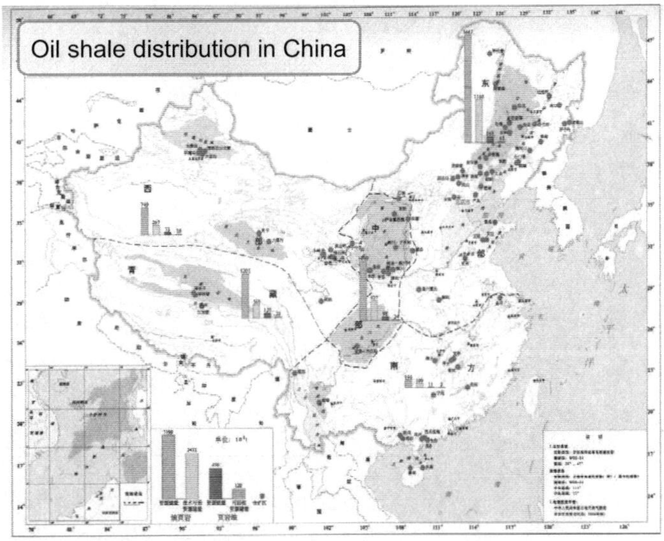

➢In China, The oil shale resources is about 720 billion tons, equivalent to 47 billion tons shale oil.

➢The basins containing oil shale resources can be classified into four categories, according to the quantities and utilization of the resources:

Class I: Have a small resource, and have been mined and processed.

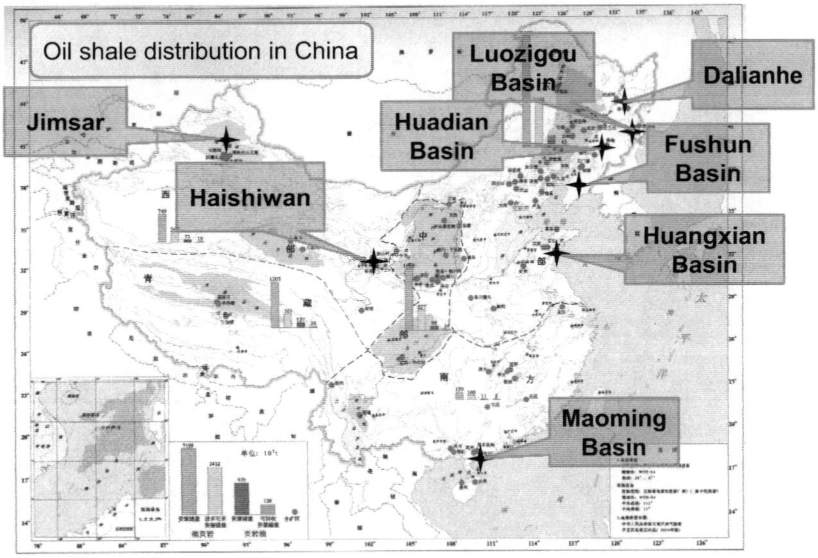

Class II: Have a large resource, yet have not been mined and processed.

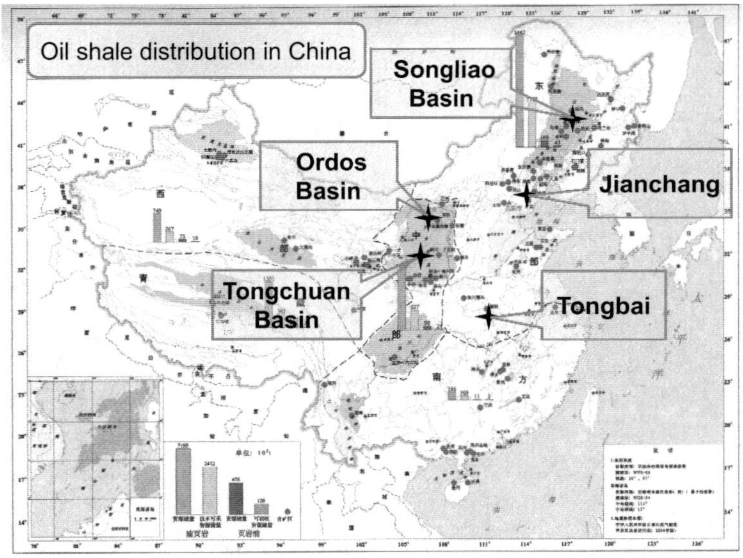

Class III: Have a small resource, and have not been mined and processed.

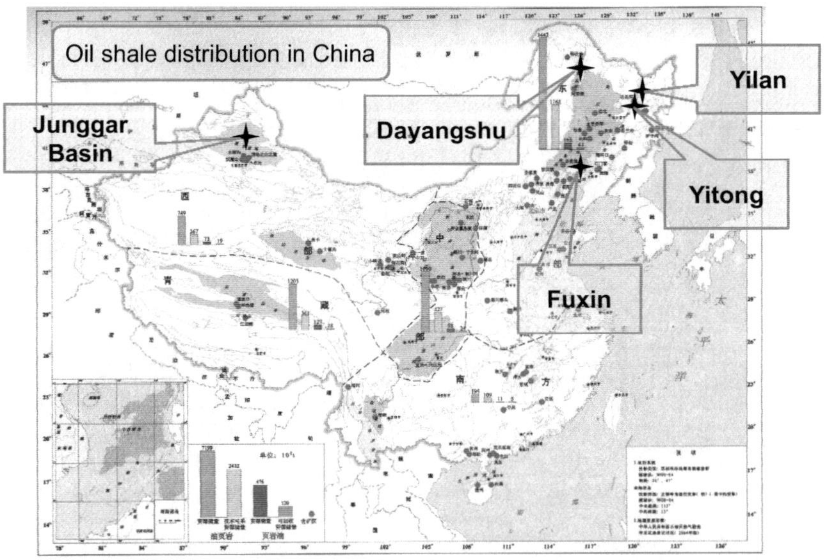

Class IV: Have not been investigated.

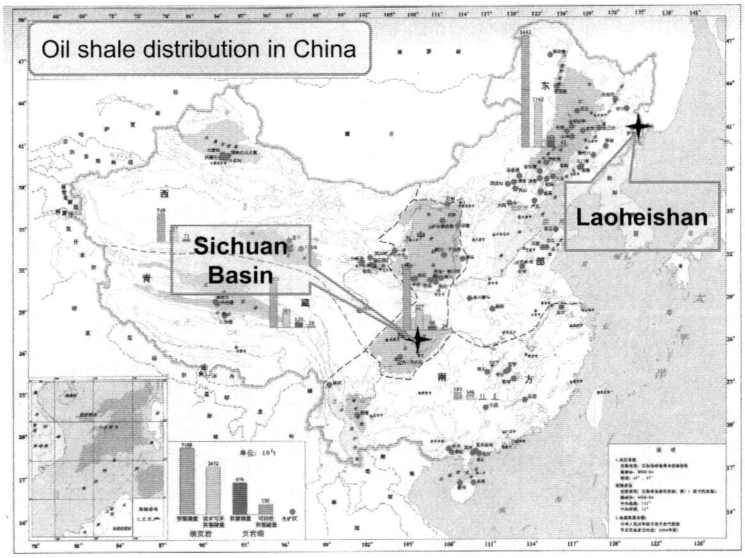

Oil shale resources in four categories of basin

	Class I	Class II	Class III	Class IV
Resource, $\times 10^9$ t	22	492	122	82
Percentage, %	3.1	68.4	17.1	11.4

2.Shale oil production in China

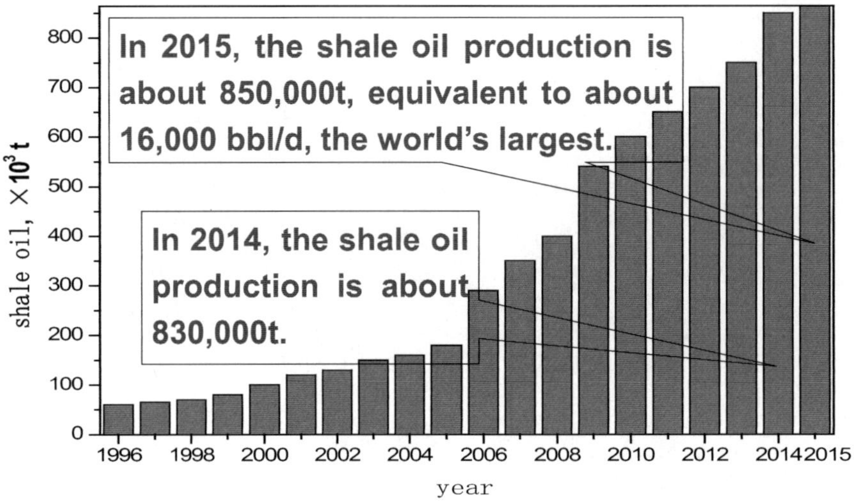

Shale oil plant distribution in China

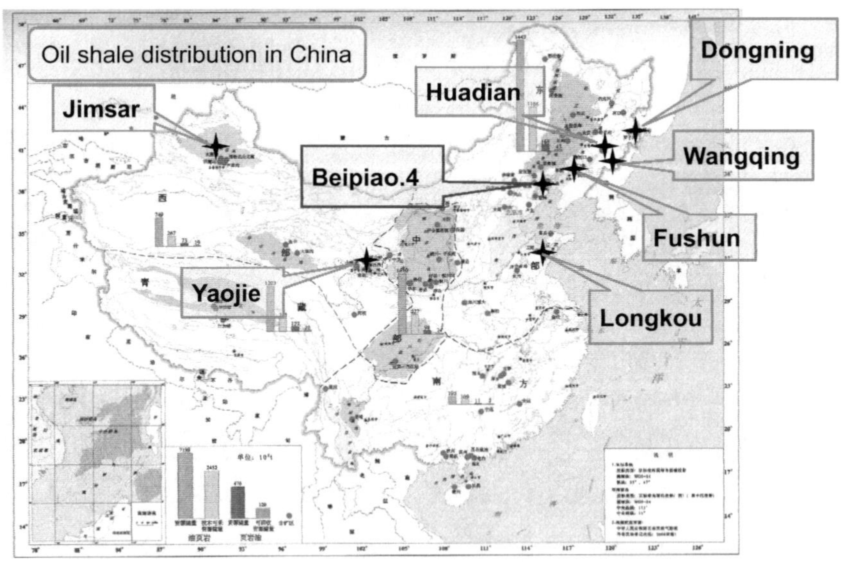

In China, there are about 11 oil shale retorting enterprises in operation.

Oil content and shale oil production in China

	Fushun	Longkou	Beipiao	Wangqing	Huadian	Barkol	Jimsar	Yaojie
Oil content %	6.5	15	4.5	6	10	7	7	7
Shale oil production	40	12	15	5	4	4	7	5

1. Fushun, Liaoning, Northeast China

- Fushun Bureau of Mines produced the largest amount of shale oil in China, about 400,000 tons in 2015.

- 220 Fushun type retorts.(100t/d)
  - The total oil shale consumed is about 10 million tons each year.

Oil content： 6%

Oil yield： 70%

1. Fushun, Liaoning, Northeast China

- The imported ATP unit operated for 40 days starting in July, 2014. From May 5,2015, ATP unit has been put into continuous operation for 90 days. About 10,048 tons of shale oil were obtained.

2. Longkou, Shandong, East China

- In Longkou Coal Mining Group Company (LCMG), there is an oil shale retorting plant with a capacity of 120,000 tons shale oil each year.

- 40 Fushun retorts divided in two blocks on average, And was put into operation in 2009 and 2011, respectively.

2. Longkou, Shandong, East China

- In 2014, LCMG Company bought Huadian Fengtai Co., whose oil shale resources are more than 40 million tons, and they have been processed by 20 Fushun retorts.
- Oil content：7%

2. Longkou, Shandong, East China

Shale Oil Price:

2400￥/t ＝（375$/t）

plans

- import Galoter technology.

3. Huadian, Jilin, Northeast China

- In Huadian, a power station has been operating for 14 years.
- There are three circulating fluidized-bed boilers(CFB) with total capacity of 195t/h steam for combusting oil shale.

- Liaoningchengda Co.,Ltd stopped producing
- 40,000 tons shale oil

4. Wangqing, Jilin, Northeast China

- This company constructed an oil shale power plant in 2009.
- There are 2 CFB boilers with the steam generation capacity of 35 t/h (6000 kw of electricity).

- Longteng Energy Company has built 40 Fushun retorts to produce about 50,000 t shale oil each year.
- Particle oil shale (<10 mm) is used for electricity generation.
- Lump Oil shale (10-75mm) is used for shale oil production.

5. Yaojie, Gansu, Northwest China

- SJ squared retort
- Yaojie coal electricity Co. Ltd has constructed 8 SJ retorts to process Yaojie oil shale.
- The capacity for each retort is 400 tons (8-60mm) per day.

Planning to develop a pilot plant of Guofu retort.

6. Beipiao, Liaoning, Northeast China

The proven reserves of oil shale are 700 million tons.

Oil content: 4.5%

Shale oil yield: 65-70%

6. Beipiao, Liaoning, Northeast China

➢ In Beipiao, 192 Fushun type retorts have been put into operation by 4 companies.

Company	Beita		Like		Panjin	Shengze
Fushun retorts	48	40	20	20	24	40
Capacity, t/d	150	100	200	100	100	120

7. Jimsar, Xinjiang, Western China

- For the first phase project, 32 retorts designed by Liaoning Chengda Company have been put into operation in October, 2013. The capacity is 3.5 million tons oil shale, which produced about 40,000 tons shale oil.
- For the second phase project, another 32 retorts have been put into operation by the end of the year 2014.

8. Hami, Xinjiang, Northwest China

In Balikun, 40 Fushun type retorts(100t/d) have been put into operation by Taimu Company.

4.New technology

Daqing, Heilongjiang, Northeast China

- Daqing Oil Company (CNPC) has finished the construction of a demonstration unit with a daily capacity of 2000 tons of oil shale using a solid heat carrier retort.
- At present, cold-testing has been finished, and adjustments are being made.

Harbin, Heilongjiang, Northeast China

- China National Coal Corporation has finished the operation of a pilot fluidized-bed retort with a capacity of 50 tons of oil shale per day.
- Size: <3mm

- This equipment has been torn down.

Shale oil chemical deep processing

Scale: 400,000t/year

- Pretreament - Catalytic pyrolysis - ethylbenzene

- FCC, Catalytic reforming, chemical product, etc.

5.New research work in China

1. Analysis and evaluation have been continually done by China University of Petroleum (CUP) both from China and abroad, such as in Myanmar and Chaoyang, Liaoning Province. In addition, research on shale oil hydrogenation, fluidized retorting, and utilization of oil shale ash have also been carried out.
2. China Academy Of Sciences and CUP have applied a national basic R&D program (973 program) on oil shale pyrolysis.

Oil shale retorting technology

contact

Editors：Jialin Qian Shuyuan Li

E-mail address：syli@cup.edu.cn

Acknowledgements

The authors would like to thank the talents start scientific research funds of China University of Petroleum-Beijing (No.2462015YJRC002), the National Basic Research Program of China (973 programs, No. 2014CB744302) and Taishan Scholar Constructive Engineering Foundation of Shandong province, China (No. ts20120518) .

Outline

1. Status update of Enefit280 plant operation
2. Construction and commissioning of the largest oil shale fired power plant in the World
3. Development of oil shale fired power plant in Jordan

Enefit

1

Status update of Enefit280 plant operation

Enefit

Main characteristics of Enefit280

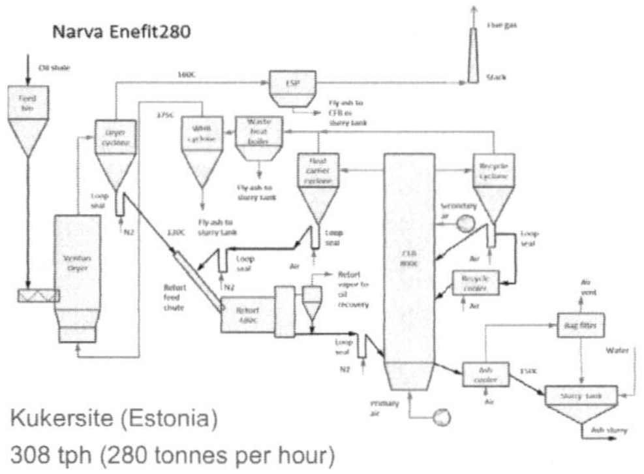

Enefit280 is combined oil, power and gas generation plant.

Design figures:

- Oil shale used: Kukersite (Estonia)
- Plant capacity: 308 tph (280 tonnes per hour)
- Annual oil shale consumption: 2,4 million tons
- Annual shale oil production: 1.7 mil. bbl or 270 th. tons
- Annual retort gas production: 2650 MMCF or 75 mil. Nm³
- Annual power production: 200 GWh
- Designed lifetime of the plant: 30 years
- Construction time: 26 months

Enefit

Enefit280 operation in numbers

Enefit280 operation in 2015:

- Total shale oil production in 2015: 650 th. bbl
- Operational capacity of the plant: 285 tpsh
- Average industrial oil yield: 30 gpt (12.5wt%)
- Average industrial retort gas yield: 1350 ft^3/t (5.5wt%)
- Thermal efficiency: >90%

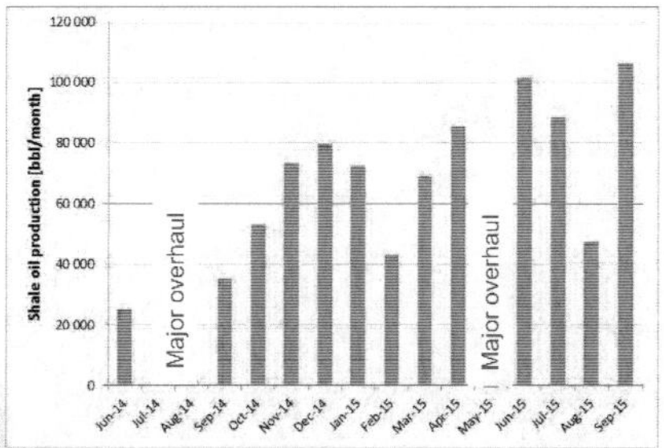

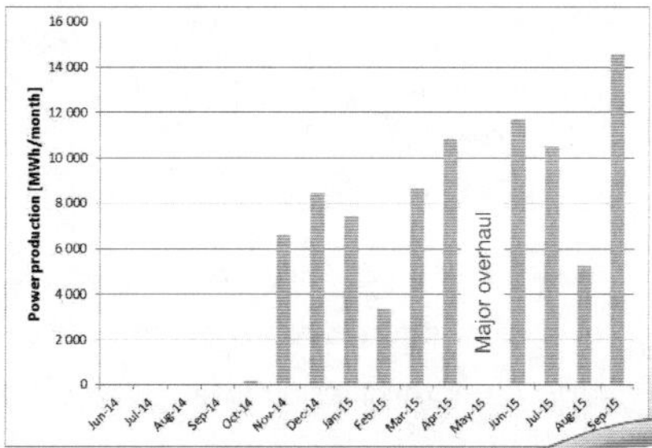

Enefit

Environmental performance

- Enefit280 plant has very low air emissions:
  - SO_2 emissions are negligible;
  - Content of incomplete combustion products (CO, TOC) in flue gases is very low;
  - Dust and NO_x emissions are below limit values.
- Enefit280 plant produces ash with very low TOC content.

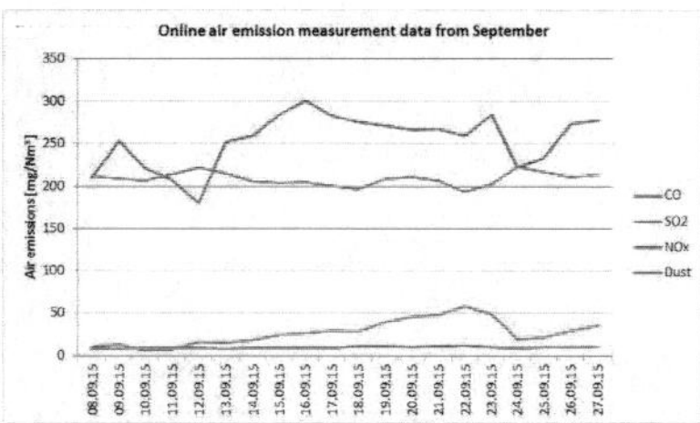

Comparison of air emissions from Enefit280 and Enefit140 plants:

	Unit	Enefit280	Enefit140
CO	mg/Nm^3	**200**	>1 000
CO_2	Vol%	12	22
SO_2	mg/Nm^3	**10**	100
NO_x	mg/Nm^3	200	150
Dust	mg/Nm^3	50	400
TOC	mg/Nm^3	**10**	>1 000
O_2	Vol%	4	0.2
H_2O	Vol%	19	36

Enefit

Progress made in operation

Comparison of operational numbers of Enefit280 and Enefit140 plants:

	Unit	Enefit280	Enefit140
Design capacity	tph	308	154
Actual capacity achieved	tph	285	150
Heating value of oil shale	MJ/kg	8.2	8.2
Industrial oil yield	gpt	**30**	28
Industrial retort gas yield	ft^3/t	**1350**	440
Pyrolysis water yield	gpt	8.4	6.3
Gross power production	MW	**25**	0
Thermal efficiency	%	**>90%**	71%

- Enefit280 should reach its design capacity in 2015 and production numbers should be close to design values in 2016.
- Enefit is planning to make a decision to build a next Enefit280 plant in 2016

Enefit

2 Construction and commissioning of the largest oil shale fired power plant in the world

Enefit

Construction and commissioning time-line

January 14, 2011	Signing the contract with Alstom
September, 2011	Civil works started
December, 2014	Mechanical completion
January, 2015	Commissioning started
April, 2015	First steam to turbine
May, 2015	First synchronization with the grid
August, 2015	Full-load achieved
August, 2015	Co-combustion of biomass tested

Enefit

Main parameters

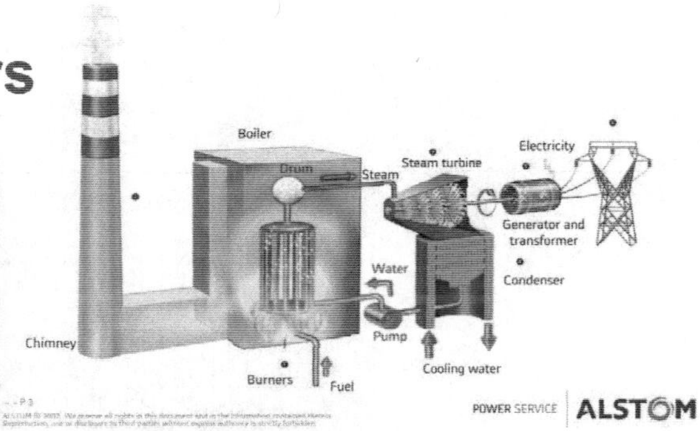

Auvere Circulating Fluidized Bed (CFB) boiler		
	Unit	Auvere PP
Design capacity	MW_{gr}	300
Main fuel		Kukersite oil shale (Estonia)
Additional fuels:		Biomass (up to 50%)
		Peat (up to 20%)
		Retort gas (up to 10%)
Steam parameters:		
- pressure	bar	172/39
- temperature	°C	540/565
- flowrate	kg/s	228/207
Net annual power production	GWh	2280

Enefit

Auvere 300 MW CFB boiler

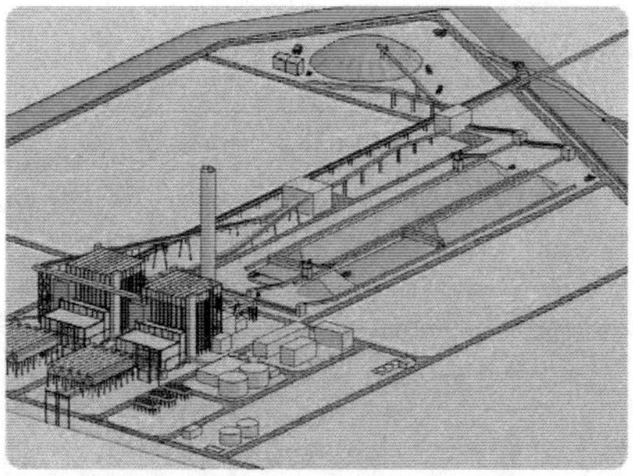

3 Development of oil shale fired power plant in Jordan

Enefit

Jordan: 553 MW Oil Shale Fired Power Plant

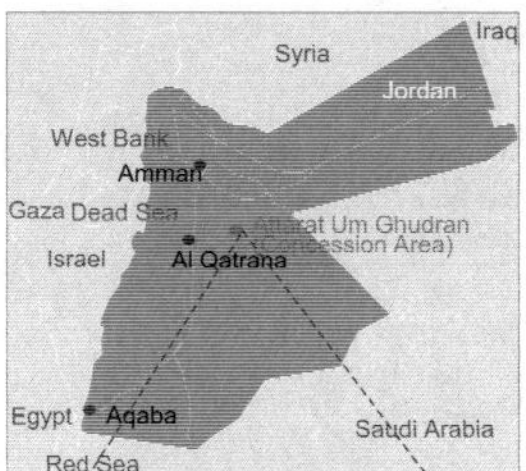

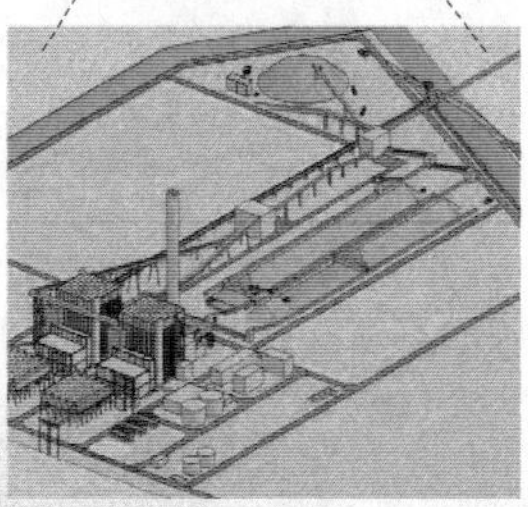

Project location: Attarat Um Ghudran, Central Jordan

Gross capacity: 554 MW (2 x 277MW CFB units)

EPC Contractor: Guangdong Power Engineering Corporation

Key equipment: Boiler: Foster Wheeler
Turbine Generator: Siemens

Mine: 10m tons oil shale per year, open cast mine
JORC compliant mineral reserve

Tariff: 30-year Power Purchase Agreement signed with the Government of Jordan

Financing: 100% financing identified
Lenders: Bank of China
Industrial Commercial Bank of China
Sinosure guarantee
Equity: YTL / Chinese Investor / Enefit

Financial Close: Expected within 2015

Enefit

Summary

- Enefit280 plant has almost reached its design capacity and significant progress has been made to increase availability of the plant;
- Enefit280 plant has the lowest environmental impact of all shale oil production plants in Estonia; *likely the world*
- The world's largest oil shale fired power plant has already reached its design capacity and is producing electricity for the Nordic Power Market. Plant will be taken into commercial operation in 2015.
- Construction of oil shale fired power plant in Jordan should start in 2016. All agreements with the Jordanian Government and construction companies have been finalized. Financing of the project should be ready in coming months.

Enefit

Thank you!

Indrek Aarna, Head of R&D

Enefit

Enefit American Oil

A Proven, Efficient and Environmentally Sound Means
to Help Utah Become Energy Independent

Enefit

Enefit's Utah Location

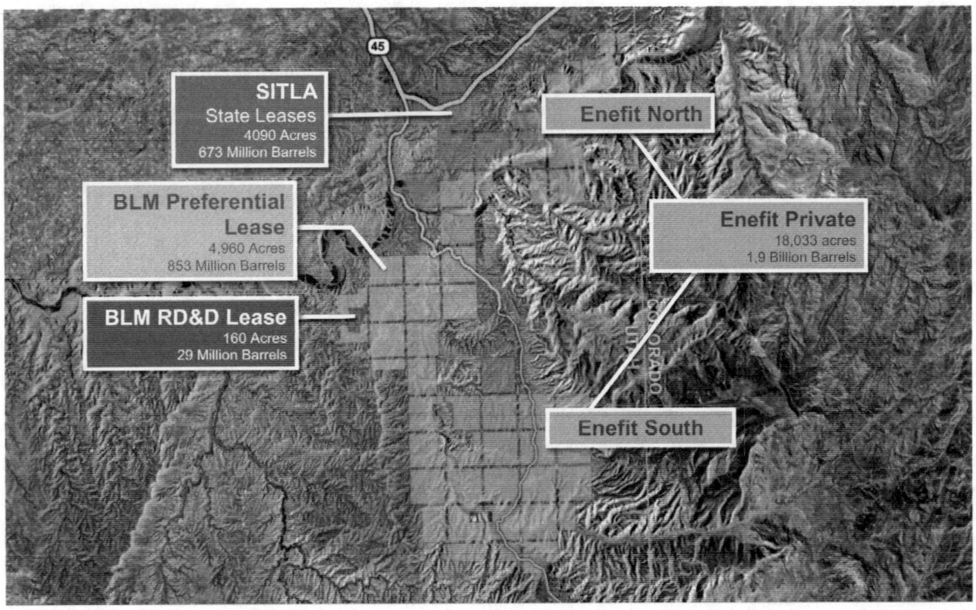

Today's Situation

- 2011 vs 2015 - Difficult global business environment
- "Perfect storm" of issues
- Opportunity to reassess business case
  - Grade: Narrower band of oil shale; more gallons per ton
  - Engineering: Improvements to Enefit process, additional revenue streams
  - Unit Size: streamlined development plan and reduction of "pre-production" investment
- Significant Accomplishments: Draft utility corridor EIS end of 2015 & Penstemon Conservation Agreement and progress

Enefit

New Business Case – What have we done?

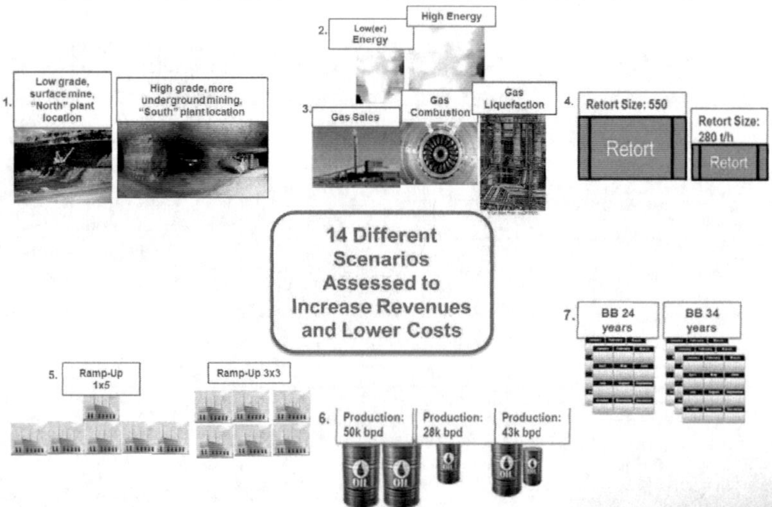

Enefit

4

Grade

- "Low" to "high" grade decreases break-even by ~30%-40%
  - Generates more revenue per mined ton
  - Decreases costs by processing less tons, ~50% less tonnage required
- Faster transition to underground mining, less surface disturbance
- Relocating plant site saves ~$100M in upfront pre-production capex

Enefit

Engineering

"High Energy" system, coupled with additional product revenue results in ~12% reduction in breakeven

Low Energy – 63 MW net el

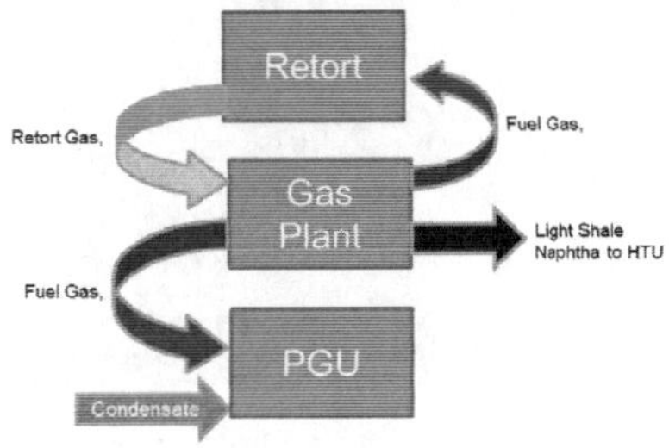

High Energy – 38 MW net el
303 MW Liquefied Products

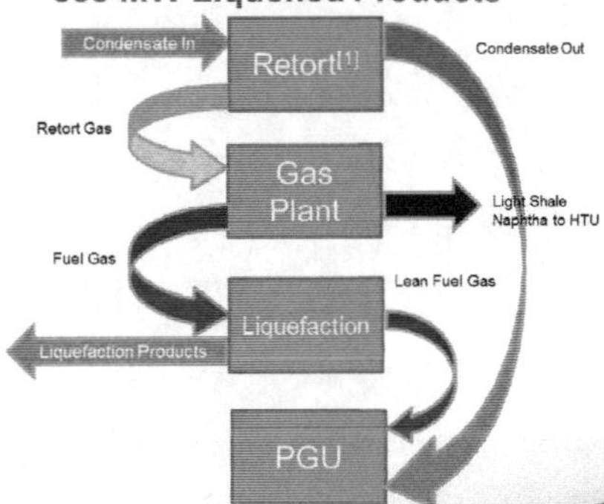

Enefit

Unit Size

Reducing the unit size from 550 t/h to 280 t/h reduces the Pre-production investment cost by more than half and the breakeven of the entire project life (when including development costs) by ~10%

- Reducing the unit size does cause loss of economies of scale
- Smaller unit size shortens development period/time to production– can move from existing testing facilities straight to commercial size unit
- Reduces pre-production investment needed for proof of scale-up facility

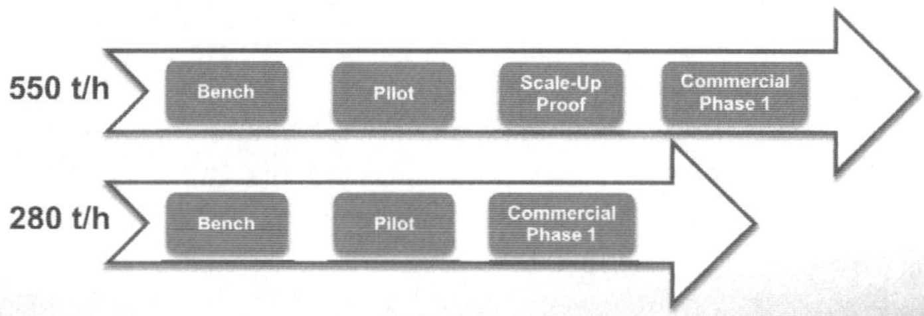

Enefit

Environmental Impact Statement

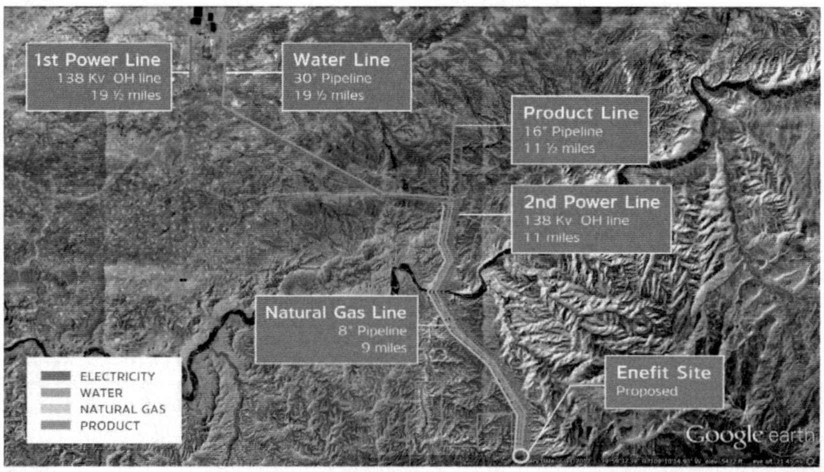

Note: The location and size of utilities contemplated in the ongoing EIS would not change due to potential changes discussed herein on the private property.

Enefit

Utility Corridor EIS Update

Project Definition and Start Up	Q4 2012
Agency and Public Scoping, Environmental Baseline	2013 - 2014
Alternative Development	Q4 2014
Data Collection	Q1 2015
Impact Assessment and Mitigation Planning	Q1 – Q2 2015
Alternative Comparison and Selection	Q2 – Q3 2015
Draft EIS	***Q4 2015***
Final EIS	Q2 2016
Record of Decision	Q3 2016
Right of Way Grant	Q4 2016

Where we are now: *Draft EIS* – *Q4 2015*

Major federal permitting milestone for oil shale project under current administration.

Enefit

Penstemon

- **May – June 2013:** EAO surveys for both Graham's and White River beardtongue on utility corridor and private property
  - 11 locations, 117 Graham's beardtongue
  - 23 locations, 413 White River beardtongue
  - No plants on BLM or state land
- **August 2013:** USFWS published proposed rule to list Grahams and White River beardtongues.
- **August 6, 2014:** USFWS withdrew proposal to list Graham and White River beardtongues and approved the Conservation Agreement.

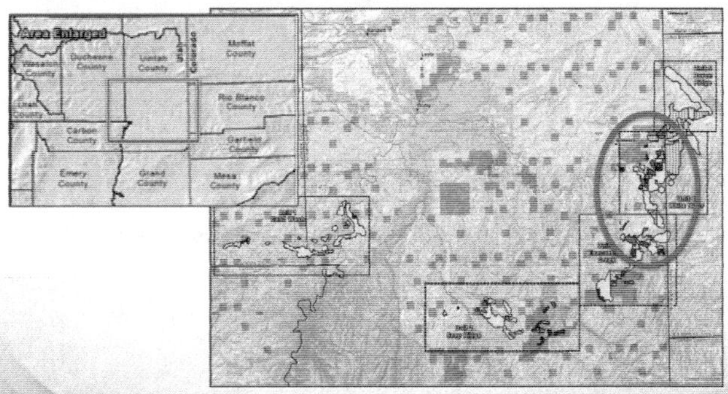

Enefit

Penstemon Conservation Activities – 2014

October 2014: Penstemon Conservation Team researchers ***successfully transplant*** 70 White River beardtongue seedlings sourced from local populations into a private conservation area on Enefit American Oil land.

Enefit

Successful Penstemon Transplants – June 2015

June 2015: Monitoring surveys of 70 transplanted seedlings on EAO property show ***approximately 90% survival*** and many plants flowering.

October 2015: Additional transplants on EAO property in disturbed and undisturbed areas are planned, with further monitoring of current transplant populations.

Enefit

Penstemon Conservation Agreement Successes

In Agreement's first year, several significant accomplishments:

- Transplantation Success:
  - Transplantation studies on EAO's property from 2014-2015 have demonstrated that plants can be transplanted – contrary to "best available science" at time of proposed listing.
- More Plants and More Populations:
  - Spring 2015 surveys identify a total ***641 and 470 previously undocumented Graham's and White River beardtongue individuals*** in/near known populations.
  - Additionally, a ***new White River beardtongue population with over 1,000 individuals*** is identified, representing a significant geographic and elevation range expansion for the species.
- The Agreement has "Teeth", Funding and Results:
  - In September 2014, Uintah County officially passed zoning ordinance to enforce the conditions of the Conservation Agreement on private land.
  - In July 2015, the official Mitigation, Weed Management, and Grazing Management Plans – conditions of the Conservation Agreement – are completed and published by the Penstemon Conservation Team.
  - Significant funding has been provided to fund these and other activities – which would not have been available had the plant been listed.

Enefit

Accomplishments – Reserve Ready Resource

- 2011: Purchase of property and leases
- 2012-2013: Drilling to 95% Measured and Indicated Resource (43-101 Compliant)
- 2012-2013: Box cut to get samples for testing
- 2012-2014: Various levels of retort and related systems testing
- 2012-2014: Environmental baseline studies complete
- 2012-current: Permitting processes underway - EIS
- 2013- current: Social License - 55% Statewide and 84% Basin support for the industry
- 2013: Sage grouse research (no birds or leks)
- 2014: Penstemon Conservation Agreement
- 2013 & 2014: All engineering to a Prefeasibility Study level (base case & "optimized" case)
- 2015: New business plan development and process re-engineering to strengthen economics

Enefit

Enefit American Oil

A Proven, Efficient and Environmentally Sound Means
to Help Utah Become Energy Independent

Visit us at: www.EnefitUtah.com

Enefit

Forward-Looking Statements

The information contained herein includes forward looking statements. By their nature, forward-looking statements involve known and unknown risks and uncertainties because they relate to events and depend on circumstances that may or may not occur in the future. Forward-looking statements are not guarantees of future performance. You should not place undue reliance on these forward-looking statements. Red Leaf Resources, Inc. makes no guarantees as to the accuracy or completeness of any information contained herein. The information contained herein may be amended. This presentation is for information purposes only and does not constitute an offer to sell or a solicitation to buy the securities referred to herein.

REDLEAF RESOURCES, INC.

Oil Shale Symposium - October 5/6, 2015

Introduction

- **Red Leaf**
  - Incorporated in Delaware in 2006
- **Concept:**
  - Economically develop the most easily accessible oil shale in the world with a low cost, environmentally responsible approach
- **Original Design:**
  - Utilize clay lined earthen structures as retorts
    - Build in mine pit and reclaim in place
    - Handle shale one time with minimum crushing
- **Summary:**
  - Red Leaf/TOTAL spent about $150MM on development of Ecoshale® technology
  - Low water use
  - One time handling of shale
  - Early reclamation
  - High quality oil, 30 – 35° API, 65% middle distillates

REDLEAF RESOURCES, INC.

Oil Shale Symposium - October 5/6, 2015 3

Corporate Overview

Assets

- Expert oil shale team
- Ecoshale® technology
  - 14 US patents issued, 15 pending
  - 119 foreign patents issued, 247 pending
  - 6 license agreements/options for licenses
- World class lab facility for oil shale testing
- Utah leases with over 500MMBL of surface recoverable oil
- Red Leaf/TOTAL JV for commercial demonstration project in Utah
  - Final permits received
  - Full funding approved

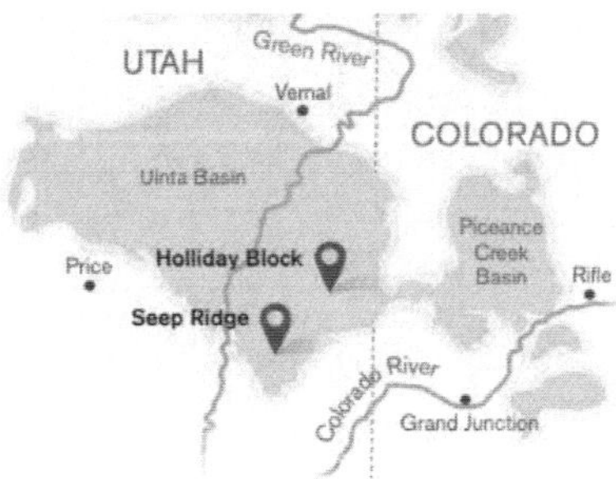

REDLEAF RESOURCES, INC.

Oil Shale Symposium - October 5/6, 2015 4

Red Leaf

Proof of Concept & Initial Funding

- Achievements 2006 – 1Q 2012
  - Patents granted for EcoShale® technology
  - Successful field pilot construction, operation 2008-2009
  - $172MM equity issued between 2007 - 2012
  - Six licenses/options for licenses issued using EcoShale® technology
  - Mineral leases acquired in Utah with 500MMB oil recoverable
  - $400MM joint venture with TOTAL for Utah oil shale projects

- Status 2012
  - Patented field proven ecoshale ® technology
  - International partner for project development
  - Funding for demonstration project and first phase of commercial
  - Brent oil price $110/bbl in Q1 2012

REDLEAF RESOURCES, INC.

Oil Shale Symposium - October 5/6, 2015 5

Red Leaf

EPS Generation 1

- Achievements 2Q 2012 – 1Q 2015
  - Agreed with TOTAL to construct commercial demonstration project (EPS)
  - Completed detailed engineering for commercial scale first generation capsule and facilities
  - Obtained mining, construction, and groundwater permits
  - Project reviewed and approved by Red Leaf/TOTAL in July 2014 – Brent Oil Price $110/bbl
  - Signed Engineering, Procurement, And Construction Management contract with WorleyParsons
  - Commenced field construction
  - Acquired additional mineral leases in Utah – 100MMB recoverable oil adjacent to existing leases
- Status
  - Mine and construction site fully developed for EPS construction and capsule floor excavated
  - Fixed-price contract in place for capsule construction with Ames Construction
  - Commercial scale engineering completed by Hatch Engineering in Canada
  - EcoShale® technology further de-risked for commercial scale operation
  - Commercial project likely viable at >$80/bbl oil
  - Brent Oil Price - $55/bbl in Q1 2015

REDLEAF RESOURCES, INC.

Oil Shale Symposium - October 5/6, 2015 6

Red Leaf

EPS Generation 2

- Achievements/Goals 2Q 2015 – YE 2018
  - Cancellation of contracts and significant slowdown of EPS project
    - Initially focus on infrastructure, pipelines, and foundations
  - Incorporate technology and cost improvements for second generation capsule design:
    - Increase oil production per capsule
    - Simplify capsule heat process
    - Reduce capsule materials and design
    - Simplify capsule construction
  - Maintain and amend permits as necessary
  - Full-scale construction of EPS generation 2 in 1H 2017
  - Complete construction in 2H 2018 and commence oil production

- Targeted Status
  - Commercial project with Ecoshale® technology fully de-risked
  - Second generation design reduces capsule costs and increases yield by 20-30%
  - Commercial project likely at $60 to $80/bbl oil

Red Leaf

Capsule Evolution

Generation 1	Generation 2
Indirect, closed loop, flue gas heating	Direct injection of hot working gas
Multiple rows of heating pipe	Single row of heat injection pipe
Significant volume of low grade material for pipe bedding/protection	Greatly reduced volume of low grade material for pipe bedding/protection
400 days for heating and 400 days for cooling	200 days for heating and cooling
Insulation: layers of dry and wetted crushed shale	Same but thinner layers
35% porosity ore	30% porosity ore
Bentonite Amended Soil (BAS) for liquid and gas containment	Considering limited Geo-membrane for liquid and gas containment

Red Leaf

Commercial Layout

- EPS2 design scaled to 1 MMBL capsules
  - ~ 200 capsules
- Capsules clusters built back-to-back and side-to-side with shared walls
- Shared ramps
- Conveyor/stacker construction

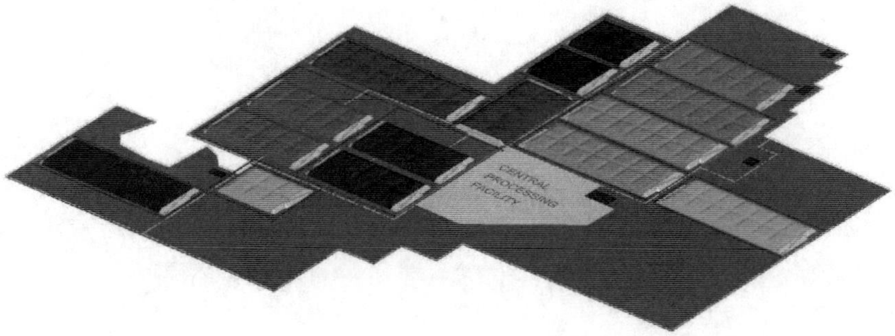

REDLEAF RESOURCES, INC.

Targeted First Commercial Project Economics

- Production rate 35,000 bbl/d
- Capsule cost < $30MM
- Recoverable oil 1MMBL
- Seep Ridge/Ambre prospect 180MMBL recoverable
- Commercial price breakeven at 10% IRR after-tax:
  - Gen1 >$80/bbl,
  - Gen2 $60-$80/bbl

	Capsule Economics
Production Life	1 year
Reserves	1,000,000 bbl
Cost	$30MM investment/$5MM O&M
Payout	< 1 yr

REDLEAF RESOURCES, INC.

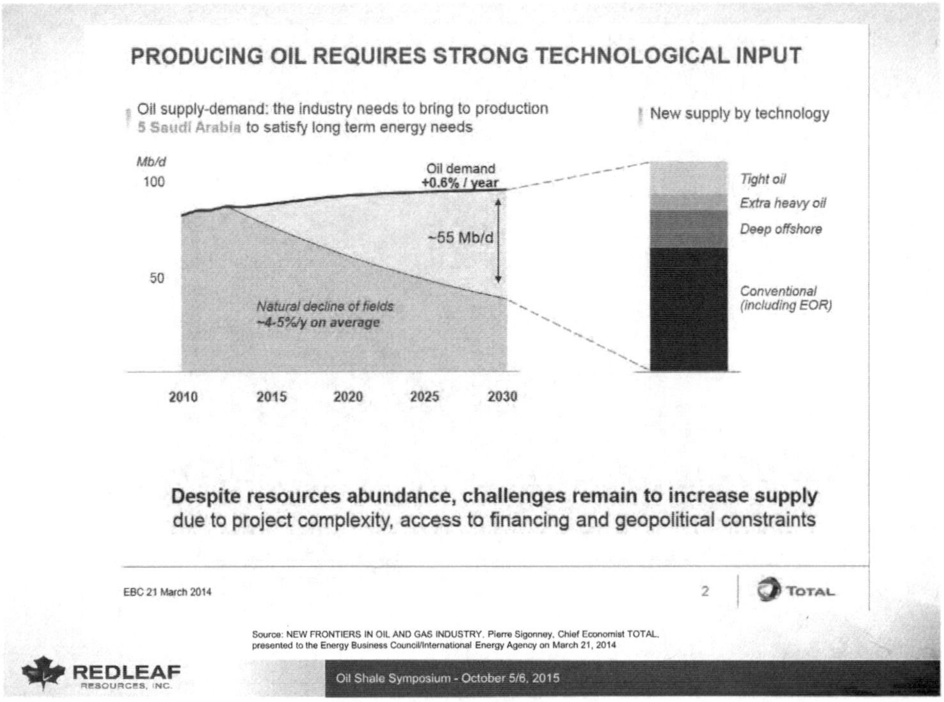
PRODUCING OIL REQUIRES STRONG TECHNOLOGICAL INPUT
Oil supply-demand: the industry needs to bring to production
5 Saudi Arabia to satisfy long term energy needs
New supply by technology
Mb/d
100
50
Oil demand
+0.6% / year
~55 Mb/d
Natural decline of fields
~4-5%/y on average
2010
2015
2020
2025
2030
Tight oil
Extra heavy oil
Deep offshore
Conventional
(including EOR)
Despite resources abundance, challenges remain to increase supply
due to project complexity, access to financing and geopolitical constraints
EBC 21 March 2014
2
TOTAL
Source: NEW FRONTIERS IN OIL AND GAS INDUSTRY, Pierre Sigonney, Chief Economist TOTAL,
presented to the Energy Business Council/International Energy Agency on March 21, 2014
REDLEAF
RESOURCES, INC.
Oil Shale Symposium - October 5/6, 2015

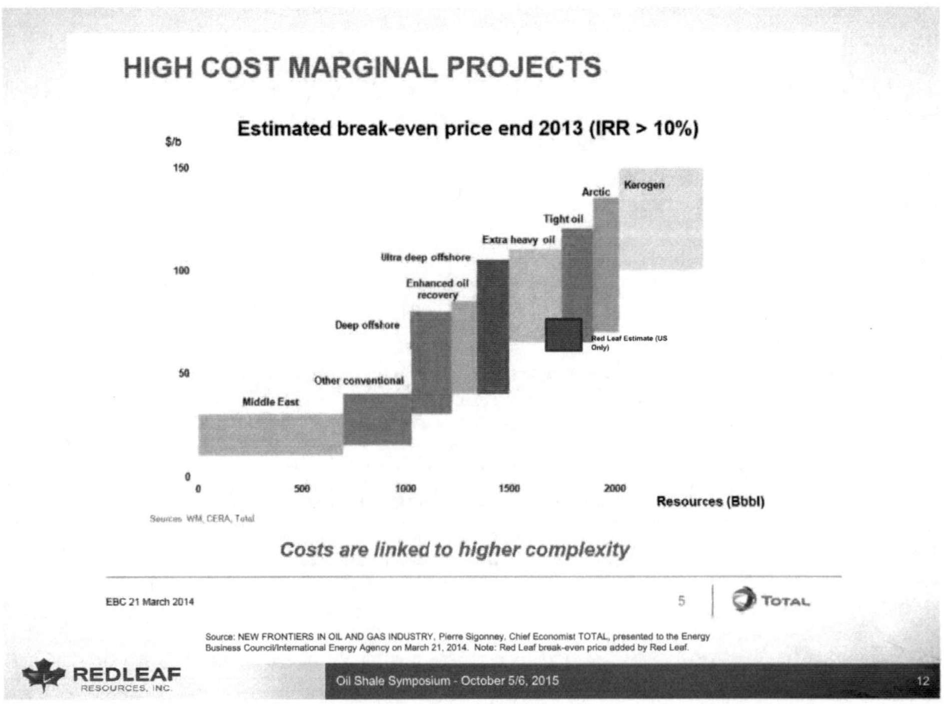
HIGH COST MARGINAL PROJECTS
Estimated break-even price end 2013 (IRR > 10%)
$/b
150
100
50
0
Middle East
Other conventional
Deep offshore
Enhanced oil recovery
Ultra deep offshore
Extra heavy oil
Tight oil
Arctic
Kerogen
Red Leaf Estimate (US Only)
0
500
1000
1500
2000
Resources (Bbbl)
Costs are linked to higher complexity
EBC 21 March 2014
5
TOTAL
Source: NEW FRONTIERS IN OIL AND GAS INDUSTRY, Pierre Sigonney, Chief Economist TOTAL, presented to the Energy
Business Council/International Energy Agency on March 21, 2014. Note: Red Leaf break-even price added by Red Leaf.
REDLEAF
RESOURCES, INC.
Oil Shale Symposium - October 5/6, 2015
12

Conclusions

- Red Leaf Ecoshale® technology
  - Environmentally responsible
  - Lowest front-end capital requirement
  - High quality oil
  - Shortest time to commercialization
- Engineering innovations accelerating
- Long-term requirement for oil from new frontiers is real
- Ecoshale® technology will be competitive
- Timing is critical

REDLEAF RESOURCES, INC.

Oil Shale Symposium - October 5/6, 2015 13

CONTACT

Alex Bocock
VP Investor Relations
abocock@redleafinc.com
801-878-8421

Red Leaf Resources
10808 South River Front
Parkway, Ste. 200
South Jordan, UT 84095
Telephone: 801-878-8100
Fax: 801-878-8101

REDLEAF RESOURCES, INC.

Oil Shale Symposium - October 5/6, 2015 14

12.3 The Oil Shale Industry's Fortunes in an Era of Dramatic – and Puzzling! – Change

Patty Limerick

Center of the American West, University of Colorado, USA

This talk will survey the changes in the world of energy in the last decade, with the goal of delivering a fresh perspective on the range of opportunities and obstacles, challenges and conundrums that face the oil shale industry. The practice of hydraulic fracturing for the development of unconventional oil and gas resources has captured the attention of the nation over this period, at once accelerating controversy over fossil fuel production and casting a spotlight on the difficulty confronted by every American claiming a stance of detachment, purity, and innocence in energy consumption. What might this mean for the oil shale industry? Returning to the Oil Shale Symposium after an absence of six years, historian Patty Limerick welcomes this opportunity to share her observations about the industry's recent experiences; reflect on the shifting configuration of optimism, pessimism, and practicality in public attitudes toward natural resources; and examine the dynamic world of federalism in Shale Country, as local, state, and federal jurisdictions compete for authority and legitimacy.

12.4 Status of the Oil Shale Industry in 2015 and Closing Remarks

Jeremy Boak[1,2]

[1]Colorado School of Mines, Golden, CO, [2]Oklahoma Geological Survey, Norman, OK, USA

The continued expansion of tight oil production in the United States and the OPEC decision to maintain its production caused a significant change in the global state of the oil industry. While numerous companies have undergone serious retrenchment, progress continues even in the face of the 50% decline in oil price, a critical control on the availability of capital for the capital-intensive oil shale industry. Genie Oil and Gas has reduced its oil shale efforts in Israel, Mongolia, and the U. S. to focus or nearer term objectives, but has not stopped efforts on oil shale. Reductions in staff by Shell in Jordan, by several companies in the United States and by VKG in Estonia, reflected slowing of efforts due to lower prices. However, the Estonian government appeared to agree that layoffs at VKG might be mitigated by tying state resource fees to oil price and that making better use of the mining quota would beneficial to the state.

A new Memorandum of Understanding with Questerre Energy Corporation and cautious progress on Enefit's power plant in Jordan are bright spots there. In Estonia, Enefit announced it was considering construction of a new Enefit280 oil shale plant, perhaps beginning in 2016, and VKG continues construction of its Petroter III plant, with a planned startup in September 2015. Overall, Estonian shale oil production in 2014 grew by nearly 12% over 2013, with most of the product being exported.

Resistance to development of oil shale continued, with lawsuits over the U. S. Fish and Wildlife Service's decision to deny protection to Graham's beardtongue, based on a 15-year conservation agreement negotiated with industry and others. On the other hand, a protest to permits temporarily approved for TomCo to develop acreage in Utah was withdrawn when TomCo agreed to await Red Leaf Resources test of the EcoShale™ process at Seep Ridge. However, Red Leaf has extended its schedule for the production system so that first heat will be deferred until the second half of 2016.

The downturn in oil prices in the last year has dealt a significant setback to the expansion of development of oil shale resources worldwide, but a few bright spots remain.

In closing, commentary on the future of the Symposium as a standalone meeting will be provided, focusing on the location(s) of future meetings and challenges for continuing an annual meeting in the U.S. on the topic.

13.1 An Overview of the Cold Extraction Method Applied to Jordan Oil Shale

Musa Resheidat

JUST, Jordan

This paper starts with a brief overview of oil shale technologies in general and on those proposed to be used in Jordan, including in-situ conversion or surface retorting. The pros and cons considering all environmental and economic aspects will be presented. Other factors will also be explained as they relate to what method will be best for utilizing Jordanian oil shale. A detailed presentation will be made regarding the new method based on solvents extraction of shale oil at room temperature, which is termed the "Cold Extraction Method" which has been developed for use in the laboratory but has not tested on a large scale. A comparative study between this new approach and other promising methods will be presented. The conclusions drawn from this study will also be outlined.

13.2 Effect of Oil Shale Alkyl Resorcinols on Physico-chemical and Thermal Properties of Polycondensation Resins

Ana Jurkeviciute[1], Larisa Grigorieva[2], Ksenia Moskvina[2]

[1]*Tallinn University of Technology Virumaa College, Oil Shale Competence Center, Estonia,* [2]*Tallinn University of Technology Virumaa College, Estonia*

Shale alkyl resorcinols are formed as a by-product in oil shale processing. They are unique raw materials for the chemical industry. Polycondensation resin production is one worthwhile application of shale alkyl resorcinols. These resins are widely applied in many branches of industry such as wood-working, metallurgic, tire, rubber products, construction, etc. The possibility of obtaining such resins using alkyl resorcinols will reduce the cost of finished products and expand the range of resins available. Synthesis of polycondensation resins using alkyl resorcinols was conducted using several methods during this investigation. In the formulations, a part of the resorcinol feedstock was replaced by fractions of shale alkyl resorcinols containing different amount of 5-methylresorcinol (40-80 mass %). Some resins were also modified with aromatic alkenes during synthesis.

Heat resistance of resins was investigated by thermogravimetric analysis (TGA) under both nitrogen and air atmospheres. TGA integral curves were obtained and processed in dynamic mode for interval of temperatures 25-830 °C. The heating rate was 5°C/min, gas flow rate - 50 ml/min. Resins potential for carbonization was evaluated by carbon residue. Physical-chemical parameters of the resins, as well as the unreacted resorcinol and 5-methylresorcinol content, were determined using a gas-solid chromatography method.

13.3 Detailed Characterization of Laminated Oil Shale and Source Rocks using Laser-induced Breakdown Spectroscopy and Infrared Microscopy

Kathryn Washburn[1], Justin Birdwell[2]

[1]Ingrain Inc., USA, [2]U.S. Geological Survey, USA

Geochemical and mineralogical analysis of shales and mudrocks at submillimeter-resolution can reveal subtle variations in the composition of these materials that are relevant to the overall understanding of oil shale and petroleum resource systems. Laminated mudrocks are of particular interest. Variations in lamina composition reflect fluctuating conditions in paleodepositional environments and can provide geochemical and geologic insights into these systems that may be useful in the examination of potential oil shale, tight oil, or shale gas plays. The application of laser-induced breakdown spectroscopy (LIBS) provides information on the elemental composition of rock samples on spot sizes of 400 μm or less allowing for characterization of individual lamina. When used with calibrated chemometric tools, LIBS can differentiate between organic and inorganic elemental components, making it possible to assess important kerogen properties such as elemental hydrogen-to-carbon ratios as well as matrix mineralogy. A complimentary tool for characterization of laminated shales is attenuated total reflectance infrared microscopy (ATR μ-IR), which provides non-destructive analysis of the covalent bonds between elements in minerals and organic matter in mudrock samples and can be applied over a range of scales including submillimeter analysis (125 μm spot size used here). Both LIBS and ATR μ-IR can be applied to intact rock pieces with minimal sample preparation. In this study, we demonstrate the utility of a combined LIBS and ATR μ-IR approach to characterize the organic geochemical and mineralogical nature of shale and mudrock samples. Variability within samples can be assessed rapidly and with quantitative accuracy similar to that of traditional analytical methods such as X-ray diffraction, total organic carbon, and programmed pyrolysis. A variety of Eocene Green River Formation oil shales, Turonian Boquillas Formation mudrocks, and Devonian-Mississippian Woodford Shale samples were examined and the results compared to bulk measurements made on crushed and homogenized aliquots of each rock. The results of this study show that submillimeter variations in mineralogy and organic matter content and quality in sedimentary units can be rapidly determined by these methods, providing higher resolution results than other techniques.

13.4 Experimental and Numerical Investigation of Dyes Adsorption on Shale Oil Ash

Mohammad Al-Shannag[1], Khalid Bani-Melhem[2], Zakaria Al-Qodah[3]

[1]The University of Jordan, [2]Hashemite University, Jordan, [3]Taibah University, Saudi Arabia

In the current study, adsorption of dyes on oil shale ash has been investigated over a series of batch experiments. The adsorption isotherm data have been fitted adequately according to the Langmuir equation. Based on the film resistance and homogeneous solid-phase diffusion, a two-resistance mass-transfer model has been proposed. This differential model has been solved using implicit Crank-Nicholsen finite-difference scheme. The study has demonstrated the excellent agreement between the numerical results and the corresponding experimental ones. Present results show that the external mass-transfer resistance is not affected by the initial dye concentration. However, it decreases notably with increasing either agitation speed or temperature. Furthermore, the internal mass-transfer resistance decreases significantly with increasing either initial dye concentration or temperature.

Research and Application of Technology to Improve the Permeability of Maoming Oil Shale in China

Qiulian Long[1], Yuanping Gao[1], Jianzheng Su[1], Bue Wang[2], Jiayuan He[1]

[1] Petroleum Exploration and Production Research Institute of SINOPEC, China ;
[2] SINOPEC, China

Abstract

Oil shale, one of the world's richest energy resources, is considered a very important alternative energy in the 21st century. In recent years, oil shale in-situ exploitation technology has advanced significantly, such as In-situ Conversion Process of Shell and in-situ heating with fluid. The property of low porosity and low permeability makes it hard to inject thermal fluid into oil shale formations. For in-situ heating with fluid, some stimulation of the oil shale formation is needed so that it can have sufficient permeability. For Maoming oil shale in China, clay minerals are the main constituent. The total content of kaolinite, hydromica and montmorillonite ranges from 50% to 70%, whereas the total content of brittle minerals ranges from 30% to 40%. This feature of mineral composition results in difficulty of getting a network of fractures using hydraulic fracturing.

In this study, the Pulse Power Fracturing Technology (PPFT) and Treatment Agent Post-processing Technology (TAPT) were designed for the stimulation of oil shale formation. The energy used in PPFT was optimized in combination with characteristics of Maoming oil shale. Treatment agents were developed and their performances were investigated. Experimental results showed that the porosity, pyrolysis rate, and oil yield rate of oil shale samples from Maoming increased by more than 25% after it reacted with the treating agent. During stimulation, PPFT broke the oil shale formation at first, and then TAPT further improved its permeability. After this process, the morphology of fracturing could meet the requirement for the injection of thermal fluid.

PPFT was applied in the Maoming block, and it broke the oil shale formation with designed energy scale which was optimized by laboratory experiments. During the process of operation, both ultrasonic testing technique and hydraulic pressure drop testing technology were used to detect and analyze the results of this treatment. Practice has proven that PPFT could break oil shale formation effectively and the retrofit distance could surpass 15 meters. This study describes effective means for increasing the permeability of oil shale, providing an important and useful support for the technical development and application of oil shale in-situ heating with thermal fluid.

Introduction

Oil shale is considered one of the world's richest energy resources and a very important alternative energy for the 21st century (James et al, 2004).

According to statistics, currently the world's shale oil resource is about 411 billion tons, whereas China's shale oil resource is about 47.6 billion tons, ranking second in the world after the United States. China's oil shale resources have three characteristics. The first is shallow depth. Oil shale at depths of 0-500 meters account for 64%, and at 500-1000 meters for 36% of the resource. The second characteristic is the lack of thick layers. The average thickness is about 20-30 meters, and many oil shale deposits have sandstone and mudstone interlayers. The third feature is the low proportion of high-quality oil shale. A considerable part (about 45.4%) of oil shale resources have oil content less than 5%, whereas oil content between 5% and 10% accounted for 37%, and oil content greater than 10% (high-grade resources) only accounted for 17.6% (Liu et al, 2006; Guo et al, 2008). Maoming oil shale has similar characteristics. Effective exploitation of oil shale requires reckoning with their resource characteristics.

At present, the exploitation methods of oil shale can be divided into two types: surface retorting and in-situ exploitation. For surface retorting, relatively mature technology, short heating period and quick return on investment are its advantages. But serious environmental pollution, extensive damage to the land, shallow exploitation depth, high water use in the residue processing and backfilling work needed after the end of exploitation limit the widespread application of surface retorting technology. The construction of large-scale mining and gas treatment facilities are not required for in-situ exploitation which can also develop the deep oil shale resources. With increasing pressure on environmental protection, in-situ exploitation has become the likely trend of large-scale commercial development for oil shale in the future (Fang et al, 2009; Feng et al, 2011).

Currently, major foreign oil companies are conduting research on oil shale in-situ exploitation and apply more than ten kinds of technology, including Shell's ICP (In-situ Conversion Process) technology (Vinegar, 2006), ExxonMobil's Electrofrac technology (Tanaka et al, 2011), Chevron's CRUSH technology (Looney, 2011) and the American Shale Oil company's CCR (Conduction, Convection and Reflux) technology (Allix and Burnham, 2011). Considering the heating method, these technologies include electric heating, heating with fluid, radiant heating, combustion heating and others. Among them, the maturity of ICP technology using electric heating is relatively high, as it has been successfully tested in Green River Formation oil shale in the United States. But a long heating period, such as 2-4 years, is needed because of slow heating speed, consumes a great deal of power and may lead to poor economic performance especially for thin layers. In September 2013, Shell announced its withdrawal from RD&D (Research, Development & Demonstration) projects in Green River oil shale.

Technology of heating with fluid has been widely considered because of fast heating speed, short heating period and taking full advantage of carbonization gas. For Maoming oil shale, it is the main in-situ exploitation technology researched. However, under normal circumstances, thermal fluid cannot be injected directly into the oil shale formation due to its low porosity and extremely low permeability. So reservoir stimulation is needed to create certain permeability. For Maoming oil shale, there are three important characteristics. Firstly, the clay minerals are the main constituents. The total content of ranges from 50% to 70%, whereas the kaolinite, hydromica and montmorillonite total content of brittle mineral ranges from 30% to 40% (Qian et al, 2008). This is lower than the requirement of 60% for forming a network of fractures when using hydraulic fracturing. Secondly, the depth of the main seam is shallow, mostly less than 300 meters, so the horizontal stress is generally greater than the vertical stress. Thirdly, horizontal bedding is well developed. These three features result in producing only horizontal fractures easily and unsatisfactory stimulation effect using hydraulic fracturing. Research on other effective means for increasing the permeability of oil shale is needed.

In this study, the Pulse Power Fracturing Technology (PPFT) and Treatment Agent Post-processing Technology (TAPT) were designed for the stimulation of oil shale formation. Treatment agents were developed and their performance was investigated. In September 2015, PPFT was applied in Maoming block, and its effect was tested.

Pulse Power Fracturing Technology (PPFT)

Technology Background

High power pulse is a technology that can produce a variety of strong electrical pulse power. Its energy is stored in a slow way (on the order of seconds). Then pulse compression and power amplification are achieved by various fast-acting switches. High power electromagnetic energy is instantaneously released to the load in a very short time (on the order of microseconds to nanoseconds) with high power (MW to GW) by the form of a single pulse or controlled repetitive pulses. High power pulse technology is developed mainly for defense applications, such as nuclear radiation simulation, high power microwave, electric guns and laser weapons.

An electric pulse device generates a strong shock wave of repeated high power pulsed voltage. When the shock wave which has a high peak pressure acts on the reservoir, it is both the action object and transfer medium of the shock wave, without any other media. If the peak pressure of the shock wave is higher than the breaking strength of the reservoir formation, the rock can be fractured. In this study, this technology is named as Pulse Power Fracturing Technology (PPFT).

Technology Theory

Strong shock wave of repeat pulse generated in the way of electrohydraulic effect can be explained by two mechanisms. One is that the plasma channel discharged in the water rapidly expands and generates a strong shock wave under the heating of discharge current. The second is that the plasma channel discharged in the water heats the water molecules around, resulting in their rapidly heating up, vaporizing, expanding and generating a strong shock wave. The peak pressure of these shock waves can be calculated by Equation 1 and 2 respectively.

$$P = \beta\sqrt{\frac{\rho\omega}{\tau T}} \qquad (1)$$

$$P = \frac{kC^{\frac{1}{2}}U}{1.4r^{5/2}} \qquad (2)$$

where P is the peak pressure of shock wave, *Pa*; β is a complex integration, the value of which is 0.7 in water, *dimensionless*; ρ is the water density, kg/m^3; τ is the leading edge of shock wave, *m*; T

is the bottom width of the shock wave, *s*; *w* is the total energy of discharge current per unit length of the discharge channel, *J*; *k* is a coefficient, *dimensionless*; *C* is the electric capacity, *F*; *U* is the voltage, *kV*; *r* is the spreading distance of the shock wave, *m*.

It can be seen from Equation 1 and 2 that the peak pressure of the shock wave is proportional to the discharge energy and energy density on the discharge channels. And it is proportional to the storage capacity and the square of the discharge voltage in the condition of constant storage energy. The electrical energy density in the energy storage component is a major obstacle limiting the application of a shock wave. In order to apply a shock wave of repeated pulses to unconventional gas development and produce stronger shock waves in the limited space, not only in-depth study of energy conversion efficiency but also compound new physical mechanisms are needed.

Test Results

Based on the results of the rock mechanical properties, the breaking strength of rock at high acceleration shock load is significantly lower than it is under static load. The peak pressure of the shock wave generated by an electric pulse device is greater than 100 MPa (Figures 1 and 2), impacting the reservoir at several thousand times the acceleration of gravity. Under this high acceleration shock wave, impact strength is much higher than the fatigue strength of rock, which can extend the existing fractures and generate micro-fractures or macro-fractures until the impact strength reduces to the fatigue strength of rock. These new fractures will increase the permeability of the reservoir.

Figure 1: Energy converter of electric pulse device (bullet in the middle).

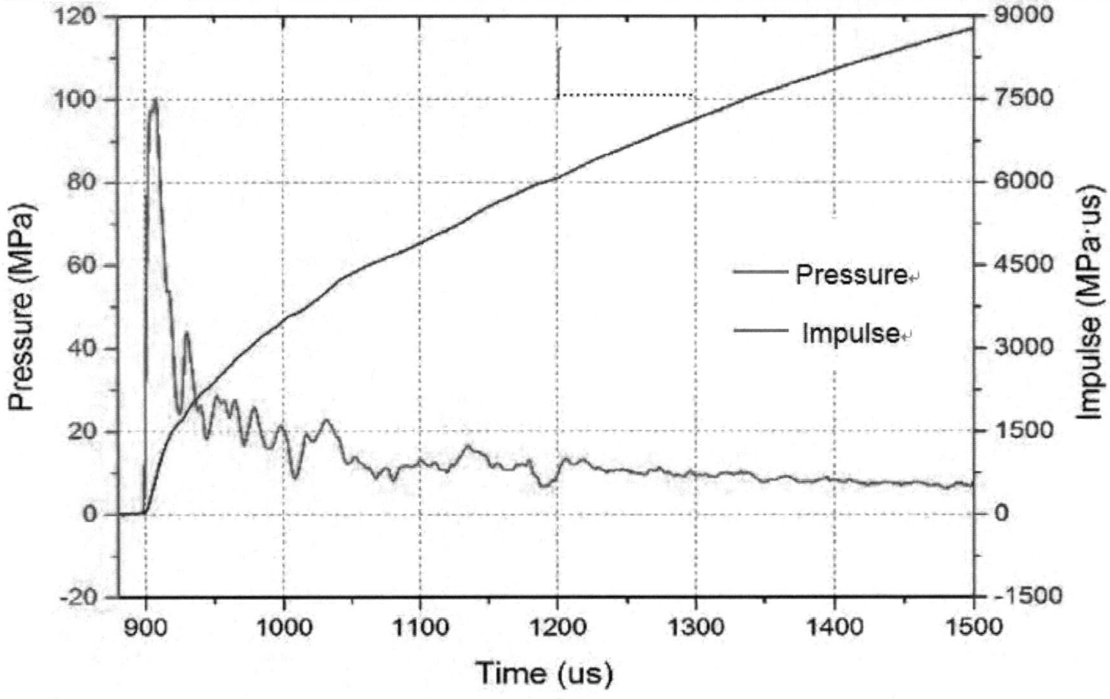

Figure 2: Energy release process of repeat pulse shock wave.

PPFT was successfully tested. Figure 3 shows the laboratory test result of the electric pulse device in Figure 1. It can be seen from Figure 3 that the sandstone sample was well fractured after the effect of shock waves generated by electric pulse device for 5 times. Every time one bullet was used, and the energy of each bullet was 30 kJ. Figure 4 shows the field test result that when the energy of the bullet increases to 90 kJ, the rock can be fractured after one shot. The test depth in Figure 4 was about 2 meters. Theoretical research and test results indicate that PPFT can be applicable to different formation, such as sandstone, shale, oil shale etc.

Treatment Agent Post-processing Technology (TAPT)

Treatment Agent Composition

The mass percentage contents of materials in the treatment agent developed are as follows: acid 5-28%, cobalt and/or nickel salt 0.1-1%, inhibitor 1-3%, mutual solvent 0.1-20% and the rest water. Among these compositions, the effect of acid is corrosion and expanding pores for oil shale formation, cobalt and/or nickel salt being a catalyst that promotes the pyrolysis of organic in oil shale, inhibitor is used for making the treatment process steady and reducing the corrosion of metal pipes.

Treatment Agent Preparation

1) Calculate the required amount of each material according to the treatment agent formula designed, and weigh each material.

2) Add the acid, cobalt salt, nickel salt, inhibitor and mutual solvent to 2/3-4/5 required amount of water, keeping stirring while adding.

3) Add the remaining water to the above mixtures. Then a treatment agent is prepared after full stirring and mixing.

In this study, three different treatment agents were designed and prepared, which were named as treatment agent A, B and C respectively.

Performance Test Process of Treatment Agent

1) Crush and screen Maoming oil shale samples. Choose particle samples whose diameter range is 3.35-4.75 mm. Weigh 30 g samples and place them in a plastic bottle.

2) Pour 100 mL treatment agent prepared into the bottle. Place it in the environment of 40°C for 2 hours. The treatment agent will react with the oil shale particle samples. Pour out the treating agent after reaction.

3) Wash the samples with water, and then place them in the drying oven at 105°C for 4 hours.

4) Weigh 5g samples dried in process 3. Observe the samples with optical microscope after they naturally cool to

Figure 3: Laboratory test on sandstone sample (30 kJ for 5 times).

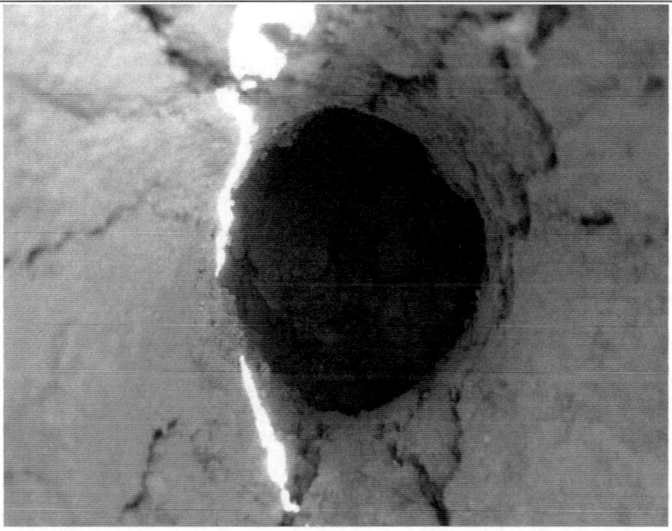

Figure 4: Field test on rock in situ (90 kJ for 1 time).

room temperature. And then test the pore size distribution and porosity of samples using 3H-2000PS1 type specific surface area and pore size distribution analyzer.

5) Weigh 15g samples dried in process 3 and place them in the vacuum reactor at 350°C for 65 hours. The organic matter in the oil shale samples will be pyrolyzed during the heating process. Collect and measure pyrolysis oil and gas. Calculate the oil recovery.

6) Observe the samples with optical microscope after pyrolysis and natural cooling, and then test the pore size distribution and porosity using 3H-2000PS1 type specific surface area and pore size distribution analyzer.

In this study, the oil shale samples treated by treatment agent A, B and C respectively are named as sample A, B and C. A contrast experiment is designed of which the test process contain 1), 3), 4), 5) and 6). In the contrast experiment, the oil shale sample is named as original sample.

Performance Test Results of Treatment Agent

The experimental results are shown in Figure 5, 6, 7 and Table 1. Figure 5 is a comparison of the oil shale samples observed by optical microscope. As can be seen from Figure 5, the original sample is very compact (Figure 5a), whereas the samples A, B and C treated by treatment agent A, B and C respectively become loose with developed micro-fractures (Figure 5b, 5c, 5d).

The comparisons of pyrolyzed oil shale samples which were treated differently are shown in Figure 6. It can be seen that the original sample after heating and pyrolysis was still compact with few fractures developed (Figure 6a), but a complex network fractures were generated after the sample A, B and C pyrolysis (Figure 6b, 6c, 6d).

Figure 7 shows a comparison of pore size distribution for different oil shale samples using 3H-2000PS1 type specific surface area and pore size distribution analyzer. The instrument uses static volumetric method (nitrogen adsorption and desorption), and its test range of pore size is 0.35 – 200 nm, which can be used to analyze the development situation of micropore, mesopore and macropore.

(a) Original sample (b) Sample A

(c) Sample B (d) Sample C

Figure 5: The comparisons of oil shale samples before and after treatment.

Figure 7 shows that compared to the sample before pyrolysis, the numbers of micropore, mesopore and macropore for all samples after pyrolysis increase, while the increasing trends are more evident for the samples treated by treatment agent A, B or C.

(a) Pyrolytic original sample (b) Pyrolytic sample A

(c) Pyrolytic sample B (d) Pyrolytic sample C

Figure 6: Effect of treatment agents on oil shale pyrolysis.

3H-2000PS1 type specific surface area and pore size distribution analyzer can test the porosity (0.35-200 nm pores). The result of this porosity before/after pyrolysis and oil recovery is shown in Table 1. The porosity and oil recovery were increased if the oil shale sample had been treated by treatment agent, wherein the effect of treatment agent B is the most obvious.

The above results show that: for Maoming oil shale, the treatment agents developed in this study can indeed promote the development of fractures, increase the porosity and permeability, and achieve better pyrolysis efficiency.

Promotion and Application of Treatment Agent

For oil shale, PPFT can break the oil shale formation and create fractures, which are useful for the injection of thermal fluid. If the treatment agent is PPFT, it will promote the development of fractures and increase the permeability of formation after use. Then the thermal fluid can be injected more easily, and the in-situ exploitation technology of heating with fluid can be achieved better. In this study, this treatment agent preparation and injection technology is named Treatment

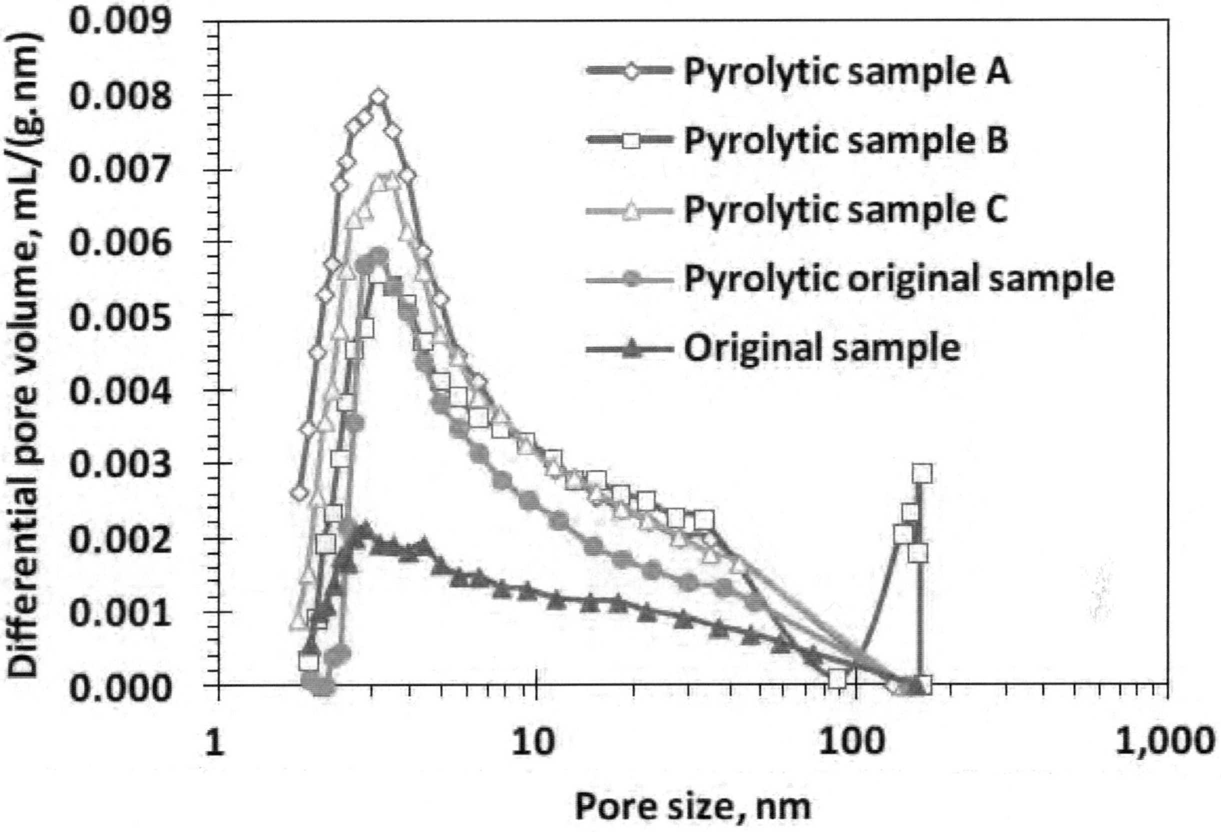

Figure 7: - The comparisons of pore size distribution for different oil shale samples before and after pyrolysis.

Agent Post-processing Technology (TAPT).

Field Application

For Maoming oil shale, the characteristics of being rich in clay minerals, shallow depth of the main seam and well developed horizontal bedding result in getting only horizontal fractures easily and unsatisfactory stimulation using hydraulic fracturing. But the effect of PPFT is different.

MYY-1 well is an oil shale coring and test well of SINOPEC in the Maoming block. In September 2015, PPFT was applied in MYY-1 well, and its effect was tested. Figure 8 shows the logging curves of MYY-1 well. Figure 8a is the

Table 1 - Effect of treatment agents on oil shale porosity and oil recovery.

	Porosity before pyrolysis /%	Porosity after pyrolysis /%	Oil recovery /%
Original sample	14.3	19.9	29
Sample A	15.1	22.7	37
Sample B	15.6	27.3	40
Sample C	15.4	24.9	39

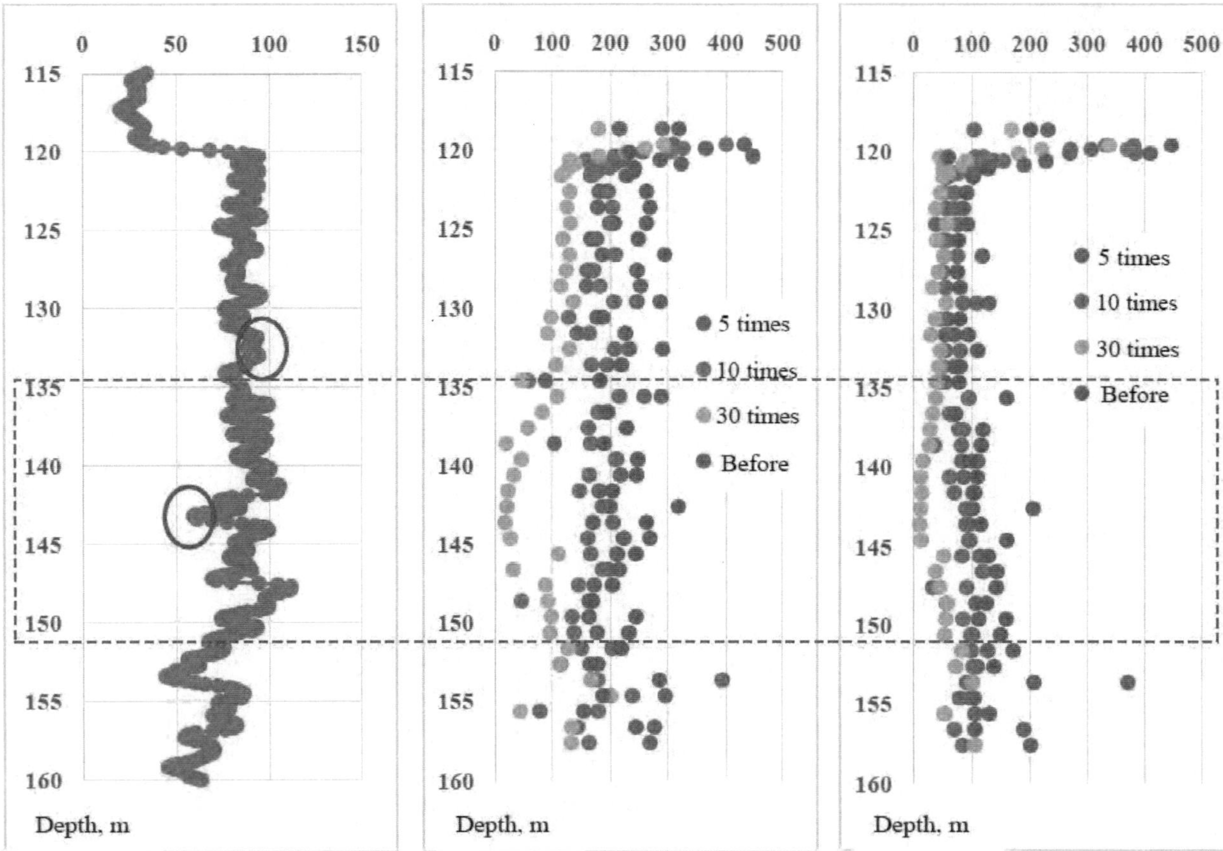

(a) Natural gamma logging curve (b) Upper ultrasonic test curve (c) Lower ultrasonic test curve

Figure 8: The logging curves of MYY-1 well.

natural gamma-ray logging curve of MYY-1 well before PPFT applied. The top depth of oil shale formation is 120 meters, and the bottom depth is 160 meters. Two acting positions of PPFT are designed.

The first one is at 143 meters depth with a relatively lower natural gamma-ray value. The second one is at 132 meters depth with a relatively higher natural gamma-ray value. For the first acting position, PPFT was applied 40 times, and 40 bullets of 120 kJ were used. For the second acting position, PPFT was applied 48 times, and 48 bullets were used (22 bullets of 120 kJ, 26 bullets of 30 kJ).

Figure 8b and 8c show the upper and lower ultrasonic test curve of MYY-1 well before and after PPFT was applied in the first position. The ultrasonic test uses a single transmitter to a double receiver, so there are two ultrasonic test curves. It can be seen from Figure 8 that the ultrasonic test curves are consistent with natural gamma logging curve. The upper and lower ultrasonic test value decreased with the increase of times applied, especially when PPFT applied 30 times, the ultrasonic test values decreased rapidly. Lower ultrasonic test value can reflect the development of fractures. This practice has proven that PPFT could break oil shale formation effectively and the penetra-

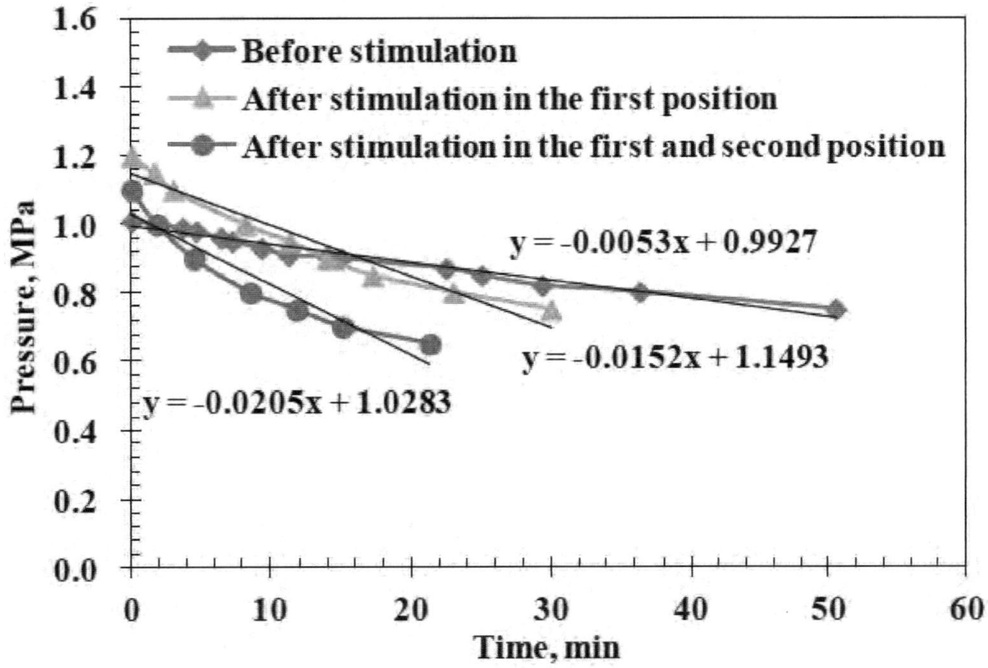

Figure 9: The hydraulic pressure drop test curves of MYY-1 well.

tion distance could surpass 15 meters. For the second position, the change law of ultrasonic test values is similar.

A hydraulic pressure drop test was also applied to explain the effect of PPFT. The results are shown in Figure 9. Before stimulation, the pumping pressure drops slowly, which indicates that the permeability of the formation is very low. After stimulation in the first position, the fall-off speed of pumping pressure increases obviously. After stimulation in the first and second position, the pumping pressure drops more quickly. These results show that PPFT makes an important contribution to the increase of formation permeability.

Conclusions

1) The theory, experiment and application of repeat pulse shock wave were researched in this study. Based on these, the Pulse Power Fracturing Technology (PPFT) was designed for the stimulation of oil shale formation.

2) Three kinds of treatment agents were developed and their performance was investigated. Experimental results showed that for Maoming oil shale, these treatment agents can indeed promote the development of fractures, increase the porosity and permeability, and achieve better pyrolysis efficiency. On this basis, Treatment Agent Post-processing Technology (TAPT) can be developed.

3) PPFT was applied in MYY-1 well in Maoming block. Both ultrasonic test and hydraulic pressure drop test were used to detect and analyze the results of this treatment. Practice proves that PPFT could break oil shale formation effectively and the penetration distance could surpass 15 meters.

4) PPFT and TAPT supplies effective means for increasing the permeability

of oil shale, providing an important and useful support for the technical development and application of oil shale in-situ heating with thermal fluid.

Reference

Allix, Pierre, Burnham, Alan K. (2011) Coaxing oil from shale. Oilfield Review (Schlumberger), 22(4): 6.

Fang, C.H., Zheng, D.W., Liu, D.X. (2009) The development and trends about in-situ mining technology of oil shale. Energy Technology and Management, 2: 78-80.

Feng, X.W., Chen C., Chen D.Y. (2011) New development of oil shale in-situ technology. China Mining Magazine, 20(6): 84-87.

Guo, Y.G., Xu X.Q., Wang H.Y. (2008) Research progress in use of oil shale as the unconventionality energy. Jiangsu Chemical Industry, 2008, 36(2): 6-9.

Vinegar, Harold (2006) Shell's in-situ conversion process. 26th Oil Shale Symposium, Colorado, 2006: 16-19.

Bunger, James W., Crawford, Peter M., Johnson, Harry R.(2004) Is oil shale America's answer to peak-oil challenge? Oil & Gas Journal, 102 (8): 16-24.

Liu Z.J., Dong Q.S., Ye S.Q.(2006) The situation of oil shale resources in China. Journal of Jilin University (Earth Science Edition), 36(6): 869-873.

Looney, Mark D. (2011)Chevron's plans for rubblization of Green River Formation oil shale (GROS) for chemical conversion. 31th Oil Shale Symposium.

Qian J.L., Yin L., Wang J.Q.(2008) Oil Shale: Supplementary Energy of Petroleum. Beijing: Sinopec Press, 2008, 71.

Tanaka P.L., Yeakel J.D., Symington W.A.(2011) Plan to test ExxonMobil's in situ oil shale technology on a proposed RD&D lease. 31st Oil Shale Symposium, Golden Colorado.

Study on numerical simulation of in-situ processing of Maoming oil shale

Jianzheng Su, Youping Wang, Yiwei Wang, Xianglong Meng, Bue,Wang

Exploaration&Production Research Institute of SINOPEC, Beijing 100083, China

Abstract

The Maoming oil shale in China is easy to recover due to the shallow rich layer and mild dip angle. The mining technology has been developed rapidly in the past few years, which brings not only significant profit but also severe environmental problems. The oil shale in-situ processing technologies show a promising future in China. The geological characters of the Maoming oil shale imply the advantages of in-situ processing. Heating the oil shale formation is an essential process during the in-situ processing and the prediction of temperature distribution is extremely significant. This paper analyzed the mechanism of heat transfer, physical and chemical change. The mathematical model for the temperature distribution was established considering temperature effect on the heat conductivity, chemical reaction, and heat loss to the upper and bottom layers. Then the heating well pattern of triangle, square and hexagon were simulated. At last the sensitivity of the main influence factors of the temperature distribution was analyzed. The results show that the heat loss, heat conductivity, and heating power play the important roles in temperature distribution. The optimization of heating parameters could shorten the heating time for the oil shale from initial temperature to required temperature, which makes a great sense to the improvement of heating efficiency for the development of Maoming oil shale.

Introduction

Oil shale is a kind of solid combustible organic mineral, from which the shale oil can be obtained by dry distillation at the low temperature. It belongs to the unconventional resources and is classified as a very important alternative energy in the 21st century because of its huge amount of resource, high abundance, and relatively concentrated distribution. In recent years with the unceasing rise of global oil price, as well as the development of conventional residual oil and low permeable reservoirs becomes more difficult, the oil shale development has been paid more and more attention worldwide. For a long time, the oil shale has been developed by the method of mining the rocks to the ground. This surface retorting has its own limitation, for example, it causes serious damage to the ecological environment. Therefore, in-situ processing of the oil shale has a promising future. In-situ technologies heat oil shale underground by injecting hot fluids into the rock formation, or by using linear or planar heating sources followed by thermal conduction and convection to distribute heat through the target area. Shale oil is then recovered through vertical wells or horizontal wells drilled into the formation.

Oil shale can cause cracks in the heating process due to thermal stress and thermal expansion occurs within the pore fluid. When the temperature reaches a certain value, kerogen within the oil shale decomposes into oil and gas, while the rock density, pore structure, and rock mineral components will change. These changes and generated hydrocarbons in turn affect the heat transfer performance of the formation. Therefore the research of the temperature distribution for in-situ processing of oil shale is more complicated. Currently, there are so many researches about the numerical simulation of oil shale, however, most of them focused on molecular motion of the fluid, involved mathematics equations are many and difficult to solve. The micro level of the research methods considers rock matrix and porous fluid as continuous flow. The research object is fluid particles or micro unit. It has to put the heat transfer and fluid flow of fine structure in the solid skeleton at the border as a boundary condition, and this quantitative description is difficult and inaccurate. Based on the mechanism of heat transfer, the mathematical model of temperature distribution for in-situ processing of oil shale has been developed in order to analyze the change rules and influence factors of temperature distribution, and to provide the basis for heating process optimization.

1. The establishment of the formation heat transfer mathematical model

1.1 Heat transfer in oil shale formation

Oil shale is a kind of multi-phase porous medium which has complicated component. The oil shale rock matrix is continuous phase, while the porous, nature fractures, and the fluids are discontinuous phases. Heat transfer processes in porous media include four parts: heat conduction in the rock matrixes, heat conduction in the porous fluids, heat convection in the porous fluids, and radiation in the rock matrix. The heat transfer in oil shale formation is the transient heat transfer. Take any element of oil shale; the heat balance equation for any point in formation should be as follows:

$$\frac{\partial}{\partial x}(\kappa_x \frac{\partial T}{\partial x}) + \frac{\partial}{\partial y}(\kappa_y \frac{\partial T}{\partial y}) + \frac{\partial}{\partial z}(\kappa_z \frac{\partial T}{\partial z}) = (1-\phi)C\rho_r \frac{\partial T}{\partial t} \quad (1)$$

Where: κ is the thermal conductivity coefficient in W/(m·K); T is the temperature in K; C is the material specific heat in kJ/(kg·K); ρ is the density in kg/m^3.

1.2 Heat convection

When there are differences in temperature or pressure between each phase within porous media, the porous fluid flows or heat and mass transfer occurs by convection heat transfer, fluid flows into the cell body per unit of time is the inflow of energy.

$$\frac{\partial}{\partial x}[\phi\sum_j (\rho_j V_j H_j) + \frac{\partial}{\partial y}[\phi\sum_j (\rho_j V_j H_j) + \frac{\partial}{\partial z}[\phi\sum_j (\rho_j V_j H_j) \quad (2)$$

Where: ρ_j is the density of j phase in kg/m^3; V_j is the seepage velocity of j phase in m/s; H_j is enthalpy of j phase in kJ/(kg·K).

1.3 Thermal radiation

It will produce thermal radiation, as long as the oil shale temperature larger than the absolute temperature. Heat transfer by radiation does not depend on the contact. Assumed that the oil shale surface is the ash surface, Heat radiation follows Stephen - Boltzmann law and Kirchhoff's law. The net heat rate of thermal radiation is:

$$\begin{cases} q_{Rx} = \varepsilon\sigma \dfrac{dT^4}{dx} \\ q_{Ry} = \varepsilon\sigma \dfrac{dT^4}{dy} \\ q_{Rz} = \varepsilon\sigma \dfrac{dT^4}{dz} \end{cases} \quad (3)$$

Where: σ is the Stephen - Boltzmann constant (5.67×10^{-8} W/($m^2 \cdot K^4$)); ε is the emissivity of radiation.

1.4 Heat loss to upper and bottom layers

During the heating process, the heat loss to upper and bottom layers is:

$$E_{loss} = \frac{A \cdot 2\kappa\Delta T}{\sqrt{\pi\alpha t}} \quad (4)$$

Where: α is the thermal diffusivity in m^2/s; κ is the thermal conductivity coefficient in W/(m·K); t is the heating time in s or hr or day; ΔT is the temperature difference between oil shale formation and upper (bottom) layer in K; A is the heat dissipation in m^2.

1.5 Pyrolysis reaction heat

When temperature of the oil shale reaches the critical temperature, kerogen within the oil shale decomposes into oil and gas. This process is endothermic reaction and the heat absorbed by the oil shale pyrolysis is:

$$E = U_R A(1-x)^m e^{-E/RT} \quad (5)$$

Where: x is hydrocarbon generation rate by pyrolysis; U_R is the pyrolysis heat in kJ/kg.

1.6 Heat transfer mathematical model

Supposed that the heat transfer between rock matrix and porous fluids is adequate, it reveals that in the basic element the temperature of the rock matrix is equal to that of porous fluid. Under this condition, local thermal equilibrium model can be used for research. From energy-balance equation, in the basic element the energy obtained by heat transfer minus heat loss is equal to the energy changes with the element.

$$\frac{\partial}{\partial x}[\phi\sum_j(\rho_j V_j H_j)+\frac{\partial}{\partial y}[\phi\sum_j(\rho_j V_j H_j)+\frac{\partial}{\partial z}[\phi\sum_j(\rho_j V_j H_j)$$

$$+\frac{\partial}{\partial x}(\kappa_x\frac{\partial T}{\partial x})+\frac{\partial}{\partial y}(\kappa_y\frac{\partial T}{\partial y})+\frac{\partial}{\partial z}(\kappa_z\frac{\partial T}{\partial z})-\frac{2\kappa\Delta T}{\rho_w L_v\sqrt{\pi\alpha t}}$$

$$+\frac{\partial}{\partial x}(\alpha\frac{\partial T^4}{\partial x})+\frac{\partial}{\partial y}(\alpha\frac{\partial T^4}{\partial y})+\frac{\partial}{\partial z}(\alpha\frac{\partial T^4}{\partial z})=\frac{\partial}{\partial t}[\phi\sum_j(\rho_j S_j U_j)+(1-\phi)C\rho_r(T-T_i)]$$

$$+\ U_R A(1-x)^m e^{-E/RT}+Q \qquad (6)$$

Initial condition:

$$T(x,y,z,0)=T_0 \qquad (7)$$

Boundary conditions:

$$\begin{cases} T(x,y,z,t)=T_1(L\in\Gamma_1) \\ T(x,y,z,t)=q(L\in\Gamma_2) \\ T(x,y,z,t)=h(T_1-T)(L\in\Gamma_3) \end{cases} \qquad (8)$$

Where: κ is the thermal conductivity coefficient in W/(m·K); C is the material specific heat in kJ/(kg·K); U_R is the pyrolysis heat in kJ/kg; Q is the source-sink term in W/m^3; T_0 is the oil shale initial temperature in K; T_1 is the fluid temperature at the boundary in K; q is the boundary heating flux in W/m^2; m is the reaction order; E is the apparent activation energy in J/mol; R is the gas constant 8.31J/(mol·K).

2. Theoretical analysis and a case study

2.1 Basic data

Top depth of Oil shale in Maoming basin is about 180m, the area 24m*24m, thickness of the oil shale about 18m, other formation parameters are as follows:

Table1 the basic data of the oil shale for simulation

Parameters	Value	Unit
Density	2400	kg/m^3
Specific heat capacity of oil shale	0.94	kJ/(kg·K)
Specific heat capacity of water	4.19	kJ/(kg·K)
Thermal conductivity coefficient of oil shale	3.17	W/(m·K)

Thermal conductivity coefficient of water	0.62	W/(m·K)
Thermal conductivity coefficient of oil	0.13	W/(m·K)
Thermal conductivity coefficient of gas	0.03	W/(m·K)
Porosity	0.05	
Initial temperature	20	℃
Heating power	2	kW/m
Heating temperature	750	℃

Using the constant temperature heat method, three types of well pattern are analyzed in this paper. The heating well employs triangle, quadrangle, and hexagon respectively and the production well is arranged in the center of well pattern, which is demonstrated in Fig.1. The triangle heating well is regarded as the basic model.

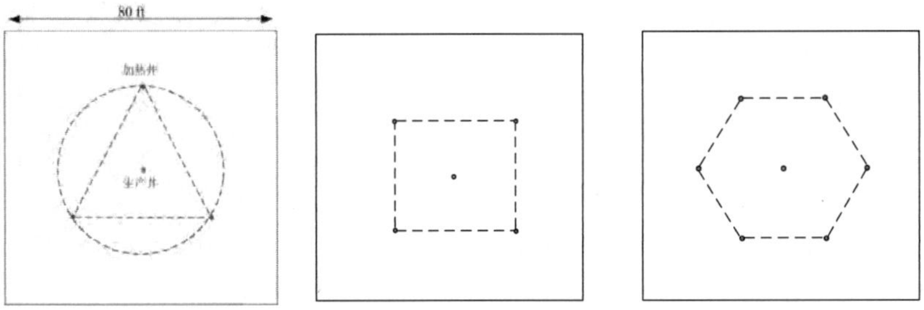

Fig.1 The schematic of the heating well pattern

2.2 Analysis of influencing factors

2.2.1Well pattern

When the constant heating power is 2kW/m and the initial condition is the same, the temperature distributions for different well patterns (triangle, quadrangle, and hexagon) are plotted in Fig.2 to Fig.4. In case of triangular heating well, the maximum temperature in the heating well is 610℃, whereas the maximum temperatures in the quadrilateral and hexagonal heating well are 733℃ and 877℃ respectively. It is important to note that the quadrilateral and hexagonal heating well can significantly increase the temperature of the heating zone and the heating total area.

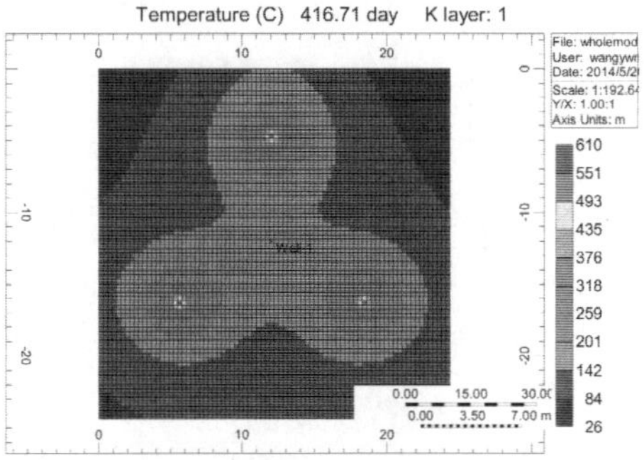

Fig.2 The temperature distribution of triangular heating well pattern

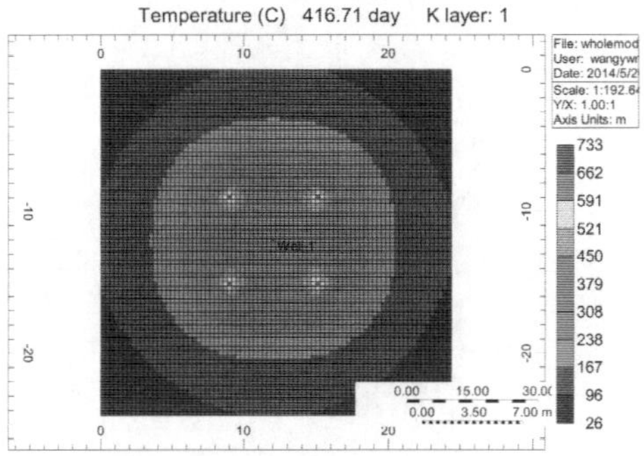

Fig.3 The temperature distribution of quadrilateral heating well pattern

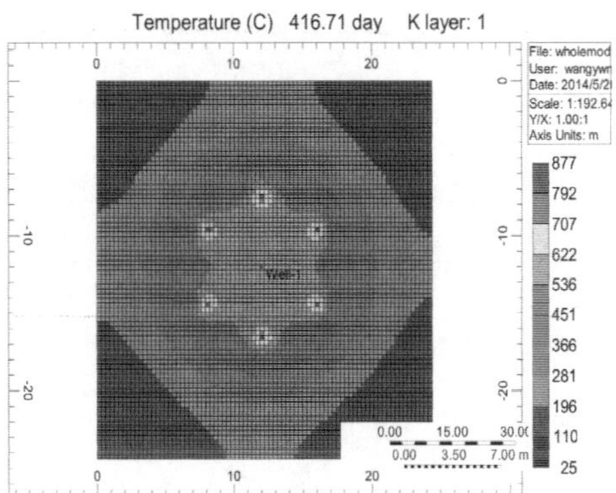

Fig.4 The temperature distribution of hexagonal heating well pattern

2.2.2 Heat loss

When the temperature of the oil shale element is higher than ambient temperature, the oil shale element loses heat to ambient, the temperature distribution that considered heat loss is illustrated in Figure 5. If the heated oil shale element is an insulator, the temperature distribution is demonstrated in Figure 6. It is verified that the influence of heat loss on temperature distribution is significant.

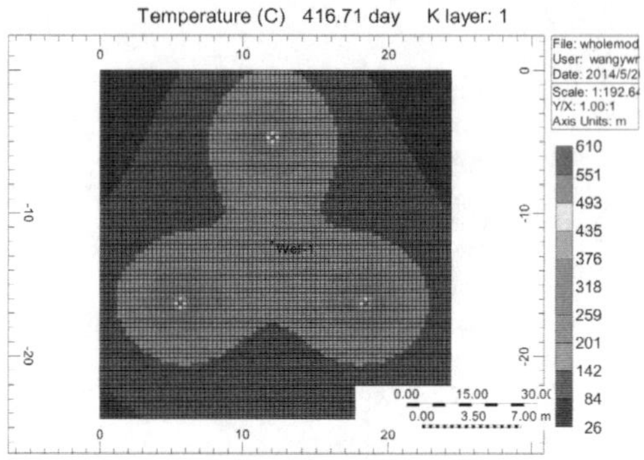

Fig.5 The temperature distribution with heat loss

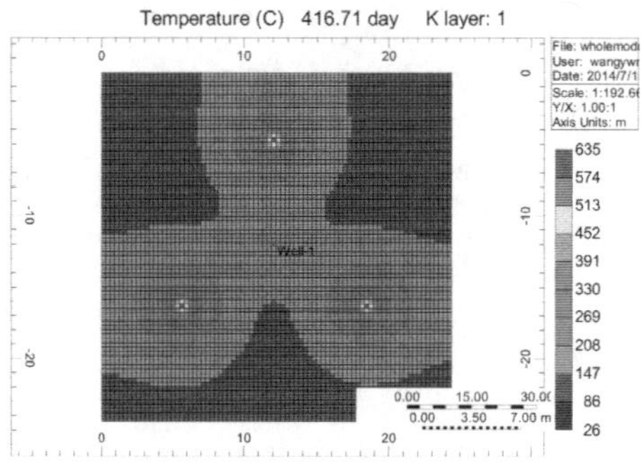

Fig.6 The temperature distribution without heat loss

2.2.3 Water saturation

The temperature distribution after 416 days heating with different water saturations (SW=0%, 30%, 60%) are plotted in Figure 7, Figure 8, and Figure 9. On the one hand, the formation water will increase the thermal conductivity of rock, thus improving the heat efficiency; on the other hand, the formation water will increase energy consumption during the heating. As a result, when the water saturation increases, the range of the heating zone becomes smaller, but the impact is not obvious.

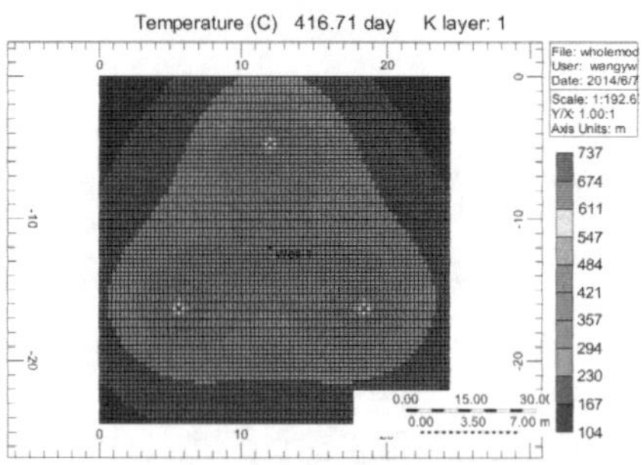

Fig.7 The temperature distribution when SW=0%

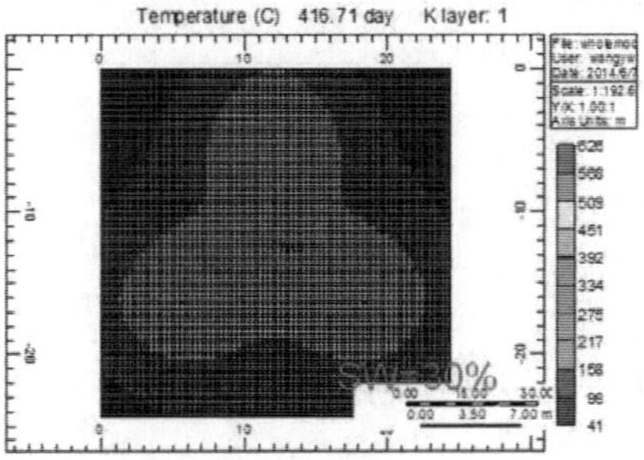

Fig.8 The temperature distribution when SW=30%

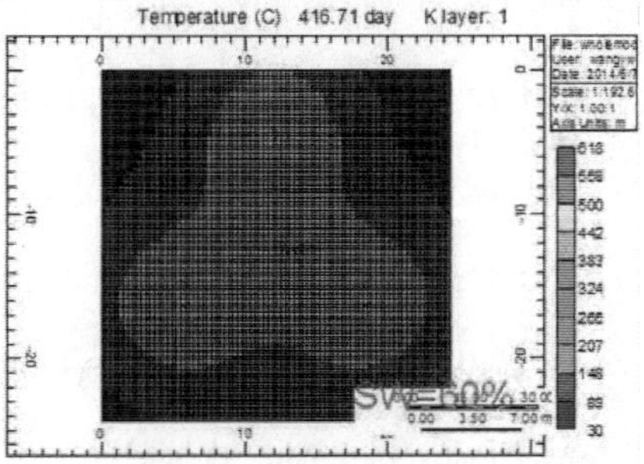

Fig.9 The temperature distribution when SW=60%

2.2.4 Heating power

Change the thermostat heating mode to constant power heating and the temperature distribution with different heating powers(1kw/m, 2kw/m, 3kw/m) is illustrated in Figure10. From the results of numerical simulation, improve the heating power can significantly increase the heating rate. However, after improving the heating power, it will make the temperature of the heater and the near wellbore temperature too high, resulting in increasing the risk of heater failure. By now the highest temperature of the heater to work long hours is about 700-800℃. For the case in our research, the heating power for oil shale formation is about 2kW/m.

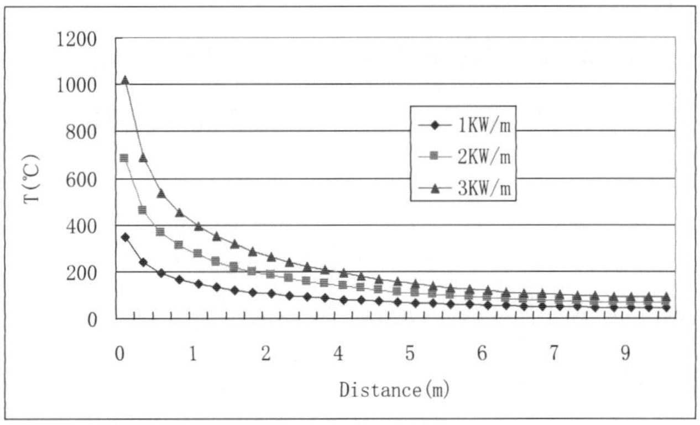

Fig.10 The temperature distribution with different heating powers

3. Conclusion

(1) The heat transfer during the in-situ processing of oil shale is an unstable transient heat transfer. The temperature distribution of oil shale is impacted by lots of factors, such as well pattern, the heat loss to upper and bottom layers, oil shale thermal conductivity, water saturation, heating power and so on. For electrical heating, heat conduction is the main method of heat transfer.

(2) Under the condition of the same heating time, the hexagonal heating well pattern is more effective than the quadrilateral pattern and triangular pattern.

(3) Under the condition of the same heating power, the heat loss to upper and bottom layers is the main factor that affects the temperature distribution of the heating zone. The more heat loss to upper and bottom layers, the more heating time are needed.

(4) The formation water will increase the thermal conductivity of rock, thus improving the heat efficiency, whereas the formation water will increase energy consumption during the heating. As a result, when the water saturation increases, the range of the heating zone becomes smaller, but this impact is not obvious.

(5) For a given oil shale formation, the controllable factor which affects temperature distribution is the heating power. For the low thermal conductivity limits the heat transfer rate of the oil shale, the heating power must match the thermal conductivity of the oil shale formation in order to avoid equipment damage due to the high temperature.

(6) For the electrical heating, the heating time for the oil shale heated to the desired pyrolysis temperature is longer, the heat loss is large, and utilization of heat is low. So improve energy efficiency technology and utilization of solar and other non-traditional energy sources are the future research direction.

References

[1] ZHANG Lei, HAN Xiangxin, WANG Zhongcun, et al. Progress of comprehensive utilization technology of oil shale[J]. China Mining Magazine, 2012, 21(9):50-53.

[2] WANG Youping, WANG Yiwei, MENG Xianglong, et al. The enlightenment of American´s oil shale in-situ retorting technology[J]. Oil Drilling & Production Technology, 2013; 35(6):55-59

[3] LIU Shanqi, LI Yongbing, TIAN Huiquan, et al. Numerical simulation on thermal conductivity of wet porous rock[J]. Chinese Journal of Geophysics, 2012, 55(12): 4239-4248.

[4] XU Mo, WANG Di, JIANG Liangwen, et al. Review on thermal conductivity coefficient of rock and soil mass[J]. Journal of Earth Sciences and Environment. 2011, 33(4):421-427.

[5] LI Xiaochuan, XU Youwei, HUANG Xiangyong. Numerical analysis of the influence of macropore on thermal conductivity of porous media[J]. Journal of North China Electric Power University, 2012, 39(6):76-79.

[6] LIU Wei, FAN Aiwu, HUANG Xiaoming. Porous medium heat and mass transfer theory and application[M]. Beijing: Science Press, 2006:81-89.

[7] Luo ji'an.Variation in the temperature field of rocks overlying a high-temperature cavity during underground coal gasification. Mining science and technology[J]. 2011, 21:709—713.

[8] Kantorovich II, Bar E Z. Heat transfer within highly porous chars: a review[J]. Fuel, 1999, 78: 279—299.

[9] MIAO Sheqiang, LI Heping, CHEN Gang. Progress of high temperature and high pressure experimental study on the thermal conductivity of the minerals and rocks[J]. Progress In Geophysics, 2013, 28(5):2453—2466.

[10] WANG Siwen, LU Shihui, HU Ke, et al. Investigation on effect of rock properties on thermal conductivity[J]. Experimental Technology and Management, 2012, 29(5):39-41.

[11] G. Roland Nguimbi, SUN Youhong and LI Qian. Pyrolysis of Huadian oil shale by electrical heating on different heating rates[J]. Global Geology, 2012, 15(2):120-125.

[12]KANG Zhiqin, ZHAO Yangsheng, YANG Dong. Physical principle and numerical analysis of oil shale development using in-situ conversion process technology[J]. ACTA PETROLEI SINICA, 2008, 29(4):592-600.

[13] Hao Y, Xiaoqiao G, Fansheng X, Jialiang Z, Yanju L. Temperature Distribution Simulation and Optimization Design of Electric Heater for In-situ Oil Shale Heating [J]. Oil Shale.2014, 31(2):105-120

[14] SHI Weiping, WANG Bingqing, YANG Xuefeng, et al. Numerical simulation of underground freeze wall generation[J]. Journal of Petrochemical Universities, 2013, 26(5):71-74.

[15] Aouizerate G, Durlofsky L J, Samier P. New Models for heater wells in subsurface simulations with application to the In-situ upgrading of oil shale[J]. Computational Geosciences, 2012, 16(2):519-533.

[16] Fan Y. Chemical Reaction Modeling in a Subsurface Flow Simulator with Application to In-Situ Upgrading and CO_2 Mineralization [D].Stanford University, 2010.

[17] Fan Y, Durlofsky L J. Numerical Simulation of the In-Situ Upgrading of Oil Shale[J]. SPE Journal, 2010, 2:368-381.

13.7 Finite Element Analysis Of Cryogenic Liquid Temperature And Flow Rate On The Underground Temperature Field

Huanan Liu, DaJun Zhao, Wei Guo, Chen Chen, Qiang Li, Mingyi Guo

Jilin University, China

During in-situ heating of oil shale, it is often necessary to seal the underground retort with a frozen wall of ice to isolate the targeted interval from groundwater and to prevent pyrolysis products from escaping. In this study, we have developed a numerical model of the underground freezing and heat transfer processes for a single row of frozen pipes to assess the impact of oil shale heating on the underground freeze wall. Temperature fields under different cooling conditions were analyzed using ANSYS analysis software and the influence of cryogenic coolant temperature and flow rate on underground the temperature field was determined. The simulation results show that lower coolant temperatures and the higher flow rates result in shorter freezing times, as expected, but also that there are optimal combinations of these parameters to minimize freezing time. Elapsed time for freezing and capital and operating costs can be effectively reduced using these results. Using an air-cooled refrigeration unit and limiting coolant temperature to between -15 and -5°C, the optimum coolant flow rate is 40 – 50 m^3/h; when using a refrigeration unit and coolant temperatures between -25 and -20°C, the optimal freezing liquid flow rate is 35 to 40 m^3/h.

13.8 Study of Pyrolysis Technology of Nong'an Oil Shale In-Situ Process and the Numerical Simulation

Pengfei Jiang, Youhong Sun, Wei Guo, Qiang Li, Sunhua Deng

Construction Engineering College, Jilin University, China

In the light of an in-situ pyrolysis pilot project for the Nong'an oil shale and its actual geological conditions, a new in-situ pyrolysis method of fracturing-nitrogen injection combination was put forward, and the corresponding surface and subsurface equipment were introduced. Pressurized nitrogen was heated at the surface and then injected to the injection well through insulated tubing. The nitrogen flowed through fractures caused by hydraulic fracturing to the production well. High temperature nitrogen heated the oil shale layer in the process of flow. At the same time, nitrogen can also carry the shale oil and gas to the surface through the production well. In this paper, a mathematical model was built, and the technology parameters related to the in-situ pyrolysis method were optimized through numerical simulation, to make the heat transfer efficiency between the high temperature nitrogen and oil shale layer reach the highest value. The simulation results suggest that the most appropriate nitrogen injection flow rate is $30m^3/h$, and the highest energy utilization ratio can reach 62.7%. The heating time is 125 days to pyrolyze the oil shale layer sufficiently between the injection well and the production well.

13.9 Characterization of Oil Shale Fracture Before and After Pyrolysis using X-ray Micro CT

Huaqing Xue[1,2], Bo Xiong[1], Shangwen Zhou[1], Gang Yan[1], Wei Guo[1]

[1]PetroChian Research Institute of Petroleum Exploration & Development-Langfang, China, [2]Key Lab for Thermal Science and Power Engineering of Ministry of Education, Department of Thermal Engineering, Tsinghua University, China

It is the common method for surface retorting and in situ conversion process that heating the oil shale to pyrolysis to produce hydrocarbon. A great number of pore and fracture were derived after pyrolysis on oil shale, which provides the space for hydrocarbon migration. Therefore, characterization of oil shale fracture before and after pyrolysis was very important to hydrocarbon producing and transportation. In this paper, the oil shale was from Erdos Basin, China and 12.7 mm diameter and 14.7mm length oil shale was drilled. We report the effect of temperature was from 350 to 500 °C) on oil shale pyrolysis by the thermal gravity apparatus and creation of fracture during thermal treatment. The oil shale was heated by different the pyrolysis temperature (350, 400, 450 and 500 °C) that resulted in higher weight loss and a corresponding increase in the oil yield. Three-dimensional X-ray micro tomography (XMT) was performed to characterize and to analyze the nature of the fracture structure before and after different pyrolysis temperature. XMT scans of the cores at 11.4 µm voxel resolution displayed distinguishable features of reaction products and source rock. The main fracture was produced in the initial of pyrolysis (350 °C) and was about 9.1 mm length. But they were almost not change, when the temperature increased. The micro fracture was produced during higher temperature and enriched in high kerogen region. It was caused by the pressure increased and the kerogen contracted in these areas, when the kerogen thermal decomposes to the hydrocarbon.

13.10 **Characterization of Type I Source Rock by Pore-Scale Simulations**

Cyprien Soulaine[1], Alexandre Lapene[1,2], Yifan Wang[1], Nicola Castelletto[1], Hamdi Tchelepi[1]

[1]Stanford University, USA, [2]Total Exploration Production Oil Shale, USA

Thermal degradation of source rock, exogenous or not, involves kerogen conversion resulting in rock structure evolution from non-permeable to permeable. Pyrolysis reactions produce oil and gas, locally increasing pressure and stress. The rock fails and cracks propagate enhancing the permeability of the medium. This represents a new challenge for porous media research as the system to be model evolves in time and exhibits strong coupling between mass and heat transfer, and rock mechanics. In this work, we propose to apply pore-scale simulations including fluid flow, heat transfer and rock mechanics on a source rock. A modelling strategy will be proposed and demonstrated on oil shale.

13.11 **Geophysical Alterations and Groundwater Flow Implications of In Situ and Ex Situ Conversion of Oil Shale in the Piceance Basin, Colorado, with Applications for Other Oil Shale Basins via Groundwater and 3D Geologic Computer Models**

Paul G Quinn[1], Jillian L Goldfarb[2], Ami Vyas[2]

[1]Ambet, USA, [2]Boston University, USA

Piceance Basin oil-shale deposits have been identified as the richest in the world. In situ and ex situ conversion of the underground shale is one potential option to recover the billions of barrels of oil located across the Piceance Basin in Colorado. Though the U.S. Department of Energy considers it a "promising" technology, its subsurface environmental impacts are not yet fully explored. The in situ and ex situ pyrolysis of oil shale will, by nature of its releasing long-stored kerogen from the shale rock, change the porosity of the remaining subsurface materials (both of extracted shale and surrounding sandstone). A series of shale samples were subjected to pyrolysis at a variety of temperatures, pressures, and reaction times to simulate in situ and ex situ conversion. The resulting surface areas and porosities were measured using nitrogen adsorption isotherms, possible crystallization and materials structure changes monitored using x-ray diffractometry, and morphological changes observed via scanning electron microscopy. Preliminary lab results show development of micro- and mesopores, which are correlated with temperature, pressure, and extraction time. To explore the impact of these morphological changes on environmental fate and transport, geophysical investigations and groundwater flow and 3D geologic computer models were created, analysed, and were correlated with potential oil migration with respect to the Eocene aged shale and associated formations.

13.12 **Aspects of Groundwater Flow and 3D Geologic Models in the Tarfaya Basin, Morroco**

Paul G Quinn[1], Zouhair Taleb[2]

[1]Ambet, USA, [2]Onhym, Morocco

Oil-shale deposits have been identified at ten localities in Morocco, the most important of which are Upper Cretaceous marinites. The two deposits that have been explored most extensively are the Timahdit and the Tarfaya deposits. This presentation explores computer modeling in relation to these environments in the present day state of diagenesis of Kerogen in the basins. Petrologic and geophysical analyses help define a geologic computer model used as the basis for developing a 3-dimensional, multi-layer groundwater flow model of the entire oil shale basin. Key characteristics of preliminary simulated groundwater flow behavior appear to confirm the assumption of low vertical hydraulic conductivity associated with confining oil shale units. This 3D and Groundwater computer(s) model is used to investigate the coupled surface water-groundwater response and spatio-temporal distribution of recharge within the basins.

13.13 Aspects of Groundwater Flow and 3D Geologic Models in the Oil Shale Basins, Jordan

Paul G Quinn[1], Omar Al-Ayed[2]

[1]Ambet, USA, [2]Al-Balqa Applied University, Jordan

Oil-shale deposits have been identified in the Jordan Basins. Geophysical and Geological Investigations have been conducted on the Oil Shale basins. The availability, natural dynamics and interaction between surface water and groundwater resources are understood and will be important to understanding integrated hydrologic impacts from development of oil shale resources on the Oil Shale Basins. Development will require groundwater to access and process oil shale deposits. The removal of this water and subsequent disposal of it will likely impact both groundwater and surface water. Understanding these impacts requires detailed knowledge of both the surface hydrology and subsurface hydrogeology of the basin system and their interaction. Natural ground water recharge within the basin aquifer is derived from precipitation whereas ground-water discharge is to either ephemeral or, perennial streams. The geohydrologic units considered in these models in geologic Formations that are unconfined and confined units. These deposits directly influence flows between surface water and groundwater. Hydraulic conductivity of the basin aquifer is related to lithology and the degree of fracturing. This presentation discusses development of a conceptual model, and inputs necessary to construct a fully integrated surface water-groundwater hydrologic model of existing conditions of the Oil Shale Basins: Lajjun, Sultani, and Attarat.

13.14 **The Change of the Permeability of Oil Shale during the Extraction by Sub-critical Water**

Li He, Sunhua Deng, Youhong Sun, Wei Guo

College of Construction Engineering, Jilin University, China

Concentrating on the characteristics of the oil shale resources in China. Jilin University put forward the concept of using sub-critical water as a heat transfer medium and extractant to the in-situ conversion of oil shale. This paper studied the change of the permeability of oil shale during the sub-critical water extraction. A physical model was proposed to simulate the in-situ extraction of oil shale by sub-critical water. THM coupling mathematical model was established to get control of computer simulation. Results showed that with the release of organic matter, the internal crack of oil shale was gradually expanding and the seepage passage was opened. The permeability of oil shale was then increasing at the same time. The result of the calculation showed that the permeability of oil shale increased with the increasing of the extraction time, the temperature and the pore pressure and with the decreasing of the volume stress. In addition, we expect to get the relationship between permeability coefficient k_i and temperature, pressure, volume stress, pore pressure and other parameters to reflect the influence of external force on the oil shale permeability combined with the experimental data.

13.15 **Effect of Original Minerals on the Pyrolysis of Oil Shale Kerogen**

Jian Shi

China University of Petroleum, China

Oil shales are fine-grained sedimentary rocks that contain organic matter which, upon heating, can be converted to liquid shale oil. The content of inorganic mineral in oil shale is more than 50%, and the composition of inorganic mineral is related to the located area. It is reported that the pyrolysis of oil shale is affected by the mineral content and composition.

The oil shale is obtained from Barkol, Xinjiang province. The effect of original mineral and acid treatment shale on the pyrolysis of oil shale is investigated with acid-treatment method and oxide addition method. Oil shale, carbonate-free oil shale, silicate-free oil shale and kerogen samples are analyzed by thermogravimetric analysis (TGA) method. The pyrolysis kinetics of acid-treatment simples is calculated using the overall n order reaction model and Friedman model. The effect of minerals on the distribution of pyrolysates is investigated through pyrolysis gas chromatography mass spectrometry (Py-GC/MS).

In terms of activation energy, the original minerals have inhibited effect on the pyrolysis of kerogen. CaO has catalytic effect on the pyrolysis process of kerogen, while MgO has inhibited effect; the effect of Fe_2O_3 is not obvious. The content order of oil fraction under C20 in the pyrolysates ranks as: silicate-free oil shale (OS-F)>Kerogen>oil shale> Carbonate-free oil shale OS-Cl.

13.16 **Recovery of Oil and Gas from Oil Shale through In Situ Pyrolysis using Radio Frequency Heating**

Preston Grounds, Jack Young, Robert Gardner

Highland Light Management, USA

The purpose of this study was to examine the feasibility of extracting of oil and gas from oil shale via in-situ pyrolysis using Radio Frequency (RF) heating. Using RF heating to increase the permeability of tight shale and to convert the kerogen found within the shale has been considered multiple times before (e.g. Radio Frequency In Situ Heating to extract Shale Oil, J.Bridges, et al, 28th Oil Shale Symposium, 2008; Slow Radio-Frequency Processing of Large Oil Shale Volumes to Produce Petroleum-like Shale Oil, A.K.Burnham, UCRL-ID-155045, 2003). This study examined the issues from a fresh perspective to determine if modern technology could make the process cost efficient. Multiple methods of in-situ heating were examined and a new paradigm for how the RF would be introduced into the rock was developed. The known problems with RF heating were explored in the context of this new design. Results of the study indicate that RF heating is a viable candidate for in-situ oil and gas production and recovery. Highlights include the fact that this new technology will provide steady production of oil and gas for 20 years from a 5,000 ft horizontal section instead of the typical well production curve; sharp spike followed by a long but rapid decline. The system generates stresses over large volumes of rock that should result in fractures. Further this new technology will allow a much higher recovery from the oil shale recovering over 50% of the oil and gas produced. A pilot program is currently being pursued to better understand the factors which govern recovery/extraction rate which is a key determinant for system economic feasibility.

13.17 **Pore-Scale X-ray Micro-CT Imaging and Analysis of Oil Shales**

Tarik Saif, Kamaljit Singh, Branko Bijeljic, Martin Blunt

Imperial College, UK

The pore structure and the connectivity of the pore space during the pyrolysis of oil shale are important characteristics which determine hydrocarbon flow behavior and ultimate recovery. We study the effect of temperature on the evolution of pore space and subsequent permeability on five oil shale samples: (1) Vernal Utah United States, (2) El Lajjun Al Karak Jordan, (3) Gladstone Queensland Australia (4) Fushun China and (5) Kimmeridge United Kingdom. Oil Shale cores of 5mm in diameter were pyrolyzed at 300, 400 and 500 °C in a nitrogen atmosphere. 3D imaging of 5mm diameter core samples was performed at 1μm voxel resolution using X-ray micro computed tomography (CT) and the evolution of the pore structures was characterized. The experimental results indicate that the thermal decomposition of kerogen at high temperatures is a major factor causing micro-scale changes in the internal structure of oil shale. At the early stage of pyrolysis, micron-scale heterogeneous pores were formed and with a further increase in temperature, the pores expanded and became interconnected by fractures (Figure 1). Permeability for each oil shale sample at each temperature was computed by simulation directly on the image voxels and by pore network extraction and simulation. Future work will investigate different samples and pursue in situ micro-CT imaging and modeling of oil shale pyrolysis to characterize the time evolution of the pore space.

AUTHOR INDEX

AUTHOR INDEX